S0-BLS-284

Random Walks and Their Applications in the Physical and Biological Sciences

(NBS/La Jolla Institute–1982)

AIP Conference Proceedings
Series Editor: Hugh C. Wolfe
Number 109

Random Walks and Their Applications in the Physical and Biological Sciences

(NBS/La Jolla Institute–1982)

Edited by
Michael F. Shlesinger
La Jolla Institute and
Institute for Physical Science and Technology
University of Maryland
and
Bruce J. West
La Jolla Institute
Center for Studies of Nonlinear Dynamics

American Institute of Physics
New York 1984

L.C. Catalog Card No. 84-70208
ISBN 0–88318–308–0
DOE CONF- 8206109

DEDICATION

These proceedings are dedicated by his friends and colleagues to the memory of

ELLIOTT W. MONTROLL (1916-1983)

Photo courtesy of John Ward

ELLIOTT WATERS MONTROLL
(1916 - 1983)

ELLIOTT W. MONTROLL (1916-1983)
(In Memoriam)

Elliott Waters Montroll distinguished himself in many fields of science and mathematics while working in high-level positions at universities, the federal government, and in private industry. Some of his most outstanding achievements and most eloquent publications have been in the field of random walks, the topic of this proceedings which we dedicate to his memory. This work is briefly summarized below.

His article in this volume traces the influence of the Vienna School of Statistical Physics from Boltzmann to present day physicists, and includes the personage of Smoluchowski, a co-inventor of diffusion which is a form of a random walk. This complements a larger work of Montroll and Shlesinger [("On the Wonderful World of Random Walks," in *Studies in Statistical Mechanics, Vol. 11*, North-Holland Publishers (1984), in press] which reviews the history of random walks.

His first random walk encounter occurred while he was Head of the Mathematics Group of the Kellex Corporation (1943-1946). His task, connected to the Manhattan Project, was to design a stable control system for the cascade separation of two uranium isotopes at the Oak Ridge gaseous diffusion plant. His endeavor was highly successful. Part of the analysis involved the behavior of a third species which he found to obey a discrete space random walk (the state space being the levels in the cascade). The project was classified so he could not publish his result as such, but its mathematical version did appear as "A Note on Bessel Functions of Purely Imaginary Argument" in *J. Math. and Physics, Vol. XXV No. 1*, 37-48 (1946). He discovered that the generating function for a lattice random walk satisfies a Green's function equation which, in turn, also represents a lattice vibration problem. His pioneering work on lattice dynamics was published in the now classic book "Lattice Dynamics in the Harmonic Approximation" Academic Press (1963) with A.A. Maradudin and G.H. Weiss. Montroll's independent discovery of the random walk Feynman-Kac path integrals for quantum and classical mechanics can be found in the article "Markoff Chains, Wiener Integrals, and Quantum Theory" in *Comm. in Pure and Applied Math. Vol. V*, No. 4, 415-453 (1952).

Montroll and K. Shuler in "The Application of the Theory of Stochastic Processes to Chemical Kinetics" [(Adv. in Chem. Phys. Vol. 1, Interscience Pub. (1958)] discussed the shock induced dissociation of a diatomic gas as a random walk up and down a ladder of quantum mechanical energy levels. Dissociation occurs with the first passage to the uppermost level. The exact solution was obtained in terms of the little known Gottlieb polynomials.

Montroll's most quoted random walk paper is "Random Walk on Lattices II" with G.H. Weiss in *J. Math. Physics* **6**, 167 (1965). Here the generating function (Green's function) method is extended to include continuous-time processes. If the mean time between events is infinite the master equation approach to this problem fails, but E. Montroll, N. Kenkre, and M. Shlesinger in *J. Statistical Physics*, **9**, 45-50 (1973), showed that the random walk is equivalent to a generalized master equation.

These random walks with infinite temporal moments for the time interval between events provided the first successful explanation of charge transport in amorphous material such as xerographic films (see "Anomalous Transit-Time Dispersion in Amorphous Solids" by H. Scher and E.W. Montroll, *Phys. Rev.*, **B12**, 2455-2477 (1975). This analysis was later extended by E. Montroll and M. Shlesinger in "On the Williams-Watts Function of Dielectric Relaxation" *Proc. Nat.*

Acad. Sci. (USA) 1984 (in press) to model electric dipole relaxation in amorphous materials as a random walk controlled reaction process with relaxation occurring when a defect (e.g. a vacancy) reach a frozen dipole.

Two recent reviews of Montroll's random walk work are "An Enriched Collection of Stochastic Processes" by E.W. Montroll and B.J. West in "*Studies in Statistical Mechanics, Fluctuation Phenomena*, Chap. 2, (1979) and the earlier mentioned volume 11 of this series which besides the history of random walks includes work on fractal (self-similar) random walk paths.

Montroll was a Distinguished Professor at the Institute for Physical Science and Technology of the University of Maryland and a member of the National Academy of Science. He was an original in whose path many have followed. The impact of his work is apparent in the growth of random walk research evidenced by the international conference "Random Walks and Their Applications in the Physical and Biological Sciences." As with Lord Rayleigh whom he greatly admired, Montroll wrote his last papers on the topic of random walks.

Michael F. Shlesinger
Bruce J. West

PREFACE

In the past decade there has been an explosive growth in the theory and applications of random walks. The need for a meeting to bring together the leaders who have contributed to this surging growth was evident, and accordingly the conference "Random Walks and Their Applications in the Physical and Biological Sciences" was organized. It was held at the National Bureau of Standards from June 29 to July 1, 1982. The sponsors were:

National Bureau of Standards
National Institutes of Health
Exxon Research and Engineering Company
Xerox Webster Research Center
La Jolla Institute
Institute for Physical Sciences and Technology
of the University of Maryland

The Organizing Committee consisted of:

Robert Jernigan (NIH)
Joseph Klafter (Exxon)
Elliott W. Montroll (IPST, U. of Maryland)
Robert J. Rubin (NBS)
Harvey Scher (Xerox)
Michael F. Shlesinger (La Jolla Institute/IPST, U. of Maryland)
George H. Weiss (NIH)
Bruce J. West (La Jolla Institute)

The program of lectures is presented subsequently.

The study of fluctuations can lead to very powerful results and insights. Einstein's 1905 treatment of Brownian motion convinced reasonable minds of the existence of molecules and atoms by allowing Avogadro's number to be calculated from the second moment of a probability distribution.[1] While Einstein's physical interpretation of his Brownian motion was novel, much of the mathematics, unbeknownst to him, had been developed in 1900 by Bachelier[2] in a doctoral thesis (Poincaré headed Bachelier's thesis examination committee) devoted to price fluctuations in the stock market. Both men were studying a random walk process, the subject of this book. In the same period Smoluchowski, whose origins were in Vienna, was making important contributions to the theory of Brownian motion. The history of the Vienna school of statistical mechanics is reviewed here in an article by MONTROLL.

The first explicit mention of the words "random walk" appear in a 1905 query to the scientific community published in Nature by the great statistician Karl Pearson[3].

"A man starts from a point 0 and walks l yards in a straight line: he then turns through any angle whatever and walks another l yards in a second straight line. He repeats this process n times. I require the probability that after these n stretches, he is at a distance between r and $r+\delta r$ from his starting point 0. The problem is one of considerable interest, but I have only succeeded in obtaining an integrated solution for two stretches. I think, however, that a solution ought to be found, if only in the form of a series in powers of 1/n, where n is large."

Today this is called a Pearson or a Rayleigh-Pearson random walk because as Rayleigh pointed out in a reply to Pearson in Nature that he had solved this problem previously. Rayleigh[4] responded:

"The problem proposed by Prof. Karl Pearson in the current number of Nature, is the same as that of the composition of n isoperiodic vibrations of unit amplitude and of phases distributed at random, considered in Philosophical Magazine X, p. 73, 1880; XLVII, p.246, 1889. If n be very great, the probability sought is

$$2n^{-1} \exp(-r^2/n)\, r dr \quad .$$

Probably methods similar to those employed in the papers referred to would avail for the development of an approximate expression applicable when n is only moderately great."

KIEFFER and WEISS treat further developments in the analysis and application of the Rayleigh-Pearson random walk. They discuss applications and develop approximations for the probability density function for end-to-end distances and projections on an axis. In 1934, Kuhn[5] characterized polymer chain configurations in terms of these random walks, but this neglects volume exclusion effects. Modern techniques using a real space renormalization group to study polymer statistics is presented here by FAMILY. An analogy is made between critical phenomena when $T \to T_c$ and polymer statistics as the number of monomers $N \to \infty$. Family categorizes the universality of polymer models and analyzes the crossover behavior between them. RUBIN reviews and further advances the investigation of the absorption of an isolated polymer chain at a solution surface. Each random polymer configuration corresponds to a self-avoiding random walk on a lattice. The random walk paths can be weighted to favor sticking to the solution surface. A critical weighting is found which leads to an absorption -desorption transition. A comparison is made to a similar model involving polymer configurations near a reflecting surface.

Of course, random walks by other names, go back far beyond Pearson, at least to Jacob Bernoulli's posthumously published treatise of 1713 *Ars Conjectandi.*[6] He described what is now called a Bernoulli process: In a sequence of independent trials, success occurs with probability p and failure with probability q=1-p. The number of successes fluctuates in an manner equivalent to a random walker on a lattice with probabilities p and q for moving forward or backward. Bernoulli's gambler's ruin problem is equivalent to a random walk on a lattice between two absorbing boundaries. This work was a generalization of Huygen's investigations of gaming odds in his 1657 treatise *De Ratiociniis in Ludo Aleae*[7], which in turn was motivated by the Pascal-Fermat correspondence of 1654 on games of chance. Bernoulli calculated the probability of m successes P_m in n trials to be

$$P_m = \frac{n!}{m!\,(n-m)!} p^m q^{n-m} \quad . \tag{1}$$

De Moivre[8] in his *The Doctrine of Chances* of 1735 showed that eq. (1) tends to a Gaussian distribution centered around n/2 when p=q. This is 76 years before Gauss' publication of the Central Limit Theorem which applies to this and also more general cases.

That the study of random walks does not end with Gaussian probability distributions was realized early on as exemplified by Daniel Bernoulli's 1731 discussion of the Petersburg Paradox (apparently receiving this nomenclature because it was published in the Comentarii of Petersburg Academy[9]). The Paradox involved calculating the "fair" entrance fee for playing a particular Bernoulli process: Flip a coin, if a head appears, win one penny, if a tail appears, flip the coin again. Repeat the process until a head appears. If a head appears for the first time on the nth trial, win 2^{n-1} pennies. The probability p_n of winning 2^{n-1} pennies is 2^{-n}, thus the expected winning is

$$\sum_{n=1}^{\infty} 2^{n-1} p_n = \infty \quad . \tag{2}$$

Now suppose you want to play this game against a bank with an infinite reserve of money, how much money would you stake just because your *expected* winning is infinite? You would hesitate to wager a large amount because your *median* winning is only one penny. Thus, the paradox is should your intuition follow the mean or the medium? It is fascinating that such questions come into play again in recent times concerning the random walk of charges in highly disordered environments such as xerographic films. Currents can be generated in the presence of an electric field because the *median* time for a charge to jump is finite, but because the *mean* time between jumps is infinite the currents are only transient.

The general theory of the limiting forms of probability distributions which are not Gaussian was discovered by Paul Lévy[10] and was called by him the theory of stable distributions. These are distributions for a sum of identically distributed random variables where the sum and any one of its terms have the same distribution except for scale factors. The Cauchy distribution is one such example. All of these distributions have their second or a lower (possibly non-integer) moment being infinite. They thus have long tails, no characteristic size in which to gauge measurements, and describe self-similar behavior. Random walks in the continuum with jump length distributions of the Lévy type (infinite second moments) are called Lévy flights. Their trajectory has a fractal dimension in the sense of Mandelbrot[2] (Lévy's protegé). HIOE'S article relates correlation functions for Lévy flights on a lattice to spin correlation functions for the spherical model of ferromagnetism with long-range interactions.

The most famous paper on lattice random walks is Polya's[11] proving that a random walker taking nearest neighbor jumps on a periodic, infinite cubic type lattice will only return to its starting point in two or less dimensions. The precise probability for not returning on various three dimensional lattices was first calculated by Montroll.[12] Starting in the 1940's Montroll has contributed much original work and many clear expositions to the theory of random walks. His Green's function and generating function techniques have made random walk theory accessible to the physics community. Many lecturers at this conference can trace their random walk roots back to Montroll. His classic 1965 work with George Weiss on continuous-time random walks[13] (CTRW) has been the cornerstone for many theories of transport in condensed matter. Lindenberg, Heminger, and Pearlstein[14] treated exciton transport in molecular solids as a random walk with randomly placed traps, using Montroll's techniques. In addition, they accounted for the possible trap configurations via the coherent potential approximation. The article by WEST and LINDENBERG treats exciton migration at finite temperatures. Previous theories were essentially infinite temperature models. West and Lindenberg add the proper dissipation term to their Hamiltonian equation of motion so the system can reach thermal equilibrium. This allows them to calculate, for the first time, the temperature dependence of quantities such as exciton lineshapes.

The growth in the interest in random walks has in part been due to the calculations of P.W. Anderson[15] showing that electronic wave functions can be localized in a sufficiently random environment. In this spirit, using the Montroll-Weiss walk to calculate a quantum-mechanical probability Scher and Lax[16] in 1973 analyzed the frequency dependent conductivity of impurity conduction of doped semiconductors. The article by LAX and ODAGAKI continues this field of research by treating more accurately the random walk in a random media. The multiple scattering theory of Lax is used to derive a coherent medium approximation for the hopping of charges in a random media. Dynamic percolation is automatically included in the calculation.

SCHER and RACKOVSKY'S article extends the earlier random walk transport model to geminate recombination of charge pairs. The Colomb field of the charges as well as the effect of an external electric field are taken into account. The random walk Green's function approach includes defective sites and allows the effects of dimensionality, disorder, tunneling, and intramolecular transitions to be analyzed exactly.

Scher and Montroll[17] were the first to treat the transient currents in xerographic films as a random walk process governed by a waiting time distribution between jumps which has an infinite mean. The connection of these processes to fractal times, Lévy distributions and renormalization groups has been explored by Shlesinger and Hughes.[18] Reaction schemes governed by this fractal time process for electron-hole recombination in an inhomogeneous environment is the subject of the article by KLAFTER and BLUMEN.

Random walks need not take place in a real positional space and the transitions need not represent actual physical jumps. Many problems can be mapped exactly onto a random walk problem, although the appropriate random walk may be semi-Markovian, non-Markovian, involve internal states, memories, a high number of dimensions, a defective lattice, difficult boundary conditions, or other complications. For example SAHIMI, in his article is able to treat flow paths through a random network as a continuous-time random walk. Percolation effects are also included, with information about the structure of the percolation backbone being derived from the dispersion of the flow.

While there has been much attention paid to the behavior of a single particle in a random potential, a more difficult question involves the dynamics of a fluid in a random potential. MUKAMEL, in his article, addresses this issue and calculates the conductivity and density function response of the fluid as these are experimentally observable quantities.

In the last article CLAY and SHLESINGER describe the passage of potassium ions through a channel in a neural membrane in terms of a random walk model whose internal states represent the contents of the channel. The kinetics of gates which can block the current are included in the analysis. Modifications of the Hodgkin-Huxley equations for the gating process are proposed. The model leads to exact results which can be used to analyze flux experiments.

We wish the reader much enjoyment with this volume. For those who find random walks irresistible we suggest the following reviews:

1. M. Barber and B. Ninham, *Random and Restricted Random Walks* (Gordon and Breach Pub. NY) 1970. This is a fairly complete review of the already extensive random walk literature up to 1970.

2. M.F. Shlesinger and V. Landman, in *Applied Stochastic Processes*, G. Adomian ed. (Academic Press, NY) 1980, p. 151-246. Treats many problems which can be cast as random walks with internal states.

3. E.W. Montroll and B.J. West, in Studies in Statistical Mechanics, **VII**, *Fluctuation Phenomena* eds. J.L. Lebowitz and E.W. Montroll, (North-Holland Pub.) 1979, p. 61-175. Good review of continuous-time random walks, boundary and defect problems, Lévy distributions, first passage time distributions, and nonlinear diffusion processes.

4. G.H. Weiss and R.J. Rubin, in Advances in Chemical Physics, **52**, eds. I. Prigogine and S.A. Rice (John Wiley & Sons Inc., N.Y.) 1983, p 363-505. A good recent review of general theory and applications to polymer physics, multistate random walks, solid state physics, and motion of micro-organisms.

5. E.W. Montroll and M.F. Shlesinger, in CCNY Physics Symposium in Celebration of Melvin Lax's 60th Birthday, H. Falk ed., (CCNY Physics Dept., NY) 1983, p. 44-147 and E.W. Montroll and M.F. Shlesinger in Studies in Statistical, **11**, *Nonequilibrium Phenomena 11* eds. J.L. Lebowitz and E.W. Montroll, (North Holland Pub) (in press). The emphasis is on the history of probability theory and the theory and applications of fractal random walks.

6. Journal of Statistical Physics **30** No. 2 (1983). Eds. G.H. Weiss and R.J. Rubin. Contains twenty-nine papers presented at the Random Walks in the Physical and Biological Sciences conference. Full of good research problems.

REFERENCES

1. A. Pais *Subtle is the Lord, The Science and Life of Albert Einstein*, 79-103, (Clarendon Press, Oxford) 1982.
2. B.B. Mandelbrot, *The Fractal Geometry of Nature*, (W.H. Freeman, San Francisco) 1982.
3. K. Pearson, Nature, **72**, 294 (1905).
4. Lord Rayleigh, Nature, **72**, 318 (1905).
5. W. Kuhn, Kolloid Z. **68**, 2 (1934).
6. Jacob Bernoulli, *Ars Conjectandi* (Basel) 1713.
7. C. Huygens, *De Ratiociniis in Ludo Aleae* (The Hague) 1657.
8. A. DeMoivre, *Doctrines of Chances: or, A Method of Calculating the Probabilities of Events in Play* (London) 1756 (3rd Edition).
9. I. Todhunter, *A History of the Mathematical Theory of Probability* (Cambridge) 1865. Chelsea Press Reprint (1949).
10. P. Lévy, *Theorie de l' addition des variables aléatoires* (Gauthier-Villars, Paris) 1937.
11. G. Polya, Math. Ann. **84**, 149 (1921).
12. E.W. Montroll, Ann. Math. Soc., XVI Sym. on App. Math. 193 (1964).
13. E.W. Montroll and G.H. Weiss, J. Math. Phys. **6**, 167 (1965).
14. K. Lakatos-Lindenberg, R. Hemenger, and R. Pearlstein, J. Chem. Phys. **56**, 4852 (1972); R. Hemenger, R. Pearlstein, and K. Lakatos-Lindenberg, J. Math. Phys. **13**, 1056 (1972).
15. P.W. Anderson, Rev. Mod. Phys. **50**, 191 (1978).
16. H. Scher and M. Lax, Phys. Rev. **B137** 4491, 4502 (1973).
17. H. Scher and E.W. Montroll, Phys. Rev. **B12**, 2455 (1975).

18. M.F. Shlesinger and B.D. Hughes, Physica **A109**, 597 (1981).

For the organizing committee,

MICHAEL F. SHLESINGER**
La Jolla Institute
and IPST
University of Maryland
College Park, MD

BRUCE J. WEST
Center for Studies of Nonlinear Dynamics*
La Jolla Institute
8950 Villa La Jolla Drive, Suite 2150
La Jolla, CA 92037

*Affiliated with the University of California, San Diego
**Present address: Office of Naval Research, Arlington, VA

PROGRAM

A history of stochastic processes.
Elliott W. Montroll, University of Maryland
From classical dynamics to continuous time random walks.
Robert Zwanzig, University of Maryland
Random walk model for 1/f noise.
Mark Nelkin, Alan Harrison, Cornell University
Diffusion in random one-dimensional systems.
W. Schneider, J. Bernasconi, Brown Boveri Research Center
Diffusion and relaxation in disordered systems.
B. Movaghar, Hirst Research Center
Multiple scattering, CPA, and CTRW treatment of hopping conductivity.
Melvin Lax, T. Odagaki, City University of New York and Bell Laboratories
Analytic continuation method for estimating effective parameters for multicomponent random walks.
George Papanicolau, Courant Institute
Random walks on inhomogeneous lattices.
P. W. Kasteleyn, W. Th. F. den Hollander, University of Leiden
Laser speckle as a two-dimensional random walk.
Richard Barakat, Harvard University
A random walk model of multiphase dispersion in porous media.
Muhammad Sahimi, L. E. Scriven, H. T. Davis, University of Minnesota
Random walks and renormalization theory: The central limit theorem as a fixed point.
P. B. Visscher, University of Alabama
Nonexponential decay in relaxation phenomena.
A. K. Rajagopal, Louisiana State University; K. L. Ngai, R. W. Rendell, S. Teitler, Naval Research Laboratory
From random to self-avoiding walks.
Cyril Domb, Bar-Ilan University
Self-avoiding walks with geometrical constraints.
S. G. Whittington, University of Toronto
Stochastic processes originating in deterministic microscopic dynamics.
Joel Lebowitz, Rutgers University
Stochastic flows in integral and fractal dimensions and morphogenesis.
John J. Kozak, University of Notre Dame
Stochastic stick boundary conditions.
Irwin Oppenheim, Massachusetts Institute of Technology; N. G. van Kampen, Rijksuniversiteit Utrecht
Stochastic aspects of biological locomotion.
Ralph Nossal, National Institutes of Health
Phase transitions in a four-dimensional random walk with application to medical statistics.
Ora E. Percus, Jerome K. Percus, Courant Institute
Protein folding as a random walk.

Nobuhiro Go, Kyushu University

Conformational space renormalization group treatment of polymer excluded volume.

Karl Freed, James Franck Institute

Cell renormalization for self-avoiding random walks and lattice animals.

Fereydoon Family, Emory University

Single and multiple random walks on random lattices: Excitation trapping and annihilation simulations.

R. Kopelman, P. Argyrakis, J. Hoshen, J. S. Newhouse, University of Michigan

Correlation factors for diffusion via the vacancy mechanisms in crystals.

Masahiro Koiwa, Shunya Ishioka, Tohoku University

Diffusion in concentrated lattice gases.

Klaus W. Kehr, Institut fur Festkorperforschung

Random walks on random lattices with traps.

V. Halpern, Bar-Ilan University

Energy transfer as a random walk with long-range steps.

Alexander Blumen, Technische Universität; G. Zumofen, ETH-Zentrum

Rotational diffusion in solid polymers.

J. T. Bendler, General Electric

On the mean motion and some statistical properties of quasi-periodic observable in a fermion-boson model.

F. T. Hioe, University of Rochester

Monte Carlo renormalization group calculation for polymers.

H. Muthukumar, Illinois Institute of Technology

Trapping of excitation in the average T-matrix approximation.

D. L. Huber, University of Wisconsin

Diffusion-controlled reactions.

R. Cukier, Michigan State University

Transport processes in disordered solids.

Kurt E. Shuler, University of California, San Diego

Equilibrium folding and unfolding pathways for a model protein.

Robert L. Jernigan, S. Miyazawa, National Institutes of Health

Physics of migration of ligands in biomolecules.

Peter Hanggi, Polytechnic Institute of New York

Master equation techniques for exciton motion, capture, and annihilation.

V. M. Kenkre, University of Rochester

Random walk to and interaction with an impurity.

Peter M. Richards, Sandia National Laboratory

Monte Carlo simulation of electronic transport in disordered media.

M. Silver, University of North Carolina; H. Baessler, G. Schoenherr, Universitat Marburg; Leon Cohen, Hunter College

Renormalization group approach to random walls on disordered lattices.

Jonathan Machta, University of Maryland

On the dynamics of excitations in disordered systems.

S. Mukamel, Weizmann Institute of Science

Approach to asymptotic diffusive behavior in strongly disordered lattices.

Itzhak Webman, Exxon Research and Engineering Co.

Kinetics of adsorption on stepped surfaces.

C. H. Wu, RCA/David Sarnoff Research Center; Elliott W. Montroll, University of Maryland

Fractal and lacunary stochastic processes.

Barry D. Hughes, University of Minnesota; Elliott W. Montroll, University of Maryland; Michael F. Shlesinger, La Jolla Institute;

Generalized average T-matrix approximations for transport on a disordered lattice.

Gregory Korzeniewski, Richard Friesner, Robert Silbey, Massachusetts Institute of Technology

TABLE OF CONTENTS

ON THE VIENNA SCHOOL OF STATISTICAL THOUGHT*

Elliott W. Montroll
Institute for Physical Science and Technology
University of Maryland
College Park, Maryland 20742, USA

ABSTRACT

Here is presented a fragment of the family tree of the Vienna School of statistical physics. Its branches have extended far beyond the Austrian national boundaries into The Netherlands, the United States, and China. Some vignettes from the lives and works of some of the Vienna School's prominent members are also given.

THE HISTORY

A remarkably large number of pioneers in the evolution of the statistical style of thought in modern physical and indeed biological science may trace their scientific family tree back to the Vienna school. The University of Vienna dates from 1365 but its prominence in our chosen field starts with the students of Christian Doppler (1803-1853); Josef Stefan and Gregor Mendel.

Doppler's birthplace, a magnificent house, still stands on Mozart Square in Salzburg. His effect (1842), the frequency alteration of sound (and light) by relative motion, was first experimentally verified in Utrecht by C.H.D. Buijs-Ballot (1845). A flat car with seven trumpeters towed by a speedy locomotive provided the principal equipment. This unusual style in science is also evident in M. Eiffel's measurement of the drag coefficient of a flat plate by dropping it (along guide wires) from the Eiffel tower. The first laboratory measurement of the Doppler shift was made by Ernst Mach (also of the Vienna School) in 1861.

Doppler became the first director of the University's Institute of Experimental Physics in 1850 and occupied that chair until his death in 1853. Gregor Mendel, the father of the statistical style in genetics "spent four terms between 1851 and 1853, at the University of Vienna, where he studied physics, chemistry, mathematics, zoology, entomology, botany and paleontology. In the first term he took work in experimental physics under the famous Doppler and was for a time an 'assistant demonstrator' in physics. He also had courses with Ettinghausen, a mathematician and physicist [later successor to Doppler's chair],. . ."[1] (Josef Stefan studied with Doppler and with the physiologist Carl Ludwig.) Ludwig Boltzmann [1844-1906] studied with Stefan and was much influenced by his colleague Joseph Loschmidt [1821-1895], who in 1865 made the first convincing estimate of the size of a molecule.

A small portion of the family tree of the Vienna school following Doppler is sketched in Figure 1. The largest branch stems from Ehrenfest and his Dutch students whose influence extended to distant parts of the world, as far as China to the East and the U.S. to the West. We continue this section with some vignettes from the lives of a sample of members of the Vienna family. I must apologize that limited space forces me to omit many distinguished names from the tree. To appreciate the difficulty I face, notice for example, that Karl Herzfeld directed the completion of more than twenty-five Ph.D. theses at Catholic University and that his student, John Wheeler, directed the completion of more

*The title of the lecture given by the author at the Random Walks Symposium was: "A History of Statistical Processes." Since a voluminous elaboration of that lecture will be published in *Studies of Statistical Mechanics* entitled, "On the Wonderful World of Random Walks," this note is limited to an extended discussion on one of the viewgraphs whose content is omitted from the forthcoming *Studies* article.

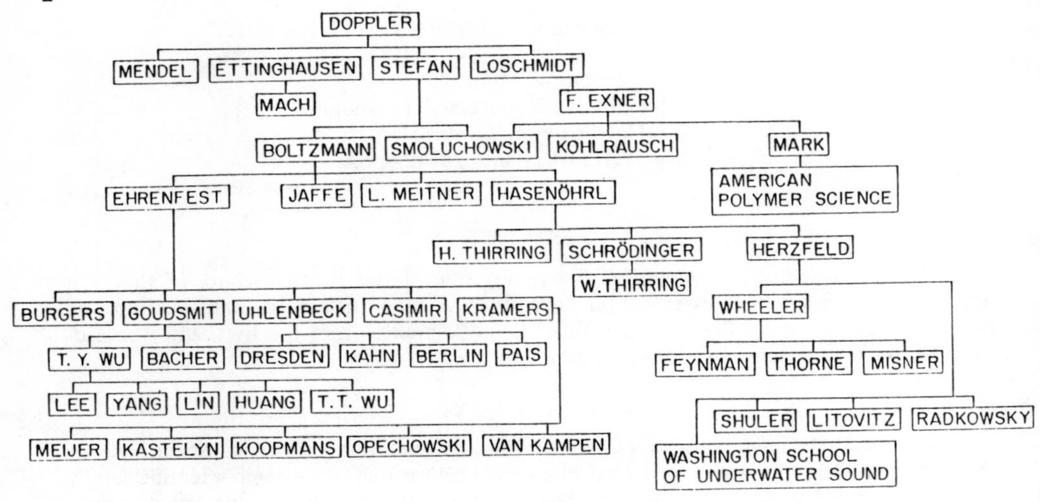

Figure 1. A fragment of the family tree of the Vienna School

than twenty-five Ph.D. theses at Princeton University. Please, Dear Reader, forgive me if you are omitted and insert your name on the appropriate branch of the tree.

Loschmidt was endowed with considerable patience and imagination. He first studied with Sigmund Exner at Innsbruck. Upon completion of his education in Vienna in 1843 he, finding no academic position, worked briefly in the chemical industry and then went in business for himself. Unfavorable economic conditions bankrupted him in 1854. Upon returning to the University of Vienna as an over-age student, his earning a teacher's certificate qualified him for a high school position which he held for several years. At the age of forty he published, at his own expense, his first scientific work, *Chemische Studien I* (II never appeared).[2]

Loschmidt's chemical treatise included structural formulae for organic compounds and was apparently the first publication to contain the standard double and triple bond symbols in such structural formulae. The book at that time aroused little interest except in Kekule. Loschmidt discussed possible benzene structures and had even thought of the Kekule one but apparently abandoned it for some reason. At a later date a controversy developed between Kekule, Loschmidt and Crum Brown over priorities and mutual influences in organic structure representation. *Chemische Studien I* only achieved a broad recognition posthumously for Loschmidt when it was reprinted as a volume in Ostwald's Klassiker.

Loschmidt's *Constitution des Ether* (Vienna 1862) stimulated more attention, but it was his calculation of the size of molecules (1865) that made him famous and procured him a chair at the university in 1869. His influence on Boltzmann's development of the kinetic theory of gases was considerable. The Vienna trio, Stefan, Loschmidt, and Boltzmann, provided a formidable foundation for a research and educational structure that encouraged a magnificent style for criticism, clarity, and innovation among its disciples.

Loschmidt brought the reversibility paradox to Boltzmann's attention. If one changes the signs of all the velocities in dynamical equations and runs the system backward in time, the equations remain invariant. This was not the case with Boltzmann's transport equation. It was in response to his criticism that Boltzmann introduced his probabilistic interpretation of entropy

$$S = k \log W,$$

W being the number of possible molecular configurations corresponding to a given macroscopic state, and developed his deeper arguments for irreversibility.

Loschmidt, a student of Sigmund Exner became, in turn, the teacher of Exner's son, Franz, who later occupied Loschmidt's chair.

Ludwig Boltzmann was of course the spiritual and intellectual leader of the Vienna school in the late 1800's and early 1900's. He gave remarkably clear, brilliant lectures and was a great inspiration to his students. If we are to believe Stefan Zweig,[3] who was a student in the philosophy and literature faculty, Boltzmann's style was not common in the University of Vienna as a whole: ". . . . in particular, in a university such as ours in Vienna, which was so overcrowded with its six or seven thousand students that fruitful personal contact between teacher and scholar was hindered from the very outset, and which had remained behind the times because of its all too great adherence to tradition. I did not see a single teacher who could make his branch of learning irresistible to me."

Boltzmann, J. Clerk Maxwell, and J. Willard Gibbs were the founding fathers of statistical mechanics. Maxwell had no school. As Cavendish Professor, he devoted most of his attention to introducing experimental laboratory procedures into the curriculum and to editing Cavendish's notebooks. He, unfortunately, died quite young shortly after becoming Cavendish Professor. Gibbs did not have the advantage of receiving students well educated and well tuned from a Vienna Hochschule; most of his students arrived at Yale ill prepared for his courses. Only about a half dozen of them continued with a scientific career, including: H.A. Bumstead, who became professor of physics at Yale; Percy Smith, who became professor of mathematics at Yale; Irving Fisher, who became a prominent Yale economist; Lyne Wheeler, who pursued a distinguished career at the Naval Research Laboratory and was the author of the definitive Gibbs biography; and Edwin Bidwell Wilson, who was probably the student closest to the Master and co-author of Gibbs-Wilson's *Vector Analysis.* Wilson also wrote a once popular advanced calculus book, edited the PNAS during the period when it evolved into an internationally prominent journal, and became the first Professor of Vital Statistics of the Harvard School of Public Health. Even though he did not do research in statistical mechanics, Wilson considered himself to be the middleman on the subject in the United States since he was Gibbs' student and the teacher of Richard Tolman. Tolman's treatises on the subject were classics of the 1920's and 1930's. Wilson, while head of the MIT Physics Department in the early 1920's, hired J.S. Slater who became a teacher of Jack Kirkwood, the man who directed more Ph.D. students and postdoctorals in statistical mechanics than any other American professor. While Kirkwood's thesis was an experimental one directed by Keyes, Slater was his theoretical mentor.

Boltzmann's first contribution to statistical mechanics was to find the energy distribution function of members of an assembly of colliding molecules in an external force field, thus generalizing the Maxwell Velocity distribution to the Maxwell-Boltzmann Energy Distribution. The great, classic paper, which gives a statistical basis to thermodynamic entropy, making entropy a measure of randomness or disorder in a physical system, was published in 1877.

If a scientist is to be graded by the number of colleagues that he keeps busy, we must give Boltzmann an A+. For over a hundred years, an army of talented physicists and mathematicians have attempted to justify, interpret, and solve the Boltzmann equation for the dynamics of the distribution function of positions and momenta of members of an assembly of interacting gas molecules. These studies remain a central issue in the investigation of nonequilibrium processes. Two other scientists, who qualify for an A+ are Navier and Stokes, in whose 150-year old equation lies buried our understanding of turbulent flows.

As is well known to students of statistical physics and chemistry, Boltzmann's central thrust was to deduce the macroscopic properties of material systems, especially gases, from constituents atomic and molecular models. Unfortunately, in his time an influential component of the scientific community, the so-called "energeticist" lead by Ostwald, Duhem, Helm, and Mach, rejected evidence for the existence of atoms and molecules and any physical theory based upon them. They preferred to introduce generalized thermodynamic arguments which they called "energetics". A clear statement of the controversy between Boltzmann and the energeticists has been given by S. Brush.[4]

The extent of the influence of Boltzmann's opponents on him is reflected in the introduction to the first volume of his treatise in the "Vorlesung über Gastheorie" (1897), where he states apologetically that he is publishing the book even though 'kinetic theory seems to have gone out of style in Germany.' After defending himself for several years, he writes in the introduction of his second volume (1904), "Even while the first part of this book was being printed the manuscript of the second and last part was almost completed . . . . Just at that time, however, the attacks against kinetic theory increased. I am convinced that the attacks rest upon misunderstandings and that the role of kinetic theory is not yet played out...In my opinion it would be a blow to science if the contemporary opposition were to cause the kinetic theory to sink into the oblivion which was the fate suffered by the wave theory of light through the authority of Newton. I am aware of the weakness of an individual against the prevailing currents of opinion. In order to insure that not too much will have to be rediscovered when people return to the study of the kinetic theory I will present the most difficult and misunderstood parts of the subject in as clear a manner as I can."

These attacks left Boltzmann in a state of despondency and induced a persecution complex. Boltzmann's suicide in 1906 has frequently been attributed to this despondency, or a combination of it with his ill health. A new view of Boltzmann's suicide emerges from the late K.F. Herzfeld's last scientific essay. We refer later to Herzfeld's own work and position in the Vienna school. Since Herzfeld's essay is difficult to obtain, being an unpublished internal report of the Center of Theoretical Studies at the University of Miami, we quote his remarks in their entirety:[5]

"I entered the University of Vienna in the Fall of 1910. The physics department was then located in a rented private house near the University. Boltzmann had committed suicide in September 1906, and many authors, in particular Broda[6], present this as induced by a depression because Boltzmann's viewpoint on a statistical explanation of thermodynamics was rejected by the majority of physicists. Urban[7], after discussion with Mrs. Flamm, Boltzmann's daughter, ascribed it to a "neurological disease," aggravated by depression. When I entered the University, Boltzmann's laboratory "Diener"—I have forgotten his name—was still in the laboratory. The "Diener" related how in the last years one could hear Boltzmann "roar with pain while in the washroom". To my knowledge there is only one condition which produces that and may lead to suicide (as it has in the case of the American bacteriologist Hans Zinsser) and that is infection (grand douleur) of the facial nerves, which has nothing to do with the psychological state. So I am convinced that this was what Boltzmann suffered from and by which he was driven to suicide."

We conclude our discussion of Boltzmann with the following quote from S. Brush:[4] ". . . . it is certainly one of the most tragic ironies in the history of science that Boltzmann ended his life just before the existence of atoms was finally established (to the satisfaction of most scientists) by experiments on Brownian motion guided by a kinetic-statistical theory of molecular motion."

Willard Gibbs, a second of the founding fathers of statistical mechanics, also failed to live to enjoy the applause eventually directed to his work on the subject or to reap the reward of its success. While Gibbs devoted many years of thought to statistical mechanics, he published no memoir on that topic before the appearance of his classic treatise, "Elementary Principles in Statistical Mechanics." His lone earlier contribution was a two-page abstract of a lecture delivered at an AAAS meeting in 1884 under the title, "On the Fundamental Formulae of Statistical Mechanics with Applications to Astronomy and Thermodynamics." From conversations with Gibbs' student, E.B. Wilson, many years ago I learned that Gibbs was reluctant to publish because of what he visioned to be inadequacies in the theory, especially as applied to the calculation of thermodynamic properties of polyatomic molecules. As Gibbs stated in the preface to the "Elementary Principles . . . .": "Even if we confine our attention to the phenomena distinctively thermodynamic, we do not escape difficulties in as simple a matter as the number of degrees of freedom of a diatomic gas. It is well known that while theory would assign to the gas six degrees of freedom per molecule, in our experiments on specific heat we cannot account for more than five. Certainly, one is building on an insecure foundation, who rests his work on hypotheses concerning the constitution of matter. Difficulties of this kind have deterred the author from attempting to explain the mysteries of nature . . . .".

Again, according to Wilson, Gibbs published the "Elementary Principles . . . ." only after some prodding by the Yale administration who wished to have contributions from prominent Yale faculty to celebrate Yale's bicentennial in 1901. The manuscript for the "Elementary Principles . . . ." was submitted in 1901. The book was published in 1902. Gibbs died on April 28, 1903, before any published response to it could appear in European journals. Incidentally, most of the early references were negative; a full appreciation of its importance and depth required several years.

Marjan (Marie) Smoluchowski (according to K.F. Herzfeld:[8] Ritter von Smolan; that is, knight from Smolan) was born in Brühl, a suburb of Vienna in 1872.[9] He attended elementary schools in Vienna and matriculated at the University, receiving his Ph.D. in 1895. His early work was experimental under the guidance of Stefan and Exner whose interest in Brownian motion infected his student. He attended lectures of Boltzmann but apparently never had a close personal interaction with the great man. Upon completion of his Ph.D., Smoluchowski went on a grand tour of important European scientific centers, listening to lectures by Poincaré and Hermite in Paris as well as conferring with Lippmann and de Bouty, then proceeding to Glasgow to work with Lord Kelvin and finally spending time with M. Warburg in Berlin, experimenting on thermal conductivity. It seems that he was most influenced by and most admired Lord Kelvin. In 1897 no position opened for him in Vienna but he found one at the Polish university at Lvow (Lemburg), later becoming a professor there in 1903. He moved on to the more distinguished post as Director of the Physics Institute at Cracow in 1913. At the height of his power in 1917, he contracted dysentery and died shortly thereafter.

Smoluchowski's imaginative research on Brownian motion, fluctuations and critical opalescence, basic aspects of stochastic processes (such as the calculation of first passage times), and his theory of coagulation of colloids give him a prominent niche in the class of pioneers in the development of our understanding of non-equilibrium processes. He also made lesser known contributions to the study of phenomenon associated with our macroscopic world, ranging from aerodynamics to meteorology to a theory of mountain folding.

Boltzmann's students Ehrenfest and Hasenöhrl did not inherit his originality or even develop into a Smoluchowski or Loschmidt. They both, in their style however, became great teachers whose students and grand-students were among the most imaginative innovators of the statistical aspects of 20th century physics.

Friedrich Hasenöhrl succeeded Boltzmann in 1907 as Professor of Physics at the University of Vienna. His lectures, modeled after Boltzmann's, became famous for their clarity. His most important research concerned the effects of radiation within a moving cavity, where he showed that the trapped radiation led to an apparent increase in the mass of the cavity, in anticipation of Einstein's general theorem of the equivalence of mass and energy.

Hasenöhrl's most famous student, Erwin Schrödinger, was apparently very much impressed by Hasenöhrl's inaugural lecture on Boltzmann's work. In his Nobel Prize lecture, Schrödinger states:

> During my four years of university in Vienna (1906-1910) the young Fritz Hasenöhrl, who had just stepped into the chair of the unfortunate Boltzmann, had the greatest influence. In a cycle of eight semesters each meeting five hours per week were presented in detail the advanced theories of mechanics as well as the eigenvalue problems of physics of continua—in the detail in which later they would become necessary for me.

Hasenöhrl's successful career was tragically cut short when, at the outbreak of World War I, he left his university spot for active duty, was killed in the early battle of Isanzo. Schrödinger's respect for Hasenöhrl is summarized in his statement that had Hasenöhrl not been killed ". . . . I have a feeling that otherwise his name would appear where mine does today."

As with Smoluchowski, Schrödinger was also very much influenced by Franz Exner.

In 1920, Schrödinger was recruited to fill Hasenöhrl's vacant chair; an offer he declined for financial reasons. He did accept a more lucrative post at the University at Zurich. That acceptance was probably the most important decision he made in his life. Moving to Zurich put him in close contact with Peter Debye and Herman Weyl. It was Debye who prodded Schrödinger to report on de Broglie's thesis to the Zurich Physics Seminar and, after the seminar presentation, suggested to Schrödinger that he find a wave equation appropriate for de Broglie waves.[11] Consultation with Weyl was extremely helpful to Schrödinger for the solving of various forms of his equation. Had Schrödinger remained in Vienna, the probability that he would have become fascinated by the de Broglie waves at just the right time (before considerable competition arose) would have been small.

Most of Schrödinger's pre-wave mechanics research was concerned with statistical mechanics, especially as applied to lattice vibrations, fluctuation theory, and quantum gases. His Handbuch der Physik article on specific heats of solids remained the classic presentation of the subject for many years. Hanle in a, as yet unpublished, Ph.D. thesis[12] has given an excellent account of Schrödinger's statistical mechanics publications. Schrödinger's post-wave mechanics monograph, "What is Life," motivated many physicists to examine biological processes and especially to become interested in the manner of passage of genetic information from one generation to another.

After Schrödinger rejected the Vienna offer, it was tendered to Hans Thirring (who had written some important papers on the theory of specific heats of solids). He thus became the director of the Vienna Theoretical Physics Institute. In the post-World War I period, funds were scarce and the Institute was not restored to its pre-war prominence. Nevertheless, it turned out a number of outstanding students. As an example, Herman Mark received his Ph.D. in 1921, having studied with Thirring and Exner, as well as with the University's physical chemists. Upon graduation, he made a career in the chemical industry and in the physical chemistry institutes in Germany, returning to Vienna as professor for the period 1932-1938. There he did research on electron diffraction and also became attracted to polymer science. With the appearance of the Nazis, he immigrated to the United States, eventually moving to Brooklyn Polytech where he established his world-famous Institute of Polymer Science. His students and postdocs have become key figures in industrial and academic polymer science in the United States.

A Hasenöhrl student who produced a distinguished line of successors was Karl F. Herzfeld.

Herzfeld and Schrödinger were luckier than Hasenöhrl in their military careers, both having survived World War I as artillery officers. Herzfeld wrote from Galicia:[13] "The Russians attacked once in force; and I saw one of the last large-scale cavalry attacks (3,000 Tartars) in history; if they had gotten through, I would not be writing this history."

At the conclusion of World War I, Herzfeld became a privatdozent at Munich in the Sommerfeld Institute. Since the Herr Geheimrat Professor was in constant demand by the European physics community and by German educational commissions, Herzfeld devoted considerable attention to substituting for Sommerfeld and eventually became ausserordentlicher professor. He was an important influence on Sommerfeld students including Heisenberg, Heitler, and Pauling, being the intermediary between the students and his distinguished boss. Herzfeld wrote a treatise "Kinetische Theorie der Wärme," which became the standard textbook from which European students learned the kinetic theory of gases and statistical mechanics, and made important contributions to *Handbuch der Physik* ("Klassische Thermodynamik" and "Absorption and Dispersion").

Herzfeld's early original research was on the old quantum theory and on the kinetic theory of gases. In the late 1920's, he was the first to give a serious discussion of the metal insulator transition (now called the Mott transition). His work on sound dispersion in polyatomic gases and liquids was extended by Litovitz and collaborators, with the subject being summarized in the Herzfeld-Litovitz book entitled, "Absorption and Dispersion of Ultrasonic Waves."

It was Herzfeld who suggested to E. Maxwell (then of NBS) that he measure the isotope effect on the transition temperature of superconductors. The positive outcome of the experiment was instrumental to the development of modern theories of superconductivity.

Herzfeld moved from Munich to the Washington-Baltimore area of the U.S. in 1926 (being at Johns Hopkins University from 1926 until 1936 and at the Catholic University of America for the remainder of his career). For many years, he was the capital area's wise theoretical physicist to whom local researchers went with their problems, which he generally clarified and frequently resolved. At Catholic University, he developed a program to respond to the needs of young staff members of the many government laboratories, specially the National Bureau of Standards and the Navy laboratories. Many part-time students from these organizations received their Ph.D.'s under his sympathetic direction. Many more had their general level of physical understanding and sophistication raised considerably by attendance at his courses.

The Catholic University acoustics program initiated by Herzfeld had a profound influence on the Navy underwater sound program, with many of the participants in that program having taken acoustics courses at Catholic University or having attended state-of-the-art workshops on the subject. Statistical mechanicers will recognize the sample Wheeler, Litovitz, and Shuler listed on the Vienna family tree. They will probably not have heard of Alvin Radkowsky who, for many years, was a chief scientist behind Admiral Rickover's nuclear Navy.

After World War I, the most virile branch of the Vienna school flourished in Leiden under the dynamic leadership of Paul Ehrenfest (1880-1922), one of Boltzmann's last students (Ph.D. 1904). Ehrenfest was considered to be the conscience of physics in the period 1900-1933, being a deep critical analyst of the foundations of the subject ranging from statistical mechanics to quantum theory. Physics was his life and passing his experience to his students was his passion. His unusual personality has been sympathetically portrayed by his biographer Martin Klein.[14]

Klein summarized the opinion of Ehrenfest students and friends by stating that[15]: "As a teacher Ehrenfest was unique. Albert Einstein described him as 'peerless and the best teacher in our profession whom I have ever known'. His lectures always brought out the basic concepts of a physical theory, carefully extracting them from the accompanying mathematical formalism. He worked closely with his students doing everything in his power to help them develop their own talents." One of Ehrenfest's most successful students, George Uhlenbeck, had on the occasion of his being awarded the Oersted medal for notable contributions in the teaching of physics given an eloquent account of Ehrenfest's teaching style, especially emphasizing his interaction with separate students. Uhlenbeck said,[16] "....I am very conscious of the fact that I have been taught how to teach, mainly by example, but also by definite advice. The man who taught me was the late Paul Ehrenfest at the University of Leiden in Holland, who was in my opinion one of the really great teachers." Readers interested in educational methods must read Uhlenbeck's reminiscences in reference 16. I will merely emphasize here the last phrase of Ehrenfest's interaction with a research student.

"....Ehrenfest recognizes that one of the main requirements for the scientific investigation is confidence in himself, or if you want, courage.... It is essential that as a student one feels at least to be equal to one's teacher! Ehrenfest used to say: 'Weshalb habe ich solche gute Studenten? Weil ich so dumm bin'!"

After Ehrenfest received his Ph.D., he and his wife Tatyana moved to Göttingen for a brief period. His scientific interests turned to statistical mechanics and especially to the approach of a system to equilibrium as described by Boltzmann's H-theorem with the entropy increasing monotonically. For this purpose he invented his ingenious Urn models or, as they are sometimes called, dog and flea models. This work was successfully presented to Felix Klein's mathematics colloquium. It impressed Klein to the degree that he invited the Ehrenfests to write the review article on the "Principles of Statistical Mechanics" for Klein's "Encyklopädie der mathematischen Wissenschaften". When published, the critical review was well accepted and with the previous work on H-theorem, made the Ehrenfests highly regarded in the theoretical physics community. From Göttingen, the Ehrenfests moved to St. Petersburg for several years where they continued to live modestly on Paul and Tatyana's small inheritances. The Ehrenfest search for a regular university appointment as described by Martin Klein is an interesting story that should be read by all those who have been job seekers themselves.

In 1912, Lorentz decided to retire from his professorship at Leiden to accept another appointment in the Netherlands. His search for a successor first lead him to Einstein who was at that time a hot item on the academic market. Einstein's acceptance of the position at Berlin left Lorentz without a candidate. Ehrenfest was then considered on the basis of Lorentz' having been much impressed with the Encyklopädie article and by Sommerfeld's response to Lorentz' inquiry to him about Ehrenfest. Sommerfeld responded that,[14] "Ehrenfest lectures like a master. I have hardly ever heard a man speak with such fascination and brilliance." The position was offered to Ehrenfest who accepted it immediately.

In his first years at Leiden, Ehrenfest attracted many outstanding students to his lectures including Dirk Struik, Hans Kramers, Jan Burgers, and A.D. Fokker. Struik later made important contributions to differential geometry and to the history of science. His classic, "Yankee Science in the Making," is a must to anyone interested in early American science and technology. Fokker is of course famous for the Fokker-Planck equation.

Kramers spent some time in the Niels Bohr Institute (where he received his Ph.D.), then returned to the Netherlands first at Utrecht and finally returning to Leiden after Ehrenfest's suicide, as professor of theoretical physics. Kramers made notable discoveries in quantum theory ranging from being the K in WKB to being one of the originators of the concept of renormalization. The Kramers-Kronig relation finds application in many branches of physics. His most important contributions to statistical mechanics were concerned with models employed in theory of magnetic systems and phase transitions, and in chemical kinetics. The Kramers-Wannier paper on the Ising model brought that model to the attention of the physics community, leading to a cottage industry that became the basis of our modern understanding of phase transitions. Through Kramers' instigation, his student Opechowski developed the theory of interacting spin waves.

Ehrenfest's first Ph.D. student was Jan M. Burgers. His thesis was entitled, "The Atom Model of Rutherford-Bohr." It contained, among other things, a detailed analysis of the manner that old quantum theory could be extended to multiply periodic systems. This gave Ehrenfest the proper background material for his paper on the Adiabatic Principle which he considered his most important work. Burgers[17] states, "But after having completed the thesis work, I became somewhat afraid of having insufficient phantasy for making fruitful advances in Bohr's theory." At that time (1918), a chair of hydrodynamics was created at the Technical University in Delft. Ehrenfest was asked to propose a candidate. He strongly recommended Burgers but added the caveat that, while Burgers was not an expert in hydrodynamics, if given the opportunity he would soon become one and would be making outstanding contributions to the field. Burgers was appointed. He developed a very active hydrodynamics-aerodynamics institute at Delft. He with Von Karman, Prandtl, and G.I. Taylor were the world leaders in hydrodynamics of their generation. His prominent place in the field was reflected in the position of his seminal review in Durand's classic six volume compendium on aerodynamics. To the general physical science community, Burgers' two most significant contributions were his theory of dislocations and his "one dimensional" model of turbulence known as the Burgers' equation.

I had the pleasure of accompanying Jan Burgers on his first airplane flight, a transatlantic trip from Europe to the U.S. in 1950, after thirty years of teaching aerodynamics. He spent an hour with the pilots and returned to the cabin with great excitement explaining that everything worked just as he had told his students it would.

According to M. Klein[14], in 1918 Ehrenfest wrote Einstein about a new passion: "In the last few months I have become completely fascinated by some very definite problems of theoretical economics and I have been reading deeply and thinking much about this whole complex of questions." The passion passed without a trace of publication. However, a few years later when his student Jan Tinbergen experienced the same passion, Ehrenfest encouraged and guided his transfer from theoretical physics to economics with the result that Tinbergen became the first Nobel Laureate recipient in economics.

Of Ehrenfest's later students, G. Uhlenbeck and S. Goudsmit became famous for the discovery of electron spin and H.B.G. Casimir and C.J. Gorter were the inventors of two fluid models of quantum fluids. Gorter eventually became Director of the Kamerlingh Onnes Laboratory and Casimir, Director of the Phillips Research Laboratory.

Ehrenfest's influence as a teacher extended far beyond the intimate circle of his research students. Ehrenfest's lectures were conscientiously attended by practically all students and postdocs in the Kamerlingh Onnes Laboratory which, for several decades, was the world's great center of low-temperature physics. The theoretical level of the experimentalists in the laboratory was raised to an unusually high point through their interaction with Ehrenfest and his students who were always available as theoretical consultants to the laboratory. Ehrenfest's great reputation as a teacher and the laboratory's pre-eminence in low-temperature physics brought many visitors to Leiden and made Leiden a mandatory stopping point on the grand tour of foreign physics visitors in Europe. Thus Ehrenfest's lectures were attended by a very broad audience. In the early days of modern quantum mechanics, Göttingen was a stopping place for young theorists the world over. Poor Max Born, while trying to keep his position in making innovative contributions to the growing subject, was swamped with foreign students demanding his attention, which he had no time to give. Many of the foreign visitors became uncertain about their ability to compete as theoreticians in the new physics. If their next stopping place was Leiden, their interaction with Ehrenfest often restored their self confidence and put them back in business. Two exceedingly important cases were those of Enrico Fermi, who stated that he found his self confidence in his Leiden period through his contact with Ehrenfest and that his was the most valuable thing he acquired in his studies abroad[16]; and Oppenheimer, who made similar statements.

In 1927, the Chairman of the Physics Department at the University of Michigan asked Ehrenfest to recommend a theoretical physicist as a candidate for a vacant professorship. Ehrenfest responded by recommending two candidates, Uhlenbeck and Goudsmit, as potential instructors, who would be available as two-for-the-price-of-one professor. His proposal was accepted and a new branch of the Vienna Family Tree sprouted in Ann Arbor, Michigan. The two young instructors, who eventually became professors at the price of two, produced many distinguished students. Ted Berlin, whose name appears on the tree, actually received his Ph.D. under the direction of Fajens in the Chemistry Department of Michigan, but his true theoretical mentor was George Uhlenbeck.

One of Sam Goudsmit's students was Ta You Wu who submitted a thesis on molecular spectroscopy, which eventually led to the publishing of an important book on the spectra of diatomic molecules. Upon returning to China, Ta You Wu fathered the brilliant Chinese branch of the Vienna Family Tree. Wu became professor at the National University of Peking which, upon the Japanese invasion of China, fled to Chang-Sa and then to Jun-Ming, with the professors and students remaining a hop and a skip ahead of the Japanese army. Many of the famous and imaginative Chinese physicists who were students of Wu obtained their education on the run. These included: T.D. Lee, C.N. Yang, C.C. Lin (the hydrodynamist), and T.T. Wu of Ising fame. K. Huang (of Born and Huang's "Dynamical Theory of Crystal Lattices") elected to remain in China where he became an important scientific leader, who like many of his colleagues suffered during the Red Guard regime, but he is now participating actively under the new administration.

I have told my story and I must confess that not a drop of Wienerblutt flows in my veins nor have any of my teachers been leaves on the Vienna Family Tree. However, my life has been very much enriched by my many friends and colleagues who are leaves identified or not identified on the tree in Figure 1.

REFERENCES

1. A.H. Sturdevant, *A History of Genetics* (Harper & Row, New York, USA, 1965), p. 9.
2. W. Böhm, *Dictionary of Scientific Biography* (C.C. Gillespie, ed., Scribners, 1978), Vol. VII, p. 507.
3. S. Zweig, *The World of Yesterday* (Viking Press, New York, USA, 1943), p. 96.
4. S.B. Brush, *The Kind of Motion We Call Heat* (North-Holland, Amsterdam, The Netherlands, 1976), Vol. 1, pp. 245, 277.
5. K.F. Herzfeld, *Questions in Statistical Mechanics: Some Reactionary Viewpoints* (Coral Gables, Florida, USA, Center for Theoretical Studies Report No. CT5-HS-71-1), p. 1.
6. E. Broda, *Ludwig Boltzmann* (F. Deuticke, Vienna, Austria, 1953).
7. F. Urban, Zum gedenken and Ludwig Boltzmann, Verlandlurgen der deutschen phys. Ges. *81*, 818 (1969).
8. K.F. Herzfeld, *Kinetische Theorie der Wärme* (Braunschweig, 1925), p. 425.
9. W. Natanson, (ed.), *Pisma Mariana Smoluchawskiego (Smoluchowski's Collected Works)*, (Cracow 1924-28), Biographical Sketch in Vol. I.
10. E. Schrödinger, *Les. Prix Nobel an 1933* (Imprimerie Royale, Stockholm, Sweden, 1935), p. 86.
11. F. Bloch, Physics Today (Dec. 1976), p. 99.
12. P.A. Hanle, *Erwin Schrödinger's Statistical Mechanics, 1912-25* (Yale Ph.D. Thesis, 1975).
13. J.A. Wheeler, Physics Today (Jan. 1979), p. 99.
14. M.J. Klein, *Paul Ehrenfest* (North-Holland, Amsterdam, The Netherlands, 1970).
15. M.J. Klein, *Dictionary of Scientific Biography* (C.C. Gillespie, ed., Scribners, 1978), Vol. V., p. 293.
16. G.E. Uhlenbeck, Amer. J. of Phys. *24*, 431 (1956).
17. J.M. Burgers, Ann. Rev. of Fluid Mechanics 7 (1), (1975).

THE PEARSON RANDOM WALK

James E. Kiefer and George H. Weiss
National Institutes of Health, Bethesda, MD 20205

ABSTRACT

The Pearson random walk consists of a sequence of straight lines each at a random angle with respect to the preceding one. This article reviews some of the techniques available for approximating the probability density function for the end-to d distances and the projection on a given axis. When there is a known upper bound on the maximum extension (i.e., the sum of the step lengths) one can write a Fourier series for the density function whose rate of convergence increases with increasing number of steps. When the step sizes are not too disparate use of the method of steepest descents leads to an approximation that is good to 0(1/n) (where n is the number of steps) over the entire range. In addition we develop a method for approximating the behavior of the density functions in the neighborhood of maximum extension. In 2-D this leads to an asymptotic series, derived by the use of a Tauberian theorem. In 3-D the technique leads to the generalization of an exact solution derived by Rayleigh and Treloar. Finally, some applications of the Pearson walk are reviewed.

INTRODUCTION

It is by now well-known in the random walk community that the first explicit statement of a random walk problem is contained in a note by Pearson[1] in _Nature_. What is appreciated to a far lesser degree is the extent to which the Pearson walk has found application in the physical and biological sciences. The original formulation of this problem is that of a 2-D random walk in which the step lengths are equal and the angles between successive steps are uniformly distributed in $(0,2\pi)$. While there are few genuinely two-dimensional phenomena, many quantities of physical interest can be represented in terms of complex numbers, i.e., are of the form

$$F = \sum_j f_j e^{i\theta_j} \qquad (1)$$

where the f_j are amplitudes and the θ_j are associated phase angles. Two typical applications of this idea occur in crystallography, in which the scattering amplitudes f_j will generally be known but the phase angles are unknown and may be assumed to be random in $(0,2\pi)$, and in acoustics, in which a scattered signal may be the sum of multiple waves, with either equal or random amplitudes and unknown phases, plus noise. The 3-D Pearson walk has played an important role in polymer physics and was first analyzed in detail by Lord Rayleigh[2] in 1919. In this article we undertake to bring together

0094-243X/84/1090011-22 $3.00 1984 American Institute of Physics

the threads that link these diverse applications, and describe some techniques useful for deriving quantitative information about these random walks as well as describing some of the applications.

EXACT ALGORITHMS

When a set of D-dimensional random vectors $\{\underset{\sim}{r}_i\}$ has the property that the tip of each vector is uniformly distributed on a D-dimensional sphere the sum

$$\underset{\sim}{R} = \underset{\sim}{r}_1 + \underset{\sim}{r}_2 + \ldots + \underset{\sim}{r}_n \tag{2}$$

has a radially symmetric distribution. Consequently the probability density function (to be abbreviated as pdf) of $\underset{\sim}{R}$, will be a function of $R = |\underset{\sim}{R}|$ only. To calculate this function when the $\underset{\sim}{r}_j$ are independent, identically distributed, random variables one uses the characteristic function to write

$$p(R) = \frac{1}{(2\pi)^D} \int_{-\infty}^{\infty} \cdots \int e^{-i\underset{\sim}{\omega}\cdot\underset{\sim}{R}} \prod_{j=1}^{n} \langle e^{i\underset{\sim}{\omega}\cdot\underset{\sim}{r}_j} \rangle \, d^D\underset{\sim}{\omega} \tag{3}$$

In 2-D, for example, when the amplitudes $\{r_j\}$ are fixed and the angle is uniformly distributed over $(0,2\pi)$ one finds

$$\langle e^{i\underset{\sim}{\omega}\cdot\underset{\sim}{r}} \rangle = \frac{1}{2\pi} \int_0^{2\pi} d\theta \int_0^{\infty} \frac{\delta(r-r_j)}{r} e^{i\omega r\cos\theta} \, r dr \tag{4}$$

$$= J_0(\omega r_j)$$

which implies that

$$p(R) = \frac{1}{2\pi} \int_0^{\infty} \omega J_0(\omega R) \prod_{j=1}^{n} J_0(\omega r_j) \, d\omega \tag{5}$$

Notice that the normalized (to 1) pdf of R is

$$g_2(R) = 2\pi p(R) R = R \int_0^{\infty} \omega J_0(\omega R) \prod_{j=1}^{n} J_0(\omega r_j) \, d\omega \tag{6}$$

which follows from a formula first derived by Kluyver[3]. Similarly, one can show that in 3-D,

$$g_3(R) = \frac{2R}{\pi} \int^{\infty} \omega \sin\omega R \prod_{j=1}^{n} \left(\frac{\sin\omega r_j}{\omega r_j} \right) d\omega \tag{7}$$

An analogous formula in terms of Bessel functions can be given in higher dimensions.

When the r_j are different one can find moments of R by differentiating the characteristic functions. This quickly becomes tedious and it then becomes expedient to calculate cumulants and convert these to moments. For example, in 2-D the generating function of R is

$$C(\omega) = \prod_{j=0}^{n} J_0(\omega r_j) \tag{8}$$

so that the cumulant generating function

$$K(\omega) = \ln C(\omega) = \sum_j \ln J_0(\omega r_j) \tag{9}$$

However, the Bessel function can be represented as an infinite product, from which it follows that

$$\begin{aligned} K(\omega) &= \sum_j \sum_{s=1}^{\infty} \ln\left(1 - \frac{\omega^2 r_j^2}{\lambda_s^2}\right) \\ &= -\sum_{m=1}^{\infty} \frac{\omega^{2m}}{m} \sigma^{(m)} \beta_m \end{aligned} \tag{10}$$

where

$$\beta_m = \sum_j r_j^{2m}, \quad \sigma^{(m)} = \sum_{s=1}^{\infty} \frac{1}{\lambda_s^{2m}} \tag{11}$$

and λ_s is the s'th root of $J_o(\lambda) = 0$. The sums that define $\sigma^{(m)}$ were first investigated by Rayleigh[5,6]. The first few of them are found to be

$$\begin{aligned} &\sigma^{(1)} = \frac{1}{4}, \quad \sigma^{(2)} = \frac{1}{32}, \quad \sigma^{(3)} = \frac{1}{192}, \\ &\sigma^{(4)} = \frac{11}{12,288}, \quad \sigma^{(5)} = \frac{19}{122,880} \end{aligned} \tag{12}$$

The j'th cumulant is given by

$$\kappa_{2m} = (-1)^m \frac{d^{2m}}{d\omega^{2m}} K(\omega) \Big|_{\omega=0}, \quad m = 1,2,3,\ldots \tag{13}$$

which is easily calculated from the series of Eq. (10). The moments, in turn, can be expressed in terms of the cumulants through the formula[7]

$$\frac{\mu_j}{j!} = \sum_{m=0}^{j} \sum_{\{n\}} \left(\frac{\kappa_{j_1}}{j_1!}\right)^{n_1} \left(\frac{\kappa_{j_2}}{j_2!}\right)^{n_2} \cdots \left(\frac{\kappa_{j_m}}{j_m!}\right)^{n_m} \cdot \frac{\square\ 1}{n_1!n_2!\ldots n_m!} \tag{14}$$

where the sum over $\{n\}$ is over all non negative n's that satisfy

$$j_1n_1 + j_2n_2 + \ldots + j_mn_m = j. \tag{15}$$

In 3-D one can find a similar series for $K(\omega)$ by using the infinite product for sin x/x. This leads to

$$K(\omega) = \sum_j \ell n \left(\frac{\sin \omega r_j}{\omega r_j}\right) = -\frac{1}{2}\sum_{m=1}^{\infty} \frac{4^m \beta_m}{m(2m)!} |B_{2m}| \omega^{2m} \tag{16}$$

where the B's are Bernouilli numbers[8].

There does not exist an expression in closed form, other than the integral shown and the Fourier series representation to be discussed, for $g_2(R)$. An explicit formula can be written for $g_3(R)$ as was first shown by Rayleigh[2], and later in more generality by Treloar[9], for the 3-D equal step size case. We derive this representation later, but it is nearly useless as a computational tool because, when n is large, it involves small differences of large numbers. However, Barakat[10] has developed an alternative representation that is well suited for calculation when n is large, provided that the maximum extension, to be denoted by ρ,

$$\rho = r_1 + r_2 + \ldots + r_n \tag{17}$$

is bounded. Barakat's derivation is for the case of equal step sizes, but this restriction is not a necessary one. The resulting representations are expressed as Fourier series, the convergence of which improve with increasing n. Barakat starts from the observation that $g(R) = 0$ for $R > \rho$ (in any number of dimensions), so that if one restricts attention to the interval $(0,\rho)$ a Fourier series can be used to represent $g(R)$. The type of Fourier series will depend on the number of dimensions as will shortly be seen.

In 3-D, for example, one starts by writing

$$G_3(R) = \frac{g_3(R)}{R} = \sum_{m=1}^{\infty} a_m \sin\left(\frac{\pi m R}{\rho}\right) \tag{18}$$

so that

$$\begin{aligned} a_m &= \frac{2}{\rho}\int_0^{\rho} G_3(R) \sin\left(\frac{\pi m R}{\rho}\right) dR \\ &= \frac{2}{\rho}\int_0^{\infty} G_3(R) \sin\left(\frac{\pi m R}{\rho}\right) dR \end{aligned} \tag{19}$$

The second line follows from the first since $G_3(R) = 0$ when $R > \rho$. To evaluate the integral defining a_m we substitute the integral representation of Eq. (7) into this last equation, and interchange orders of integration. In this way we find

$$a_m = \frac{4}{\pi\rho} \int_0^\infty \omega\phi(\omega)d\omega \int_0^\infty \sin \omega R \sin(\frac{\pi m R}{\rho}) dR \qquad (20)$$

where we have set

$$\phi(\omega) = \prod_{j=1}^{n} \left(\frac{\sin\omega r_j}{\omega r_j}\right) \qquad (21)$$

The integral over R in Eq. (20) is readily evaluated, leading to the result

$$\begin{aligned} a_m &= \frac{2}{\rho} \int_0^\infty \omega\phi(\omega)\delta(\omega - \frac{\pi m}{\rho}) d\omega \\ &= 2 \frac{\pi m}{\rho^2} \phi(\frac{\pi m}{\rho}) \end{aligned} \qquad (22)$$

Thus, the coefficients of the Fourier series in Eq. (18) are simply related to the characteristic function in Eq. (21). Notice that as n increases $\phi(\omega)$ decreases for all $\omega \neq 0$ so that the convergence of the Fourier series improves, in contrast to other exact expressions which simply become increasingly complicated. A similar technique can be used to show[11] that in 2-D

$$g_2(R) = \frac{2R}{\rho^2} \sum_{n=0}^{\infty} \frac{\phi_2(\gamma_n/\rho)}{J_1^2(\gamma_n)} J_0(\frac{\gamma_n R}{\rho}) \quad 0 < R < \rho \qquad (23)$$

where γ_n is the n'th root of $J_o(\gamma_n) = 0$ and

$$\phi_2(x) = \prod_{j=1}^{n} J_0(xr_j) . \qquad (24)$$

One can derive similar expansions in higher dimensions by expanding in a series of higher order Bessel functions but since no applications are known for this generalization, we omit the development of such formulae. It is interesting to note, by comparing the expansions in Eqs. (18) and (23), that the convergence of the 3-dimensional series is faster than that of the 2-dimensional series. (This property of the convergence also carries over to higher dimensions.) Equations (18) and (23) constitute much more convenient exact representations of the $g_D(R)$ than those otherwise available, i.e., they are a more sensible starting point than the use of numerical integration on Eqs. (6) or (7), and easier to calculate, for large n than Rayleigh's[2] or Treloar's[9] exact representations when these are applicable.

The functions $g_D(R)$, whose calculation we have just described, are density functions for the end-to-end distance. Another quantity whose study is interesting in several applications is the projection of the random walk onto an arbitrary axis. This quantity was first discussed (not quite correctly) by Kuhn and Grün[12] in three dimensions and later by a number of authors in the contexts of sound scattering in the ocean and of crystallography. The technique for deriving a pdf for the projection of Kuhn and Grün is an interesting one and leads to an approximation developed by them which is valid for large n. However, their argument is somewhat ad hoc, making it very difficult to assess the validity of the resulting approximation. A more satisfying approach to the problem is to start from an exact expression in terms of characteristic functions and use the method of steepest descents to evaluate the resulting integral. Consider first the two-dimensional case. Let the angle between an arbitrary step and the given axis be θ. The projection of this step onto the designated axis is then $r \cos \theta$ which implies that the characteristic function for the projection is

$$C(\omega) = \frac{1}{2\pi} \int_{-\pi}^{\pi} e^{i\omega r\cos\theta} \, d\theta = J_0(\omega r). \tag{25}$$

Consequently the pdf of the sum of projections in 2-D is

$$h_2(x) = \frac{1}{2\pi} \int_{-\infty}^{\infty} e^{-i\omega x} \prod_{j=1}^{n} J_0(\omega r_j) \, d\omega \tag{26}$$

Similarly one can show that in 3-D of the sum of projections onto an arbitrary axis is

$$h_3(x) = \frac{1}{2\pi} \int_{-\infty}^{\infty} e^{-i\omega x} \prod_{j=1}^{n} \left(\frac{\sin\omega r_j}{\omega r_j} \right) d\omega \tag{27}$$

A Fourier series representation similar to that shown in Eq. (18) can be developed for the h's.

When the r's are not fixed but are themselves random variables the product terms appearing in the last three equations must be replaced by appropriate averages. Unless the step lengths are uniformly bounded, one does not have the convenience of a generalized Fourier series representation because of the impossibility of defining the maximum extension ρ.

So far we have been interested in the statistical properties of configuration parameters, i.e., the end-to-end distance, and the projection on an arbitrary axis. Another class of problems arises in various areas in which one considers the properties of a time-dependent sum

$$z(t) = \sum_{j=1}^{n} r_j \exp[i(\omega_j t + \alpha_j)] = \sum_{j=1}^{n} r_j \exp(i\theta_j) \tag{28}$$

where the parameters r_j, ω_j, and α_j may be deterministic quantities, and in which one inquires about certain asymptotic properties of $x(t) = \mathrm{Re}\, z(t)$ as $t \to \infty$. This class of problems arises in the astronomical theory of orbits, as has been pointed out by Montroll[13] whose valuable article is partially summarized in the following paragraphs. The question of interest is whether and under what circumstances the phase of $z(t)$ is asymptotically of the form $\phi(t) = it\Omega + O(t)$ where Ω is a constant frequency. The solution to the problem can be shown to imply that a set of planets in concentric coplanar orbits that are in conjunction at $t=0$, will be arbitrarily close to conjunction at an infinite set of later times. More generally this property holds no matter what the initial condition provided that the condition of Eq. (29) is valid. The phase problem was first solved by Weyl[14] in 1938. The solution to the problem says that $\phi(t)$ approaches the desired functional form when the ω_j's are incommensurate, i.e., when

$$\sum_j m_j \omega_j \neq 0 \tag{29}$$

for all sets of integers $\{m_i\}$. This condition is required for the validity of the equidistribution theorem first proved by Kronecker and Weyl[15]. The theorem states that if α is irrational then the set of numbers $n\alpha-[n\alpha]$, where "[ ]" means integer part, is uniformly distributed over the unit interval. This theorem has the implication that, when $\theta = (\theta_1, \theta_2, \ldots, \theta_n)$, the θ_j being defined in Eq. (28), the vector θ becomes uniformly distributed over the n torus in which each θ_j goes from 0 to 2π. Consequently it can be shown to imply for any Riemann integrable function, that as $T\to\infty$,

$$\frac{1}{2T}\int_{-T}^{T} f(\theta(t))dt = \frac{1}{(2\pi)^n}\int_{-\pi}^{\pi}\cdots\int f(\theta_1,\theta_2,\ldots,\theta_n)d^n\theta \tag{30}$$

When Eq. (29) is violated the curve $\theta(t)$ will be periodic rather than equidistributed. Equation (30) provides the link between the purely deterministic problem posed earlier and the Pearson random walk. As an example, if in Eq. (30) we set $z(t) = x(t) + iy(t)$ it then follows that the joint characteristic function of $(x(t),y(t))$ is

$$\langle e^{i(ax(t)+by(t))}\rangle = \lim_{T\to\infty}\frac{1}{2T}\int_{-T}^{T}\exp\{i(ax(t)+by(t))\}dt$$

$$= \frac{1}{(2\pi)^n}\int_{-\pi}^{\pi}\cdots\int \prod_{j=1}^{n}\exp\{ir_j(a\cos\theta_j+ib\sin\theta_j)\}d^n\theta$$

$$= \prod_{j=1}^{n} J_0(r_j\sqrt{\alpha^2+\beta^2}) = \prod_{j=1}^{n} J_0(r_j\zeta) \tag{31}$$

which is just the characteristic function in Eq. (8). Thus a description of the deterministic process $z(t)$ can be described in terms of Pearson random walks. Montroll has shown that one can find the parameter Ω that appears in the asymptotic formula for the phase $\phi(t) = it\Omega + O(t)$.

A similar technique can be used to determine the asymptotic average number of times that $x(t) = \mathrm{Re}\, z(t)$ attains a target value, call it q. That is to say, one wants to determine the average number of times that $x(t) = q$ in the interval $0 < t < T$, for T large. A rather elegant analysis by Mazur and Montroll[16] suffices to demonstrate that this number is

$$N_T(q) = \frac{T}{2\pi^2} \int \cdots \int_{-\infty}^{\infty} \frac{1}{\eta^2} e^{i\beta q} \left\{ \prod_{j=1}^{n} J_0(\beta r_j) - \prod_{j=1}^{n} J_0\left(r_j \sqrt{\beta^2 + \omega_j^2 \eta^2}\right) \right\} d\beta\, d\eta \tag{32}$$

as was first shown by Kac[17]. Various asymptotic properties of this formula can be found. For example, when n, the number of terms contained in the definition of x(t), is large, one can find an expression for $N_T(bn^{\frac{1}{2}})$, when b is a constant. Define the constants

$$\overline{r^2} = \frac{1}{n} \sum_{j=1}^{n} r_j^2 \qquad \overline{\omega^2} = \frac{1}{n} \sum_{j=1}^{n} r_j^2 \omega_j^2 \tag{33}$$

Then, under some further minor restrictions it can be shown that

$$N_T(b\sqrt{n}) \sim \frac{(\overline{\omega^2})^{1/2}}{\pi (\overline{r^2})^{1/2}} \exp[-(b/\overline{r^2})] \tag{34}$$

An approximation is also available when q is close to maximum extension. When $r_1 = r_2 = \ldots = r_n$ then

$$N_T(q) \sim \frac{(n\overline{\omega^2})^{1/2}}{\pi \Gamma(n/2 + 1/2)} \left(\frac{n-q}{2\pi}\right)^{(n-1)/2} \tag{35}$$

These ideas have been used in a discussion of Poincaré cycles for the autocorrelation function of coupled harmonic oscillators[16].

APPROXIMATIONS

The expressions in Eqs. (6) and (7), are clearly awkward to deal with, even by numerical integration. While Fourier series representations are convenient to evaluate with a computer they are not otherwise revealing. Hence one wants a more useful approximation to the integrals defining the pdf's, valid in the limit of a large number of steps. The first candidate for such an approximation is, of course, the Gaussian limit that results from the application of the central limit theorem. This, however, has the deficiency of leading to an unbounded relative error for random walks with finite maximum extension. A physical consequence of having a bounded maximum extension is that if one thinks of the random walk as a model for a polymer chain, the average restoring force when a single chain is stretched, should become infinite as R, the end-to-end distance, approaches the maximum extension ρ. The force can be expressed in terms of the entropy S, and temperature T, as $F = -TdS/dR$. The entropy of a single chain is, up to an additive constant,

$$S = k \ln p(R) \tag{36}$$

where $p(R)$ is derivable from $g_3(R)$ in Eq. (7). When $p(R)$ is a Gaussian it is evident that the force is proportional to R with no discontinuity at $R = \rho$. In this section we discuss the approximation obtained by using the method of steepest descents for evaluating the relevant integrals. As a byproduct we will derive a method for approximating various pdf's when the end-to-end distance is close to maximum extension.

There have been several earlier attempts to find useful approximations to the pdf of a Pearson walk[18-24]. These have mainly appeared in the statistical literature, although because of the importance of the problem in the context of noise it has also been analyzed in engineering journals. None of the results obtained in the investigations just mentioned have produced approximations as accurate as that found by the steepest descents method[25-29]. A useful summary of some of the earlier work, mainly that of statisticians, is found in an article by Johnson[24]. Much of the work discussed by him deals with Gram-Charlier expansions, which is a refinement of the Gaussian approximation. Another variety of approximation suggested by Johnson is that in which the pdf is taken to be

$$P(y) = A\, y^{\alpha-1}(1-y)^{\beta-1} \tag{37}$$

where A is a normalizing constant, α and β are determined from the exact first and second moments, and $y = R^2/(NL)^2$ where N is the number of steps and L is the step length. While the result just mentioned has the advantage of being in a convenient form, it is not particularly accurate, hence we restrict our attention to the steepest descents approximation.

What we will call the Gaussian, or central-limit approximation can be obtained by noting that a term such as

$$\prod_{j=1}^{n} J_0(\omega r_j)$$

attains its maximum value at $\omega=0$. Consequently, an approximation to $g_2(R)$ in Eq. (6), is found by expanding the product to order ω^2 around $\omega=0$:

$$\prod_{j=1}^{n} J_0(\omega r_j) = \exp\left(\sum_{j=1}^{n} \ln J_0(\omega r_j)\right) \sim \exp \sum_{j=1}^{n} \ln\left(1-\frac{\omega^2 r_j^2}{4}\right)$$

$$\sim \exp\left(-\frac{\omega^2}{4} \sum_{j=1}^{n} r_j^2\right) \tag{38}$$

Inserting this expression into Eq. (6) we have

$$g_2(R) \sim R \int_0^{\infty} \omega J_0(\omega R) \exp\left(-\frac{\omega^2}{4} \sum_{j=1}^{n} r_j^2\right) d\omega \tag{39}$$

$$= \frac{2R}{\mu_2} \exp\left(-\frac{R^2}{\mu_2}\right)$$

where

$$\mu_2 = \sum_{j=1}^{n} r_j^2$$

In the expression in Eq. (39) there are no restrictions on R other than it be non-negative, whereas $g_2(R)$ should be equal to zero for $R > \rho$. Similarly we find that in 3-D

$$g_3(R) \sim 3\left(\frac{6}{\pi \mu_2^3}\right)^{1/2} \exp\left(-\frac{3R^2}{2\mu_2}\right) \tag{40}$$

where μ_2 has the same meaning as before. Similar formulae can be written for $h_2(x)$ and $h_3(x)$ by using the same approach. One can show that (for $r_1 = r_2 = \ldots = r_n = r$) the next term in the approximation for the g's or h's is of order $n^{-1/2}$ for large n. It is less well known that the use of the method of steepest descents leads to an error that is $O(1/n)$, as was first shown by Daniels[25]. Several authors have applied the steepest descent technique for the three-dimensional Pearson random walk[26-28] (or freely-jointed chain, as it is known in the polymer literature) with equal step sizes. The steepest descent approximation not only has the advantage of better convergence with increasing n, but it also leads to functional forms that have compact support whenever the maximum extension is finite. Domb and Offenbacher[30] have recently calculated asymptotic properties of several random walks using this method.

In order to see how this method works, let us first consider the calculation of an approximation to $g_3(R)$ whose exact representation is given in Eq. (7). As a first step, put $\omega = iv$, which transforms Eq. (7) into

$$g_3(R) = \frac{R}{\pi i} \int_{-i\infty}^{i\infty} v e^{-vR} \prod_{j=1}^{n} \left(\frac{\sinh vr_j}{vr_j}\right) dv \tag{41}$$

The approximation is obtained by writing the product in logarithmic form and expanding around the root, v_o, of

$$\sum_j r_j \; L(v_0 r_j) = R \tag{42}$$

where L(x) is the Langevin function

$$L(x) = \coth x - \frac{1}{x} \tag{43}$$

When the r_j's are all equal Eq. (42) reduces to the known

$$L(v_0 r) = \frac{R}{nr} = \frac{R}{\rho} \tag{44}$$

The resulting approximation to $g_3(R)$ is

$$g_3(R) \sim A v_0 R \exp\Big(\sum_{j=1}^{n} \ln\left(\frac{\sinh v_0 r_j}{v_0 r_j}\right) - v_0 R\Big) / (\mu(v_0))^{1/2} \tag{45}$$

in which

$$\mu(v_0) = \frac{1}{n} \sum_{j=1}^{n} r_j^2 \left(-\operatorname{csch}^2(v_0 r_j) + \frac{1}{(v_0 r_j)^2}\right) \tag{46}$$

The constant A that appears in Eq. (45) is chosen, following Daniel's suggestion[25], so that the normalization condition

$$\int_0^\rho g_3(R)\,dR = 1 \tag{47}$$

is preserved. Numerical comparisons of the values of $g_3(R)$ generated by the steepest descents method were made by Dvořák[27] and Daniels[28] in the case of equal step sizes.

Before presenting our own comparisons that include the effects of unequal step sizes we note that Eq. (41) allows us to derive the generalization of Treloar's[9] exact expression for the pdf in a simple fashion. For this purpose set

$$R = \rho - t = \sum_j r_j - t \tag{48}$$

which transforms Eq. (41) to

$$g_3(r) = \frac{R}{\pi i}\int_{-i\infty}^{i\infty} ve^{vt}\prod_{j=1}^{n}\left(\frac{1-e^{-2vr_j}}{2vr_j}\right)dv \tag{49}$$

$$= \frac{R}{i\pi 2^n r_1 r_2 \ldots r_n}\int_{-i\infty}^{i\infty}\frac{e^{vt}}{v^{n-1}}\prod_{j=1}^{n}(1-e^{-2vr_j})\,dv$$

When $g_3(R)$ is written in this form it is obvious that it is proportional to the inverse Laplace transform of

$$\frac{1}{v^{n-1}}\prod_{j=1}^{n}(1-e^{-2vr_j}) = \frac{1}{v^{n-1}}\Big(1 - \sum_j e^{-2vr_j} + \sum_{i\neq j}\sum e^{-2v(r_i+r_j)} - \sum_{i\neq j\neq k}\sum\sum e^{-2v(r_i+r_j+r_k)} + \ldots\Big) \tag{50}$$

But the inverse transform of $\exp(-\lambda v)/v^{n-1}$ is

$$\frac{1}{(n-2)!}(t-\lambda)^{n-2}H(t-\lambda)$$

where $H(x)$ is the Heaviside step function, $H(x) = 1$, $x>0$, $H(x) = 0$, $x<0$. Hence we have

$$g_3(R) = \frac{R}{2^{n-1}r_1\ldots r_n}\frac{1}{(n-2)!}\Big[t^{n-2} - \sum_j(t-2r_j)^{n-2}H(t-2r_j) + \sum_{i\neq j}\sum(t-2r_i-2r_j)^{n-2}H(t-2r_i-2r_j) - \ldots\Big] \tag{51}$$

This expression reduces to that of Treloar[9] when all of the r_j's are equal.

The pdf of projection length, $h_3(x)$, can be analyzed in the same way, leading to a steepest descent approximation analogous to Eq. (45) (in fact, equal to the formula shown, divided by v_oR) and an exact expansion similar to Eq. (51). A modification of these techniques can be used to derive properties of the joint probability density of the projection along two perpendicular axes, which is a function of $R=(x^2+y^2)^{1/2}$ only. This function is given in Eq. (5). One cannot use the method of steepest descents directly, but $p(R)$ can be related to $h_2(x)$. For this purpose we will use Eq. (26) to provide an expression for the product of Bessel functions:

$$\phi_2(\omega) = \prod_{j=1}^{n} J_0(\omega r_j) = 2\int_0^\infty h_2(x)\cos\omega x\,dx \tag{52}$$

We can therefore express p(R) as

$$p(R) = \frac{1}{\pi}\int_0^\infty h_2(x)\,dx \int_0^\infty \omega J_0(\omega R)\cos\omega x\,d\omega$$
$$= \frac{1}{\pi}\int_0^\infty h_2(x)\left\{\frac{d}{dx}\int_0^\infty J_0(\omega R)\sin\omega x\,d\omega\right\}dx \qquad (53)$$
$$= -\frac{1}{\pi}\int_0^\infty (h_2'(x)/(x^2-R^2)^{1/2})dx$$

where the third line follows from the second by an integration by parts. This may further be simplified by letting $x=R\cosh\theta$, in which case we find

$$p(R) = -\frac{1}{\pi}\int_0^\infty h_2'(R\cosh\theta)d\theta \qquad (54)$$

This expression is easily shown to be properly normalized. Thus, a good approximation to $h_2(x)$ will lead, through Eq. (54), to a good one for p(R). To find this approximation define the function

$$\Omega(t) = \frac{1}{2\pi i}\int_{-i\infty}^{i\infty} e^{vt}\prod_{j=1}^{n} e^{-vr_j} I_0(vr_j)dv \qquad (55)$$

Then $h_2(x)$ can be expressed as

$$h_2(x) = \Omega(\sum_j r_j - x) \qquad (56)$$

This observation allows us[29] to derive an approximation to $h_2(x)$ valid for large n. Let $\bar{v}$ be the solution to

$$\sum_j r_j \frac{I_1(\bar{v}r_j)}{I_0(\bar{v}r_j)} = x \qquad . \qquad (57)$$

The steepest descents approximation to $h_2(x)$ is

$$h_2(x) \approx \frac{1}{(2\pi\sum_j r_j^2\psi(\bar{v}r_j))^{1/2}} e^{-\bar{v}x}\prod_{j=1}^{n} I_0(\bar{v}r_j) \qquad (58)$$

in which

$$\psi(x) = 1 - \frac{1}{x}\frac{I_1(x)}{I_0(x)} - \left(\frac{I_1(x)}{I_0(x)}\right)^2 \qquad (59)$$

Further improvement is attained by using the normalized form

$$h_2(x) \sim h_2(x) / \int_{-\Sigma r_j}^{\Sigma r_j} h_2(y)\,dy \ . \tag{60}$$

When $|x/\sum_j r_j|$ is small, the solution to Eq. (57) can be expanded in a power series

$$\bar{v} \sim \frac{2x}{\Sigma r_j} + \frac{\Sigma r_j^4}{(\sum_j r_j^2)^4} x^3 + \ldots \tag{61}$$

The use of just the first term in this series is equivalent to a Gaussian approximation, corrections to which are obtained by using higher order terms in the series. However, a further bonus is available in the two-dimensional case. One can also find an approximation to $h_2(x)$ when x is close to maximal extension, i.e., when

$$x \sim \Sigma r_j$$

or, equivalently when t in Eq. (55), is small. Since $\Omega(t)$ is an inverse transform, we can use a Tauberian theorem for Laplace transforms[31] to infer the behavior of $h_2(\Sigma r_j - t)$ for small t from the large $|v|$ behavior of the transform

$$\prod_{j=1}^{n} e^{-vr_j} I_0(vr_j)$$

The expansion of $\exp(-x)I_o(x)$ for large x is

$$\exp(-x) I_0(x) \sim \frac{1}{(2\pi x)^{1/2}} \left(1 + \frac{1}{8x} - \frac{9}{128x^2} + \ldots\right) \tag{62}$$

so that

$$h_2(\sum_j r_j - t) \sim \frac{1}{\prod_{j=1}^{n} (2\pi r_j)^{1/2}} \frac{t^{n/2-1}}{\Gamma(n/2)} \Big[1 + \frac{1}{4}(\sum_j r_j^{-1}) \frac{t}{n} + \tag{63}$$

$$+ \frac{1}{32} [(\sum_j r_j^{-1})^2 - 10 \sum_j r_j^{-2}] \frac{t^2}{n(n+2)} + \ldots \Big]$$

valid when t is small. Higher order terms in this expansion can be calculated recursively. If we write

$$\prod_{j=1}^{n} e^{-vr_j} I_0(vr_j) = \frac{1}{\prod_j (2\pi r_j v)^{1/2}} \left(1 + \frac{a_1^{(n)}}{v} + \frac{a_2^{(n)}}{v^2} + \ldots\right) \tag{64}$$

and

$$e^{-vr} I_0(vr) = \frac{1}{\sqrt{2\pi rv}} \left(1 + \frac{b_1}{v} + \frac{b_2}{v^2} + \ldots \right) \tag{65}$$

where $b_1 = 1/(8r)$, $b_2 = -9/(128r^2)$, ..., then

$$a_j^{(n+1)} = \sum_{s=0}^{j} a_s^{(n)} b_{j-s} \tag{66}$$

In writing this formula we have used the convention that $a_o^{(n)} = b_o = 1$. The same analysis in 3-D was found to lead to the generalization of Treloar's exact solution. This technique can also be used to develop end point corrections ($t \sim 0$) in higher dimensions.

How well do these various approximations perform? A comparison of the performance of the Gaussian and steepest descents approximations was first given by Dvořák[27] and Daniels[28], the former for n = 200 or more steps and the latter for n of the order of 4-10. A similar comparison for the two-dimensional pdf was first given by Weiss and Kiefer[29]. We summarize results related to the different approximations. The main results shown are for the pdf's of projections, $h_2(x)$ and $h_3(x)$. In all cases, we have chosen a Pearson walk with five steps. In Fig. 1a we show a plot of $h_2(x)$ for the case (1,1,1,1,1). Figure 1b shows curves of the relative error:

$$E = (h_{exact}(x) - h_{app}(x))/h_{exact}(x) \tag{67}$$

for the steepest descents (SD), Gaussian (G), and end point (EP) approximations for this case. It is easy to see that the steepest descents approximation is uniformly good over the domain of interest, whereas the Gaussian approximation is very poor at the end of the interval. Daniels' estimate of error for the steepest descents method leads to a maximum relative error of the order of 0.1 which is seen to be quite accurate in the present case. The relative error of the end point correction using five terms is essentially equal to zero on the scale shown. In practice, there might conceivably be some difficulty in deciding on the domain of validity of this approximation, but the correction terms will give some indication of this domain. When one takes into account the possible variations of step length, then it can hardly be surprising to note that neither the steepest descents approximation nor the Gaussian approximation is particularly good, because they require large n and rough equality between step sizes, i.e., no single step or small group of steps should predominate. Such a case is illustrated in Fig. 2 for which the step sizes are (1,1,1,1,5). Bimodality of the pdf is to be expected since the step of length 5 predominates. The pdf of the projection on a single axis, of a random vector of step size L is

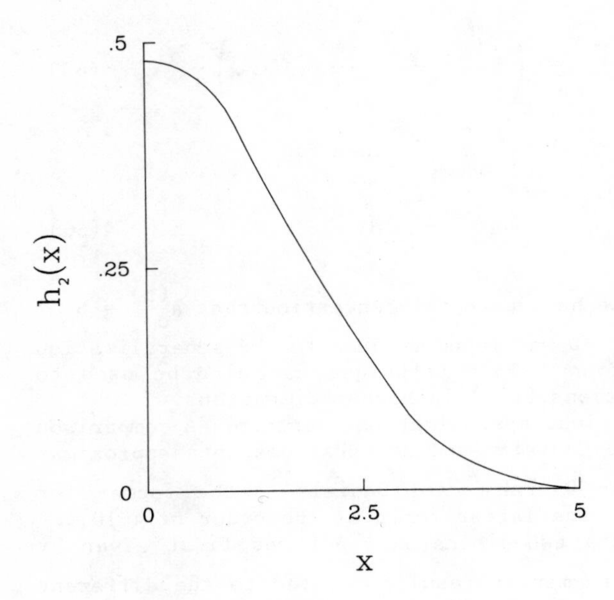

Fig. 1a. The function $h_2(x)$ for a five step walk with step lengths (1,1,1,1,1). The curve is symmetric around x=0 so that only the x>0 section is drawn.

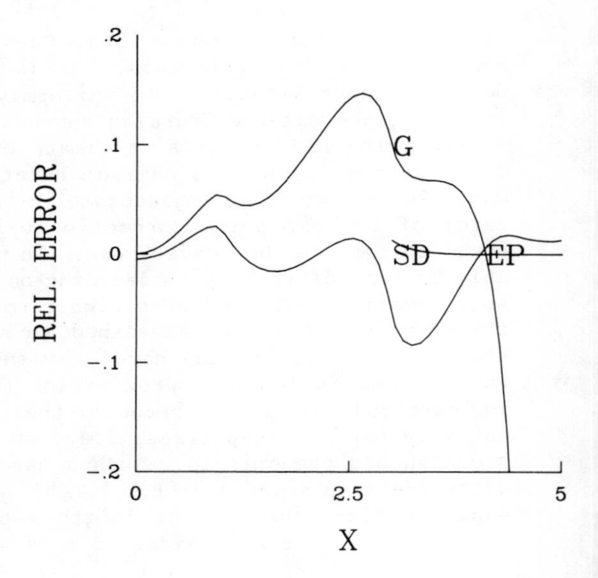

Fig. 1b. Curves of the relative error in approximations to $h_2(x)$. The curve labelled G is the relative error of the Gaussian approximation, SD refers to the steepest descents approximation, and EP refers to the approximation obtained using a Tauberian theorem argument. The EP approximation was obtained using five terms in the power series for t. The coefficients of successive powers were: 1.5(-2), 7.6(-3), 3.5(-3), 1.7(-3), 9.3(-4), 5.4(-4).

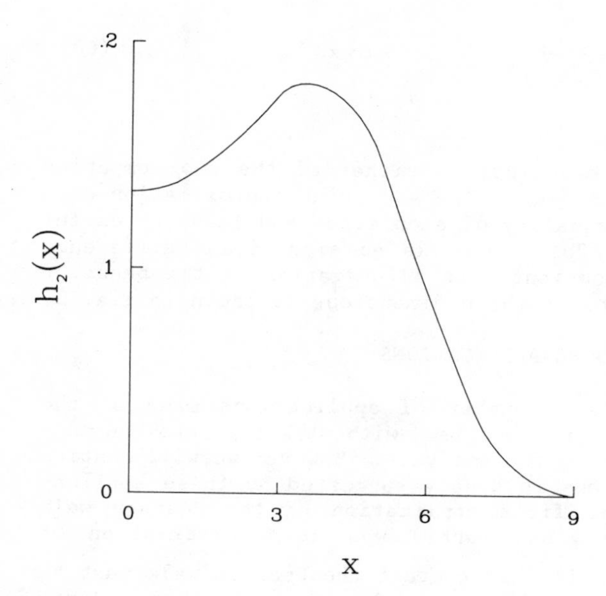

Fig. 2a. The function $h_2(x)$ for a five step walk with step lengths (1,1,1,1,5). Since the step of length 5 is much greater than the others, the curve is bimodal (only the positive half being shown).

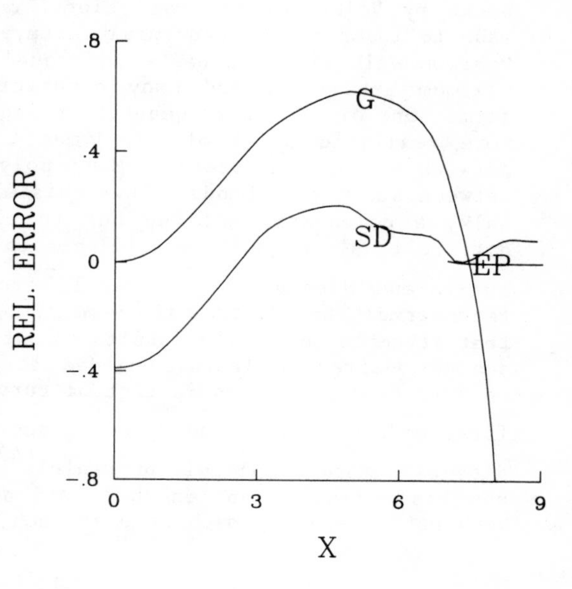

Fig. 2b. Curves of the relative error corresponding to the case of Fig. 2a. The endpoint approximation leads to a substantial error in the neighborhood of x=0, but is otherwise well behaved. The Gaussian approximation is not useful over a wide range because it cannot reproduce bimodal behavior.

$$h_2(x) = \frac{1}{\pi(L^2-x^2)^{1/2}} \quad , \quad -L<x<L \tag{68}$$

$$= 0 \quad \text{otherwise}$$

which is infinite at its endpoints. Neither of the two competing approximations is uniformly good. The end point approximation does not depend critically on equality of step sizes and leads to useful results as shown in Fig. 2b. There is no significant difference between two and three dimensions. An illustration of the behavior of different approximations in three dimensions is shown in Fig. 3.

SOME APPLICATIONS

There have been a large number of applications made of the Pearson walk. We can mention only a few, without going into the many problems or details attending the analysis. However we will mention a few of the generalizations in theory suggested by these applications. Historically, the first application of the Pearson walk (outside of Pearson's original paper) was as a description of polymer configurations[32]. In that context the Pearson walk must be considered as a somewhat unrealistic model without further modification because it fails to take into account known structural information such as fixed bond angles. Nevertheless, the Pearson walk has been useful in a qualitative way, since it does have the important features of a random walk, and can be used to test computational methods that may be useful on more realistic models. Good accounts of the Pearson walk known as the freely jointed chain in the polymer literature) in the context of polymers are to be found in the books by Volkenshtein[32] and Flory[33]. Various attempts have been made to incorporate structural features into the formulation of the Pearson walk since it has a pdf equal to zero beyond the point of maximum extension, and many quantities are calculable in closed form. One of the first quantities requiring modification to make a more realistic model of a polymer is the distribution of angles between successive steps. Many polymers have fixed bond angles between successive bonds. When this element is added to the Pearson walk, a calculation of the pdf in closed form becomes much more difficult, although it is still possible to calculate moments of the end-to-end distance[33]. Several attempts[34-41] have been made to reconstruct the pdf from the moments and there is a large literature that attempts to use the results so obtained to fit measured data. A second desireable feature needed to provide realism in modelling polymer configurations is that of curvature. This modification was first made by Kratky and Porod[42], and its mathematical implications discussed more completely by Daniels[43] and others[44]. In this model one has a fixed step length l, and a fixed bond angle θ, and lets both parameters approach zero in such a way that $2l/\theta^2 \to a$ where $\underline{a}$ is

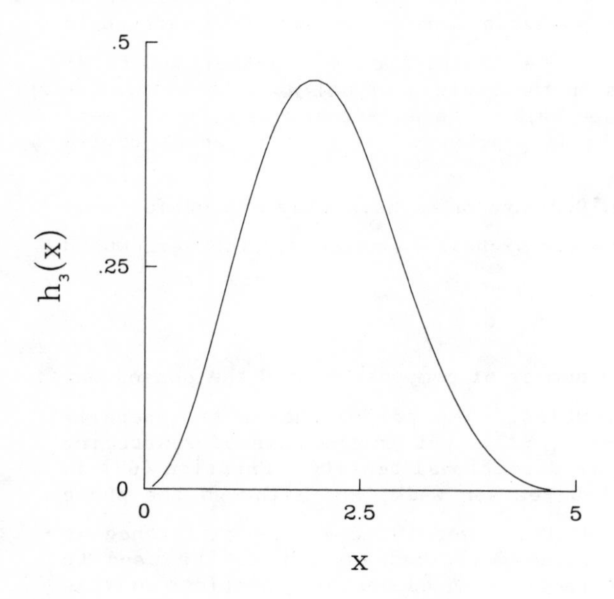

Fig. 3a. A curve of $h_3(x)$ for (1,1,1,1,1). In contrast to the 2-D case, the pdf goes to zero at x=0.

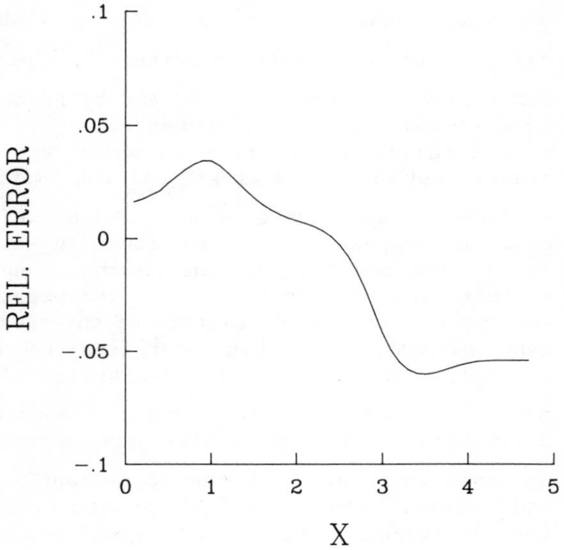

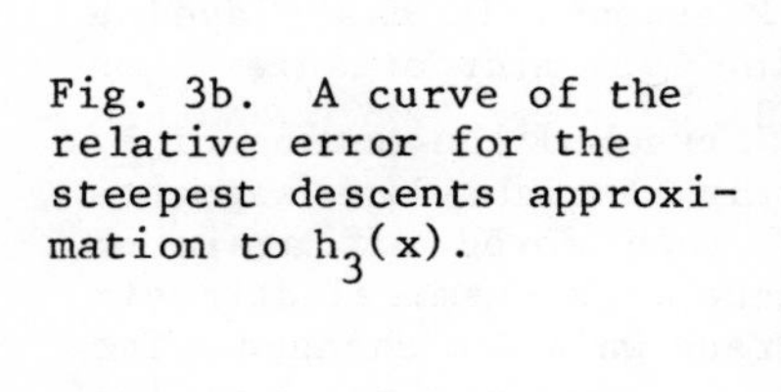

Fig. 3b. A curve of the relative error for the steepest descents approximation to $h_3(x)$.

known as the persistence length. The mathematics of these models is by no means trivial, and the analysis is most conveniently carried in terms of path integrals[45,46]. A description of further modifications of random walk models in the context of polymers leads too far from the notion of the Pearson walk to be undertaken here, but a more extensive description and bibliography is given in a recent review article[47].

A second area in which Pearson walks have played a useful role is in the analysis of narrowband signals in noise[48]. The narrowband signal is usually modelled as

$$S = \Sigma a_j \cos\theta_j \tag{69}$$

where the amplitude, a_j, the number of components, and the phases may all be taken as random variables. The pdf of phases is generally assumed to be uniform in $(-\pi/2, \pi/2)$, but in the case of scattering from a surface may have a more directional density. Equation (69) is clearly a projection of a 2-D Pearson walk, and although the phase density[49] differs from the uniform over $(0, 2\pi)$, the difference is insignificant and the techniques described earlier can be used to analyze statistical properties of S. A number of variations of this model were analyzed by Lyon[50] and Dyer[51] in the context of acoustic scattering in the ocean, by Barakat[48] in the context of laser speckle patterns, and by Barakat and Cole[48] as a paradigm of the interaction of random waves, with application to many areas of engineering. An extension of the Pearson walk required by these applications is that of finding the distribution of S + N where S has been previously defined, and N is Gaussian noise. This problem is analyzed by Rice[52] for a single component in S, and by Barakat and Cole[48] for the case of a random number of components.

A third subject area in which the Pearson walk has played a significant role is that of the motion of microorganisms on surfaces. Experiments by Berg and Brown[53] tracked the motion of _E. coli_ and suggested that they moved in nearly straight line segments for random amounts of time, turned, and then moved off again on another straightline track. In the presence of a chemical attractant the statistical parameters of this Pearson walk are changed. Two types of change have been identified; the time spent moving along the straight line may have a distribution that depends on whether the motion is towards or away from the source of attractant[54], and the distribution of turn angles may depend on the location of the bacteria in relation to the attractant[55]. Both of these responses, and possibly others, exist in nature. A complete theory of such (mainly two-dimensional) random walks would be of considerable experimental value because of the possibility of using the theory to relate observed behavior to more fundamental biological determinants. An early discussion of random walks in biological problems was given by Patlak[56], and two useful reviews of random walks specifically applied to the motion of microorganisms have recently been written by Nossal[57,58]. A further general reference of some interest is a paper by Alt[59].

There are other applications of the Pearson walk, particularly in the area of crystallography[60,61] which have lead to useful tools for the reduction and interpretation of data. These, however, do not appear to raise any new analytical problems and will not be discussed here. The interested reader is directed, for further references, to the monograph by Srinivisan and Parthasarthy[60]. It should be apparent from these examples, and the related theoretical developments that in spite of its age the simple Pearson walk is still alive and well as a source of engaging problems, with application to many subject areas.

REFERENCES

1. K. Pearson, Nature 72, 294 (1905).
2. Lord Rayleigh, Phil. Mag. 37, 321 (1919).
3. J. C. Kluyver, Proc. K. Akad. van Wet. de Amst. 8, 341 (1906).
4. G. N. Watson, A Treatise on Bessel Functions (Cambridge University Press, Cambridge, 1966), 2nd ed.
5. J. W. Strutt, Proc. Lond. Math. Soc. V, 119 (1874).
6. D. H. Lehmer, Math. Tables and Other Aids Comp. 1, 405 (1946).
7. M. G. Kendall, A. Stuart, The Advanced Theory of Statistics, vol. I (C. Griffin, London, 1977), 4th ed.
8. M. Abramowitz, I. Stegun, Handbook of Mathematical Functions (U. S. Government Printing Office, Washington, D. C., 1964).
9. L. R. G. Treloar, Trans. Far. Soc. 42, 77 (1946).
10. R. Barakat, J. Phys. A6, 796 (1973).
11. R. Barakat, Optica Acta 21, 903 (1974).
12. W. Kuhn, F. Grün, Koll. Zeit. 101, 248 (1942).
13. E. W. Montroll, J. Soc. Ind. Appl. Math. 4, 241 (1956).
14. H. Weyl, Am. J. Math. 60, 889 (1938); 61, 143 (1939).
15. G. H. Hardy, E. M. Wright, An Introduction to the Theory of Numbers (Oxford University Press, Oxford, 1960), 4th ed.
16. P. Mazur, E. W. Montroll, J. Math. Phys. 1, 70 (1960).
17. M. Kac, Am. J. Math. 65, 609 (1943).
18. F. Horner, Phil. Mag. 37, 145 (1946).
19. M. Slack, J. Inst. Elec. Eng. 90, 76 (1946).
20. R. D. Lord, Phil. Mag. 39, 66 (1948).
21. J. A. Greenwood, D. Durand, Ann. Math. Stat. 26, 233 (1955).
22. D. Durand, J. A. Greenwood, Ann. Math. Stat. 28, 978 (1957).
23. M. A. Stephens, Biometrika, 49, 463, 547 (1962).
24. N. L. Johnson, Technometrics 8, 303 (1966).
25. H. E. Daniels, Ann. Math. Stat. 25, 631 (1954).
26. H. Yamakawa, Modern Theory of Polymer Solutions (Harper and Row,

San Francisco) 1971.
27. S. Dvořák, J. Phys. A5, 78, 85 (1972).
28. H. E. Daniels, Proc. Sixth Berkeley Symp., vol. 3, 533 (1972).
29. G. H. Weiss, J. E. Kiefer, J. Phys. A (to appear).
30. C. Domb, E. Offenbacher, Am. J. Phys. 46, 49 (1978).
31. G. Doetsch, Theorie und Anwendungen der Laplace Transformation (Dover reprint, New York, 1945).
32. M. V. Volkenshtein, Configurational Statistics of Polymeric Chains (Interscience, New York, 1966).
33. P. J. Flory, Statistical Mechanics of Chain Molecules (John Wiley & Sons, New York, 1969).
34. K. Nagai, J. Chem. Phys. 38, 924 (1963).
35. R. L. Jernigan, P. J. Flory, J. Chem. Phys. 50, 4165, 4178, 4185 (1969).
36. D. Y. Yoon, P. J. Flory, J. Chem. Phys. 61, 5358, 5366 (1974); Polym. 16, 645 (1975).
37. R. L. Jernigan, G. H. Weiss, Polym. Prepr. ACS, Div. Polym. Chem. 14, 273 (1973).
38. M. Fixman, R. Alben, J. Chem. Phys. 58, 1553 (1973).
39. M. Fixman, J. Skolnick, J. Chem. Phys. 65, 1700 (1976).
40. J. Freire, M. Fixman, J. Chem. Phys. 69, 634 (1978).
41. J. Freire, M. M. Rodrigo, J. Chem. Phys. 69, 6376 (1980).
42. O. Kratky, G. Porod, Rec. Trav. Chim. 68, 1106 (1949).
43. H. E. Daniels, Proc. Roy. Soc. Edinb. A63, 290 (1952).
44. J. J. Hermans, R. Ullman, Physica 18, 951 (1952).
45. N. Saito, K. Takahashi, Y. Yunoki, J. Phys. Soc. Jap. 22, 219 (1967).
46. K. F. Freed, Adv. Chem. Phys. 22, 1 (1972).
47. G. H. Weiss, R. J. Rubin, Adv. Chem. Phys. 52, 363 (1982).
48. R. Barakat, J. E. Cole III, J. Sound Vib. 62, 365 (1979).
49. R. Barakat, J. Opt. Soc. Am. 71, 86 (1981).
50. R. H. Lyon, J. Acoust. Soc. Am. 48, 145 (1970).
51. I. Dyer, J. Acoust. Soc. Am. 48, 337 (1970).
52. S. O. Rice, Bell Syst. Tech. J. 24, 46 (1945); 27, 109 (1948).
53. H. C. Berg, D. A. Brown, Nature 239, 500 (1972).
54. R. M. McNab, D. E. Koshland, Jr., Proc. Natl. Acad. Sci. 69, 2509 (1972).
55. R. Nossal, S. H. Zigmond, Biophys. J. 16, 1171 (1976).
56. C. S. Patlak, Bull. Math. Biophys. 15, 311, 431 (1953).
57. R. Nossal, Biological Growth and Spread (W. Jager, H. Rost, P. Tautu, eds., Springer-Verlag, Heidelberg, 1980), p. 410.
58. R. Nossal, J. Stat. Phys. (to appear).
59. W. Alt, Biological Growth and Spread (W. Jager, H. Rost, P. Tautu, eds., Springer-Verlag, Heidelberg, 1980), p. 353.
60. R. Srinivisan, S. Parthasarathy, Some Statistical Applications in X-ray Crystallography (Pergamon Press, London, 1976).
61. A. J. C. Wilson, Technometrics 22, 629 (1980).

POLYMER STATISTICS AND UNIVERSALITY: PRINCIPLES AND APPLICATIONS OF CLUSTER RENORMALIZATION*

Fereydoon Family
Department of Physics, Emory University
Atlanta, Georgia 30322

ABSTRACT

Principles and applications of a direct position space renormalization group for lattice models of polymers—called the Cluster Renormalization (CR)—are reviewed in detail. Particular emphasis is placed on the study of crossover phenomena and determination of universality classes in polymer models. The first part is largely a pedagogical description of the scaling concept and the idea of fractal dimensionality for polymers, and discusses the relation between polymer statistics and critical phenomena. The generalized lattice animal model for polymers is then introduced and it is shown that within a grand canonical ensemble polymers can be described as critical objects. This description enables us to apply CR to polymer models. Next, essentials of the CR are presented. Discussion of the applications begins with a single-parameter CR study of the scaling properties of models of linear polymers, branched polymers and polymer networks. Finally various applications of two-parameter CR to crossover phenomena are discussed. These include a study of the crossover from a random walk to a self-avoiding walk, which contains a detailed discussion of a renormalization group approach for random walks, and a discussion of the effects of branches and loops on the universality classes of polymers. It also deals with the effects of screening in solutions of branched polymers and the crossover from lattice animals to percolation.

INTRODUCTION

The recent theoretical discovery [1] that there exists an analogy between polymer statistics and the phase transition problem in condensed matter physics has made an entirely new approach to the study of polymers possible [2]. This new unusual connection implies that many of the modern techniques developed to analyze critical phenomena [3,4] can be applied to polymers [5].

In particular the Wilson renormalization group [6,7] (RG) which has been instrumental in increasing our understanding of thermal phase transitions, has now been developed into an important theoretical and computational tool in the study of the scaling properties of polymers. In recent years both position space RG [8-16] (PSRG) and field-theoretic RG [1,17,18] have been used to calculate the critical exponents for a number of polymer models.

* Supported by grants from Research Corporation, Emory University Research Fund, and by NSF grant no. DMR-82-08051.

However, in studies of thermal phase transitions [7] it has been found that the real merit of RG, in addition to serving as a technique for calculating the critical exponents, lies in its ability to search for all possible types of critical behavior and to describe crossover phenomena. More specifically, the RG method provides a framework for a systematic study of the effects of perturbations which lead to changes in the universality class of a system and to multi-critical points [7].

Investigations of the crossover between universality classes are particularly important in polymer physics [2]. In contrast to thermal phase transitions [3,4] in which the scaling behavior occurs in the limit where $T \to T_c$, in polymer systems the scaling behavior [1,2] (i.e. the critical point) occurs in the limit as the number of monomer units N in a polymer tends to infinity. Therefore an accurate knowledge of the possible types of scaling behavior, i.e. universality classes, in polymers and how they can be changed will be very useful in the interpretation and analysis of experimental results.

In this chapter we describe how a direct PSRG approach [11,19-24] — which we call the Cluster Renormalization (CR) — can be used to study the crossover phenomena and universality classes in models of linear polymer [11,20,22], branched polymer [11,19,22,23] and network polymer [19,21] systems. The direct PSRG has two main advantages over the other RG methods. First, being based on a position space description of the system, it is well suited for studies in physically relevant dimensions of d=1,2,3. Second, the direct PSRG does not rely on the existence of a model Hamiltonian and therefore can be applied to a much wider variety of models.

The organization of this review is as follows: In Section I we present a pedagogical introduction to the scaling concept in polymers, and in Section II we show how this concept is used to treat polymers as fractals. Lattice models of various polymer systems are described in Section III. The generalized lattice animal model and its special cases are introduced in Section IV where it is shown that within a grand canonical ensemble polymers can be described as critical objects. This description enables us to apply the CR to polymers. We present the essentials of CR in Section V. Simple applications of single parameter CR to models of linear polymers, branched polymers and network polymers are discussed in Section VI. Finally in Section VII we present the applications of CR to crossover phenomena. We begin with a discussion of the excluded volume effect in linear polymers in Section VIIA. We first present a renormalization group approach for random walks and then use it to study the crossover from random walks to self-avoiding walks as the excluded volume effect is increased. The effects of branching and loops are discussed in Section VIIB and VIIC, respectively. In Section VIID we discuss the effects of screening in branched polymer solutions and discuss the crossover from lattice animals to percolation.

I. SCALING CONCEPTS IN POLYMERS

Polymers are macromolecules which form by the chemical bonding of many monomeric constituents [5]. The number of reactive chemical bonds, or functionality, of a monomer is important in determining the topological structure of the resulting polymer. If all the monomers are bifunctional, then only linear polymers can form, as shown in Figure 1a. On the other

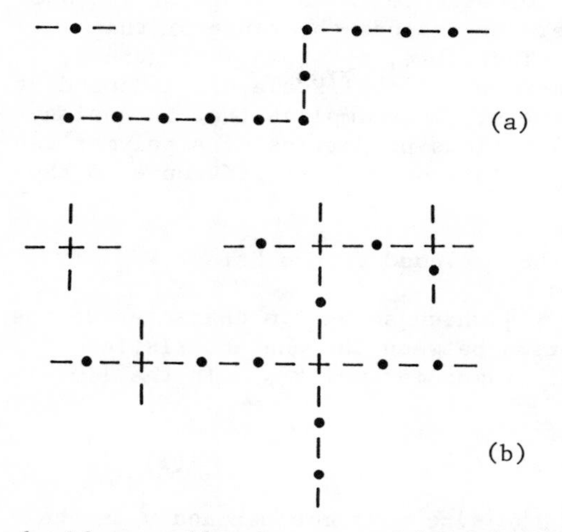

Figure 1:
Topological structure of (a) linear polymers, and (b) branched polymers. Linear polymers are formed by chemical bonding of bifunctional monomers only. The branched polymer shown is composed of both tetrafunctional and bifunctional monomers.

hand, if one adds monomers with higher functionality, a <u>branched polymer</u> of the type shown in Figure 1b may result. In the formation of branched polymers if the reaction between monomers can proceed far enough, a <u>network polymer</u> or gel is formed [2,5]. An important physical characteristic of the gelation process is the existence of a sharp gel point. The gel point is a well defined stage in the polymerization process at which the solution transforms suddenly from a viscous liquid or sol phase to an elastic network or gel phase. Prior to the gel point only finite size polymers can exist. Beyond the gel point there exists a connected polymer network which extends across the entire sample.

There are generally two basic length scales over which polymers are studied [2]. In the first approach one investigates the polymer properties on scales of the monomer size which is typically of the order of a few Å. As shown in Figure 2, in this regime polymers do not look much different from small molecules. On this local scale one learns only about the properties of the monomers in the macromolecule.

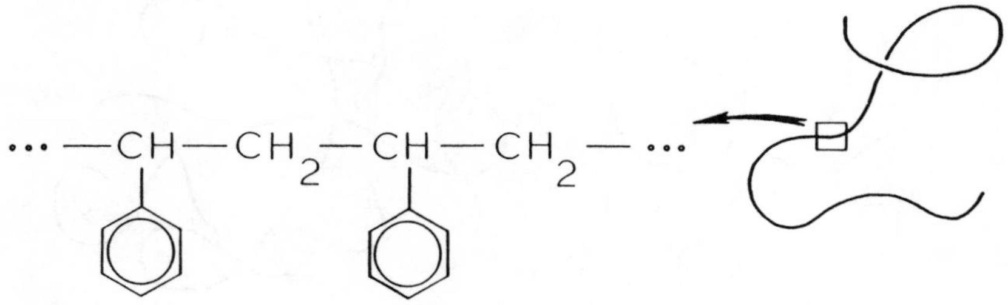

$\cdots - CH - CH_2 - CH - CH_2 - \cdots$

Figure 2: A small segment (a few Å long) of a polystyrene chain viewed under high magnification.

The second approach [2] is to study a polymer on a length scale that is much larger than the monomer size, for example in the range of 100's of Å. In this scale the size of a polymer far exceeds the range of the forces holding the momomers together. Therefore, as shown in Figure 3, the characteristic features of a polymer on a global scale are independent of the detailed properties of its monomers. This implies that in analogy with thermal critical phenomena, [3,4] various properties of a polymer can be described by some universal scaling relations without reference to the details of the interactions.

A. Critical Exponents and the Excluded Volume Effect

One fundamental scaling relation [2] which serves to characterize the conformation of a polymer is the relation between the characteristic length of a polymer, R, and the number of monomer units N. In the limit $N \to \infty$, it is found that [2]

$$R = aN^{\nu} \qquad (1)$$

where a is a constant of the order of the size of a monomer and ν is the exponent describing the power law divergence of this length. The characteristic length R could be any measure of the extent of a polymer. The radius of gyration, the end-to-end length, and the spanning length have all been used to characterize R.

One of the simplest models of a linear polymer is an N-step random walk on a lattice in which each successive step is uncorrelated with all previous ones [2]. This description is very appealing because all properties of the walk can be calculated exactly. For instance, for very large N, the mean end-to-end distance R for an N-step random walk of step size a varies as [2]

$$R = aN^{\frac{1}{2}} \qquad (2)$$

in any spatial dimension d. In comparing (2) with (1) we see that $\nu = 1/2$ for the random walk model of a linear polymer.

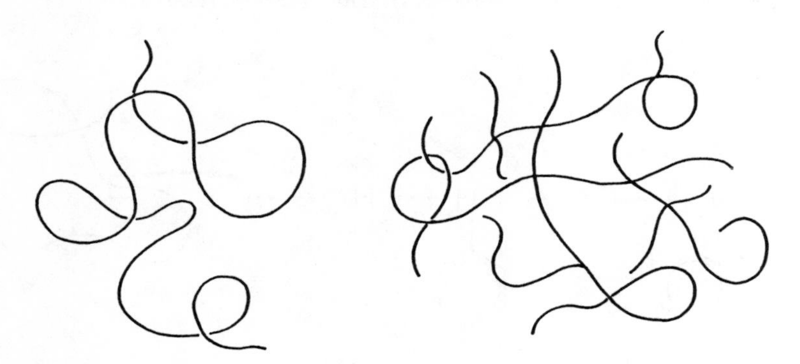

Figure 3: Conformation of linear and branched polymers in solution on a global scale, i.e. length scales much larger than monomer size.

Table 1

Dimensional dependence of the critical exponent ν for linear polymers, branched polymers and polymer networks. The upper critical dimension d_c, above which ν remains constant, is 4, 8 and 6 for linear polymers, branched polymers and network polymers, respectively. Data for branched polymers and network polymers are from [54].

Dimension	Linear polymer	Branched polymer	Network polymer
1	1	1	1
2	3/4 [63]	0.64	48/91
3	0.59 [64,65]	1/2	0.40
$4-\varepsilon$	$\frac{1}{2}+\frac{1}{16}\varepsilon$ [1]	-	-
4	1/2	0.45	0.31
5		0.40	0.27
$6-\varepsilon$		-	$\frac{1}{4}+\frac{5}{168}\varepsilon$
6		0.32	1/4
7		0.28	
$8-\varepsilon$		$\frac{1}{4}+\frac{1}{36}\varepsilon$	
8		1/4	
Flory theory [27-31]	3/(d+2)	5/[2(d+2)]	2/(d+2)

A simple model of a randomly branched polymer is the branched Gaussian chain model, first studied by Zimm and Stockmayer [25]. In this model, a random walk is allowed to have branches, but each branch is described by a random walk. Thus, there are no spatial correlations between the steps in any of the walks or branches. The radius of randomly branched Gaussian chain of N statistical units of length a varies as [25,26]

$$R = aN^{\frac{1}{4}} \tag{3}$$

in all dimensions d. Thus, $\nu = 1/4$ for the randomly branched Gaussian chain model. This result is clearly different from the random walk result and indicates that branching reduces the overall size of the polymer by reducing the exponent ν.

Although the random walk and the branched Gaussian chain are both simple and useful models from a pedagogical point of view they do not describe the statistics of real polymers in solution [2,5]. The main shortcoming of these models is that they ignore the "excluded volume effect," [5] which prohibits two different monomers from occupying the same spatial position. This simple geometrical constraint leads to non-trivial long-range correlations among monomers in a polymer. An immediate consequence of the excluded volume effect is to expand the polymer relative to its size when the excluded volume effect is not present. This implies that the exponent ν is generally larger than 1/2 and 1/4 for linear polymers and branched polymers, respectively.

The values for ν for models of linear polymers, branched polymers and network polymer systems where the excluded volume effect is present are listed in Table 1. These results are for polymer models which we discuss in Section III. Rational numbers indicate (presumably) exact results, numbers with a decimal point are numerical extrapolations with an estimated error typically of the order of one unit in the last digit given. The closed form expressions given in the last line of the Table are the results for ν obtained from the generalized Flory theory [27-31].

B. Upper Critical Dimension

An inspection of Table 1 shows that ν, in analogy with all other critical exponents, [3,4] has two important characteristics: first of all, ν depends strongly on d, and second, there exists a dimension d_c at which ν takes on its value corresponding to the models with no volume exclusion. Above d_c, which is called the upper critical dimension, [4] ν does not change any longer.

Clearly as the dimension of space increases the excluded volume effect becomes less and less important and the decrease of ν with d is obvious. However, the existence of the upper critical dimension is less obvious, but can be determined with the help of a simple argument due to de Gennes [29].

Consider a polymer with a radius R and monomer density N/R^d. The excluded volume effect leads to a repulsive energy per unit volume that is proportional to the mean of the square of the particle density. Let us make a simplifying (usually called a mean-field approximation) and replace the mean-square density by the square of its average value, i.e., let the repulsive energy be $(N/R^d)^2$. Thus, the total repulsive energy is N^2/R^d. We can estimate the maximum repulsive energy by letting $R \sim N^{\nu_o}$, where $\nu_o = 1/2$ for a linear polymer and $\nu_o = 1/4$ for a branched polymer. The repulsive energy is thus proportional to $N^{2-d\nu_o}$. The dimension above which this repulsive energy, i.e., the excluded volume effect, is negligible is $d_c = 2/\nu_o$. This implies that for linear polymers $d_c = 4$ and for branched polymers $d_c = 8$, as verified by field-theory results [1,18].

For a network polymer [29,30] in which an infinite branched polymer coexists with a collection of finite clusters, the repulsive energy is "screened" [29]. The degree of screening is determined by the weight average degree of polymerization of polymers in solution. Assuming this to be proportional to $N^{1/2}$, for a solution of branched molecules [30] at the gel point where a network polymer is present, the screened repulsive energy becomes $N^{3/2-d\nu_o}$. This repulsive energy becomes negligible for dimensions larger than $d_c = 6$, in agreement with field-theory results for the percolation model of gelation [29].

II. FRACTAL DIMENSION OF MACROMOLECULES

Fractal is a word coined by Mandelbrot [32] to describe mathematical sets or concrete objects whose form is extremely irregular or ramified at all length scales. Large macromolecules in solutions have a high degree of structural irregularity. Therefore, on the scales that are large compared to monomer size, but smaller than the mean radius of the chain, polymers may be parameterized by an effective or fractal dimensionality. This concept has received considerable attention recently [23,24,32-36].

The fractal (or Hausdorff-Besicovitch) dimension D of a particular structure embedded in a d-dimensional space is determined in the following way [32]. First, the size of the object is determined by covering it with (hyper-) spheres of diameter a. Let us assume that N is the minimum number of spheres needed to cover an object. If in the limit as $a \to 0$, N varies as

$$N \propto a^D \tag{4}$$

then D is the fractal dimension of the object.

The relation which most directly describes the degree of ramification of a macromolecule is the asymptotic dependence of the mean length of a polymer R on the number of monomer units N in the molecule, i.e., relation (1). In order to determine D [24], we consider a polymer of size N and cover it with hyperspheres of radius a. In the limit as $N \to \infty$, from (1) we have that $N \propto a^{1/\nu} R^{1/\nu}$, and therefore by definition,

$$D = 1/\nu \tag{5}$$

The values of the fractal dimension for models of linear polymers, branched polymers and network polymers can be obtained from the data in Table 1 and are given in Table 2. Note that the fractal dimension of linear polymers is bounded by D = 2 and those of branched polymers by D = 4.

Table 2

The fractal dimension D of linear polymers, branched polymers and netw

polymers in d-dimension. Data from Table 1.

Dimension	Linear Polymer	Branched Polymer	Network Polym
1	1	1	1
2	4/3	1.6	91/48
3	1.7	2	2.5
4-ε	$2 - \frac{1}{4}\varepsilon$	-	-
4	2	2.2	3.2
5		2.5	3.7
6-ε		-	$4 - \frac{10}{21}\varepsilon$
6		3.1	4
7		3.6	
8-ε		$4 - \frac{4}{9}\varepsilon$	
8		4	
Flory theory	(d+2)/3	2(d+2)/5	d+2/2

III. MODELS OF POLYMERS

In the study of polymer statistics it is often useful to treat lattice models in which monomer units are restricted to lie on sites of a lattice, and reactive endgroups at lattice bonds. It is believed that such models display the same scaling behavior as continuum systems [37,38]. Since there exists a one-to-one correspondence between a connected cluster and a linear graph, study of clusters on lattices first arose in the graph theoretic literature [39,40] where they were called lattice animals. Lattice animals represent all the possible "multicellular organisms" that one can "create" out of a fixed number of "cells" on a lattice.

A lattice animal is called a site (bond) animal if its size is determined by the number of lattice sites (bonds) in the cluster. From a purely mathematical problem in combinatorics site lattice animals have been studied under the term of polyominoes [41]. Six examples of lattice animals on a square lattice are shown in Figure 4.

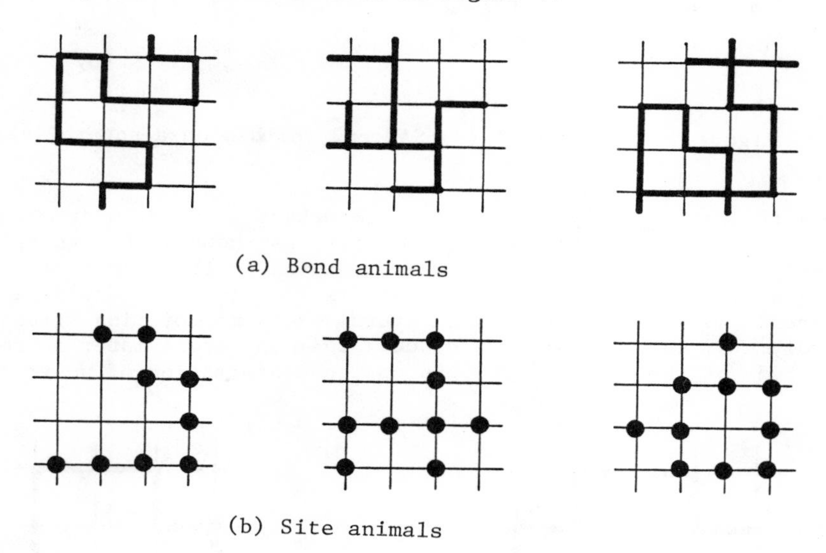

Figure 4: Six examples of lattice animals on a square lattice.

• Self-avoiding Random Walks and Neighbor-avoiding Random Walks: In analogy with polymers, topological structure of a lattice animal is determined by the valence or functionality of its elements. Linear lattice animals are composed exclusively of bifunctional elements, i.e., a site (bond) in a cluster cannot have more than two sites (bonds) attached to it. Such clusters are equivalent to the path of a random walk on a lattice with the constraint that no lattice site is visited more than once. Because of this self avoidance condition linear bond animals are called self-avoiding walks and have been extensively studied as models of linear polymers with excluded volume [42]. Similarly, linear site lattice animals are called neighbor-avoiding walks [43] because in addition to the self-avoiding constraint, the walk may not visit the nearest-neighbor of a previously visited site. Neighbor-avoiding walks have also been studied as a model of linear polymer in a solution [43,20]. Examples of self-avoiding walks and neighbor-avoiding walks on a square lattice are shown in Figure 5. The exponent ν (and the fractal dimension D) of self-avoiding walks and neighbor-avoiding walks are expected to be the same [44]. The values of ν and D for these models are given in the first columns of Tables 1 and 2, respectively.

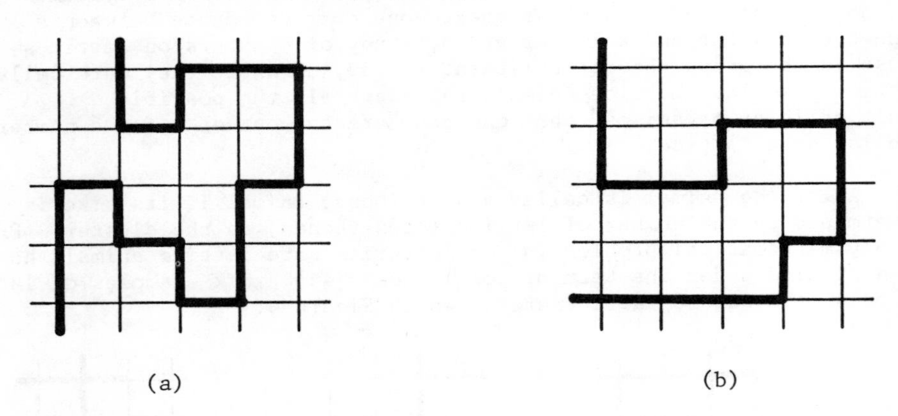

Figure 5: Examples of (a) SAW and (b) NAW on a square lattice.

• Self-avoiding Polygons: Another important class of macromolecules are the ring polymers [2,5]. A ring polymer, as shown in Figure 6, can be viewed as a self-avoiding walk which returns to the origin, i.e., a closed polygon [45]. Study of ring polymers is a subject of considerable interest [46-48] because of the natural occurance of ring structures in biopolymers, such as DNA. We discuss them in detail later in relation to the study of the effects of loops on the conformation of polymers.

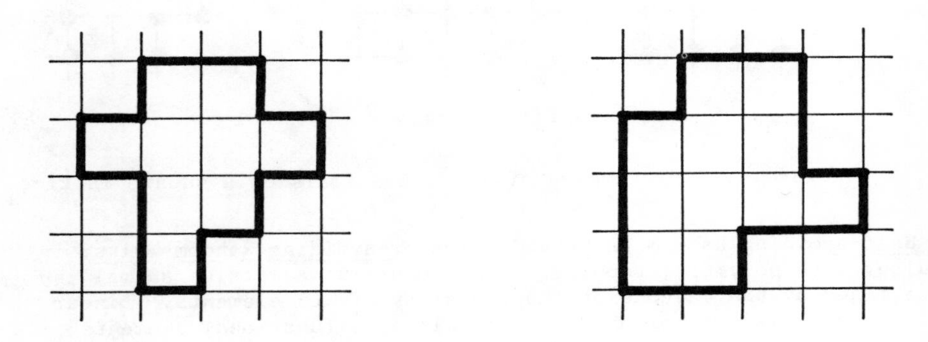

Figure 6: Examples of self-avoiding polygons on a square lattice.

• Branched Lattice Animals and Trees: Non-linear or branched animals are formed when the valence of some of the elements is greater than two. Branched lattice animals are further divided into two types depending on whether the animal has a closed loop or not [11,18]. A loop is a continuous line of elements which begins and ends on the same element. A branched animal without loops is a tree because two elements belonging to it are connected through one and only one path in the animal. Examples of branched animals with and without loops are given in Figure 7.

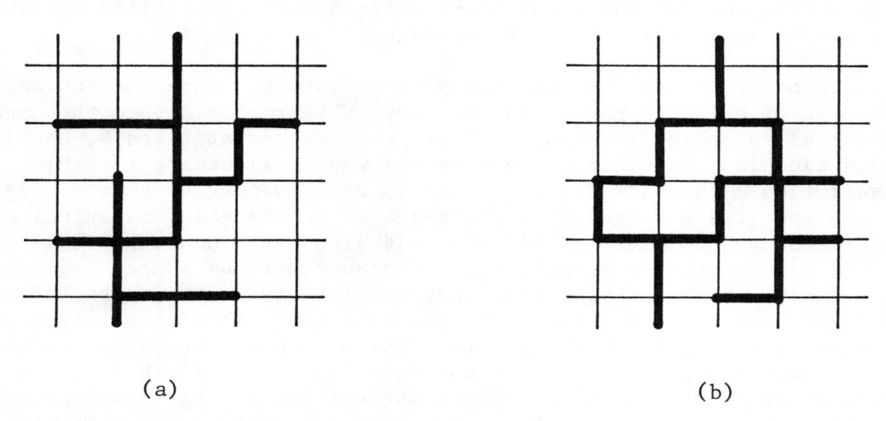

Figure 7: Examples of (a) lattice animals without loops (trees) and (b) lattice animals with loops, on a square lattice.

• Percolation: A simple lattice model of gelation [49,50] is percolation [34,51-53]. Consider a lattice where each site represents one polyfunctional monomer with a functionality that depends on the coordination number of the lattice. Let us assume that between two nearest-neighbor sites a bond is formed randomly with probability p. Clearly at small p we have only small clusters, but when p exceeds a well defined value p_c, we get an infinite network that percolates across the system [34,51]. An example of the formation of a percolating network on a square lattice is shown in Figure 8.

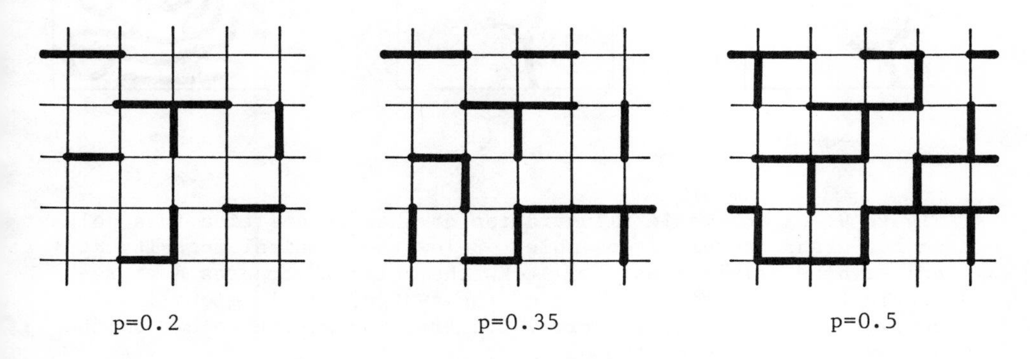

Figure 8: Formation of a percolating network as occupied bond concentration is increased from p=0.2 to 0.5. Bond percolation threshold is at p_c=0.5 for square lattice.

IV. POLYMERS AS CRITICAL OBJECTS: GENERALIZED LATTICE ANIMALS

The scaling relation (1) implies that the $N \to \infty$ limit in a polymer is analogous to the limit $(T-T_c) \to 0$ in a system undergoing a second-order phase transition. This analogy [1,2] implies that, unlike the case of thermal critical phenomena, there is no finite temperature-like parameter corresponding to the critical point of a polymer. Therefore the exponent ν is not defined in the usual sense where a temperature-like variable in the system approaches its critical value.

In order to cast the polymer problem into the more conventional language of critical phenomena, consider the problem not in the usual canonical ensemble, in which polymers are studied at fixed N, but in a grand canonical ensemble where N is allowed to fluctuate [1,2,9]. In this description a fugacity K is assigned to each monomer so that a polymer of size N receives a weight K^N. The averages in the grand canonical ensemble are taken over polymers of all possible sizes at a constant fixed fugacity K. At a fixed fugacity, polymers can assume various shapes and sizes up to a maximum size. Beyond this characteristic size the number of possible configurations dies off exponentially with the size of the polymer. At a critical fugacity K_c there appears a polymer that extends across the entire sample. Below this critical fugacity, i.e., for $K < K_c$, only polymers of finite length can occur whereas for $K > K_c$, a macroscopic fraction of the volume of the system is occupied by a polymer. At the critical point K_c the polymeric system undergoes a phase transition from a state of local connectedness to one of global connectivity. Thus, at K_c a polymer is a self-similar or critical object. This phase transition which may be called a geometrical phase transition, is schematically illustrated in Figure 9.

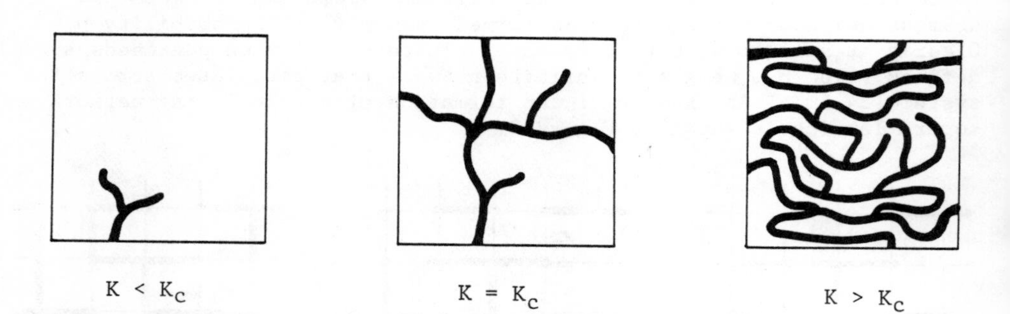

Figure 9: A schematic illustration of phase transition in a polymer in the grand canonical ensemble. Below the critical fugacity K_c only finite polymers exist. At K_c the polymer occupies a microscopic fraction of the volume of the system, but is globally connected. Above K_c, a macroscopic fraction of the volume of the system is occupied by the polymer.

A. Generalized Lattice Animals

Recently the generalized lattice animal model, which is a generalization of percolation and lattice animals, has been introduced [19]. In this model each cluster is characterized by two independent parameters: a fugacity K for each occupied element (site or bond) in the cluster and an independent fugacity q for each perimeter site or bond. The generating function for generalized lattice animals is written as [19,21,54]

$$G(K,q) = \sum_N D_N(q)K^N \qquad (6)$$

where the perimeter polynomials $D_N(q)$ are defined by

$$D_N(q) = \sum_{N_p} C(N,N_p)q^N \qquad (7)$$

Here $C(N,N_p)$ is the number of clusters per site of size N and perimeter size N_p. To study a particular polymer model one specifies the type of configurations that are included in $C(N,N_p)$.

Special cases of (6) that are of particular interest are:

• Self-avoiding Random Walks (SAW) and Neighbor-avoiding Random Walks (NAW): The generating function for SAW (NAW) is defined by [2]

$$G(K) = \sum_N A(N)K^N \qquad (8)$$

where A(N) is the number of different SAW (NAW) of size N that emanate from the origin. In order to obtain this generating function from (6) and (7) we restrict the configurations included in $C(N,N_p)$ to (bond or site) linear animals only (i.e., to SAW or NAW). In addition q is set equal to unity so that all animals of the same size are weighted equally (by K^N) regardless of their perimeter shape.

• Lattice Animals and Trees: The generating function for lattice animals [55,56] (also called random animals) is analogous to G(K) defined in (8), except that for lattice animals there are no restrictions on the type of clusters included in A(N). That is, A(N) is the total number of animals (site [55] or bond [56]) of size N per site. If the configurations included are bond (site) animals then the generating function is that of bond (site) animals. By restricting the configuratins to animals without loops, i.e., trees, one obtains the generating function for tree-like animals [57].

• Random Percolation: The usual random percolation [34,51] generating function is obtained when K = p = 1 - q in (6) and (7). In this case G(K,q) reduces to

$$G(p,q) = \sum_N D_N(q)p^N \qquad (9)$$

where p = 1-q, and $D_N(q)$ is defined in (7).

B. Relation between K and N

The generating function G(K,q) exhibits singular behavior at $K_c(q)$, corresponding to the radius of convergence of the power series in K. It is not difficult to show [2,21] that the path $(K_c-K) \to 0$ corresponds to the polymer scaling limit $N \to \infty$. Consider a function of N, say f(N). Th generating function for f(N) is defined by

$$f(K,q) = \sum_N D_N(q) N^m f(N) K^N / \sum_N D_N(q) N^m K^N \quad (10)$$

For appropriate values of m the sums in both the numerator and the denominator of (10) exhibit singular behavior at $K_c(q)$, the radius of convergence of the series in K. If the asymptotic behavior of f(N) is

$$f(N) \propto N^h \quad (11a)$$

then that of F(K,q) is

$$f(K,q) \propto | K-K_c(q) |^{-h} \quad . \quad (11b)$$

By combining (11a) and (11b) we find

$$N \propto |K-K_c(q)|^{-1} \quad (12)$$

Therefore N and K are conjugate variables [2] and the limit $K \to K_c$ corresponds to the scaling limit as $N \to \infty$. In particular, if f(N) is the mean-squared cluster radius, then F(K,q) is proportional to the square of the correlation length ξ . Since the asymptotic form of R is given by (1), then

$$\xi \propto |K-K_c(q)|^{-\nu(q)} \quad (13)$$

It is the divergence of ξ as $K \to K_c$ that enables us to apply PSRG to the generalized lattice animal model. Although ν and other exponents [54] in the generalized lattice animal model are functions of q, according to universality, they cannot depend continuously on q. In fact, renormalization-group results [19,58] imply that ν is a discontinuous function of the form

$$\nu(q) = \begin{cases} \nu_{animals} & q_c \le q \le 1 \\ \nu_{percolation} & q = q_c = 1-p_c \end{cases} \quad (14)$$

Recently [54] relations between exponents ν and θ, where θ is the cluster number exponent, have been obtained for various models of isotropic and directed polymers. This implies that all exponents in the generalized lattice animal model are discontinuous functions of q.

V. CLUSTER RENORMALIZATION APPROACH FOR POLYMERS

In the position-space version of the Wilson RG [59] one investigates the manner in which the physical properties of a system change upon repeated length rescaling. To carry out the length rescaling we first choose cells of linear dimension b on a regular lattice and under the RG transformation replace them by a cell of linear dimension 1. This rescaling reduces all lengths in the system by a factor of b. A schemat representation of this rescaling procedure is shown in Figure 10.

There are two steps in any PSRG [59]. The first step is to determine the essential physical property of the system under investigation, and the second step is to define a proper rescaling procedure under which this property is invariant.

The essential characteristic property of a polymer is its geometrical connectivity. The reason for this is that the short-range interactions in a polymer are transformed into long-range correlations through the connectivity of the polymer. Therefore, geometrical connectivity is the main characteristic property of a polymer and must remain invariant under an RG transformation. However, in defining an RG transformation for polymers it is important to note that the details of the short-range connectivity are unimportant in comparison to the overall connectivity of the polymer.

In recent years a wide variety of PSRG approaches have been developed for polymers [8-16]. The particular approach that is adopted in the Cluster Renormalization [10-12,19-24] (CR) is the "connectivity rule" first proposed by Reynolds et al. [60] for percolation. In this approach [60] a cell on the original lattice is replaced with a new cell on the rescaled lattice if and only if there exists a cluster in the original cell that connects one side of the cell to the other side. Therefore, under this transformation the global connectivity of the cluster is preserved, but the short-range details are averaged over.

Although we use the general connectivity rule of percolation [60] for polymer models [11,19-24], there is one basic difference between the two systems. In the study of polymers [2] we are generally interested in the dilute limit where there is only a single polymer in the system; whereas in percolation [34,51] there exists a collection of distinct clusters. Therefore in the CR study of the dilute limit of linear [10-12,20,22] and branched polymers [11,19,22-24] the only cell configurations that are considered are those containing a single cluster. Thus, under the CR only a single cluster is renormalized under the lattice rescaling procedure. In addition, the lattice animal generating function is a sum over the statistical weight of all clusters originating at a fixed site on a

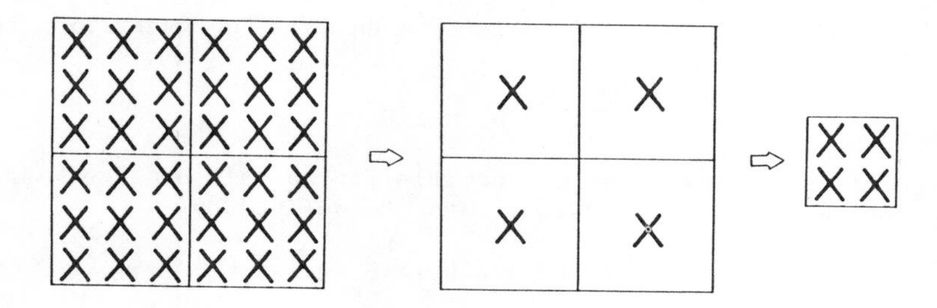

Figure 10: Illustration of length rescaling procedure in position-space renormalization-group approach. Here, the rescaling length b is 3.

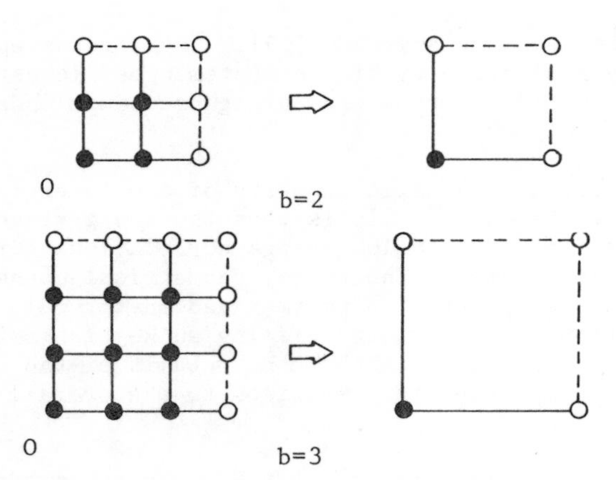

Figure 11: Examples of 2x2 and 3x3 cells on a square lattice, and their renormalization after an RG transformation.

lattice. Thus, in the CR approach only single connected clusters starting from a fixed origin in the cell are renormalized under the RG transformation. Although it is possible in principle to choose any site on the cell as the origin, we choose the lower left hand corner of the cell as the origin, as shown in Figures 11-13.

In general one can define d+1 connectivity rules for a d-dimensional cell. [For a discussion of connectivity rules in two dimensions, see reference 60]. We denote these by r_i [23], where i = 0, 1, ... d. In r_0 a cell is rescaled to a new cell if it contains a connected cluster that extends in any of the d possible directions across the cell; whereas r_i ($i \geq 1$) requires that a cluster span in i specific directions. It is expected that all these rules converge to the same results in the large cell limit [60].

A. Recursion Relations

The RG recursion relation is defined by requiring that the generating function for the "connected" clusters is invariant on the original and rescaled levels. For generalized lattice animals the generating function is [19,21]

$$R_i(K,q;b) = \Sigma\, D_N(q)K^N \tag{15}$$

where $D_N(q)$ are the perimeter polynomials for animals of size N spanning a cell of size b according to rule r_i and are defined by

$$D_N(q) = \Sigma\, C_i(N,N_p)q^N \tag{16}$$

Here $C\ (N,N_p)$ is the number of animals of size N and perimeter size N_p spanning according to rule r_i on a cell of side b. In the <u>site</u> animal problem [23] each cell is replaced by a single site. The RG recursion relation is of the form

$$K'(q) = R_i(K,q;b) \tag{17}$$

where the renormalized fugacity K'(q) is the generating function of a single site at fixed q on the rescaled lattice. Examples of RG transformations for site animals are shown in Figure 12.

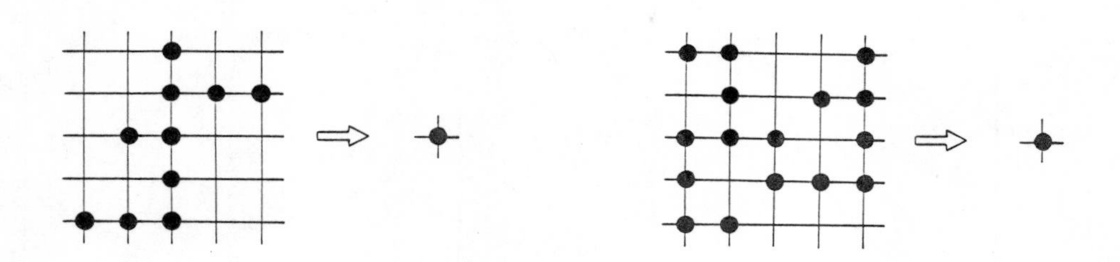

Figure 12: Examples of RG transformations for site animals.

The natural connectivity rule for bond animals [11] is rule r_1. Under this rule a bond animal spanning in a specific direction on a cell is replaced with a single bond in the same direction on the rescaled lattice [11]. The recursion relation for bond animals is then similar in form to (17) except that for bond animals $i = 1$ and K' is the renormalized fugacity of a single bond. Examples of bond animals and their RG transformation is shown in Figure 13.

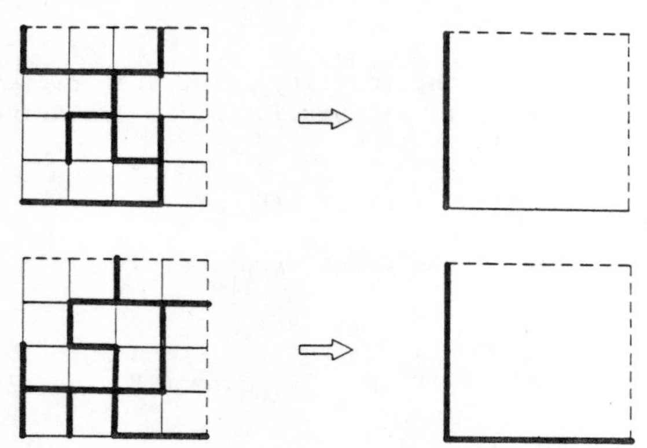

Figure 13: Examples of RG transformations for bond animals.

The general form of the recursion relation (17) is shown in Figure 14. These recursion relations [11,19-24] have three fixed points (at a constant value of q) where the parameter K remains unchanged under the RG transformation, i.e., $K' = K$. There are two trivial fixed points located at $K' = K = 0, \infty$, where the polymer length is zero. The third fixed point at an intermediate value of $K = K^*$ is unstable and corresponds to a

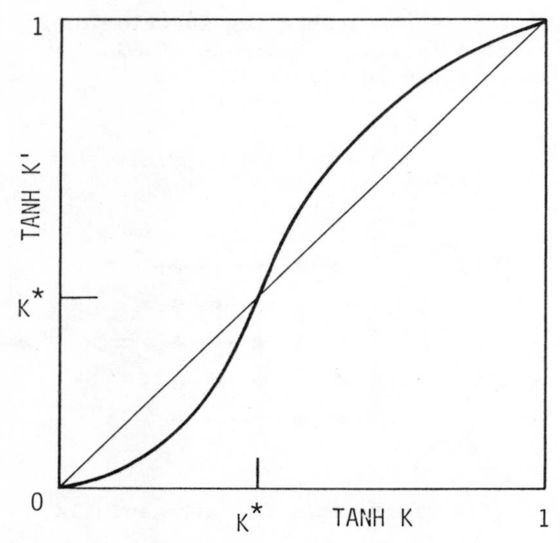

Figure 14: The general form of the recursion relations in CR for polymers. K^* is the critical (unstable) fixed point of the RG transformation.

critical fixed point at which the polymer length is infinite. For all $K < K^*$, polymers are finite in length and their length shrinks upon each length rescaling. For all $K > K^*$ a macroscopic fraction of the lattice is occupied by the polymer [See Figure 9].

B. Critical Exponents and Fractal Dimension

Since the RG transformation rescales the system by a factor of b, the new (rescaled) correlation length is

$$\xi' = \xi/b \tag{18}$$

Assuming that the value of the fugacity K^* at the critical fixed point is an approximation to the critical fugacity K_c we linearize $R_i(K,q;b)$ about K^* and write

$$K' - K_c = \lambda(K - K_c) \tag{19}$$

where λ, the eigenvalue of the transformation, is given by

$$\lambda = \partial K'/\partial K|_{K^*} \tag{20}$$

Since $\xi \propto |K-K_c|^{-\nu}$ and $\xi' \propto |K'-K_c|^{-\nu}$, from (18) and (19) we get

$$\lambda^{\nu} = b \tag{21}$$

which immediately gives the exponent ν by

$$\nu = \ell n b/\ell n \lambda \quad . \tag{22a}$$

Using relations (5) and (22a) we find

$$D = \ell n \lambda/\ell n b \ . \tag{22b}$$

Therefore, using CR one can directly determine the fractal dimension of the clusters in a model. This approach has recently [61] been used to determine the fractal dimension of the clusters in various kinetic models of aggregation.

VI. APPLICATIONS: SINGLE-PARAMETER PSRG

In this section we present applications of CR to models of linear polymers, branched polymers, and network polymer systems. For the sake of clarity we shall only discuss RG calculations where a single-parameter — the monomer fugacity K — is renormalized. In order to improve the accuracy of the small-cell, one-parameter calculations, one can enlarge the parameter space to include additional fugacities for longer-range connections between monomers [13]. This approach is not convenient because the number of parameters quickly becomes large [13]. To improve the accuracy of the calculations, while maintaining the simplicity of the one-parameter approach, we consider larger cells and extrapolate to the infinite cell limit. Applying an extrapolation scheme [60,62] to the results for a sequence of cell sizes, we obtain estimates that are less uncontrolled than those based on other RG procedures.

A. Self-avoiding Walks and Linear Polymers

For simplicity we only consider the SAW model of linear polymers in two dimensions [10-12]. For a discussion of NAW see reference 20. The 2x2 cell on a square lattice (Figure 11) has 8 bonds on which a SAW may be placed. Walks that starting from the lower-left hand corner extend across the cell vertically (horizontally) rescale to a single vertical (horizontal) step [10-12]. The 4 walks that rescale to a vertical step are shown in Figure 15. The horizontally spanning walks are the same as these because of the symmetry. The resulting recursion relation [10-12]

$$K' = K^2 + 2K^3 + K^4 \tag{23}$$

has a non-trivial fixed point at $K^* = 0.4656$ and $\nu = 0.7152$.

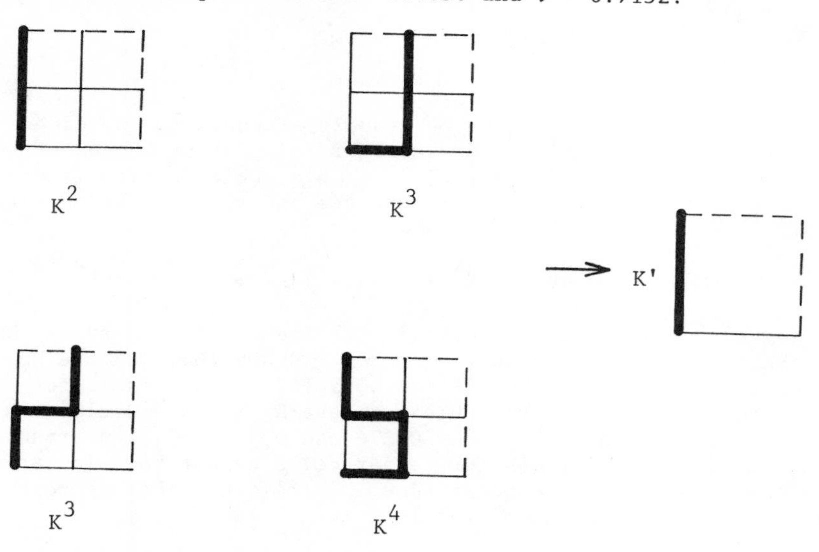

Figure 15: The 4 SAW on a 2x2 cell that rescale to a vertical step.

In order to improve the accuracy of the b = 2 result, but maintain the one-parameter approach, it is natural to consider using larger cells In fact, by using a computer the exact recursion relations for SAW [12] a square have been calculated up to b = 6 [12]. From previous RG studie [60,62] it appears reasonable to assume that the error in finite b resul vanishes as $b \to \infty$ in the following form:

$$\nu^{-1}(b) = \nu^{-1} + c_1(\ell n b)^{-1} + c_2(\ell n b)^{-2} \qquad (24$$

Thus the error decreases as b increases in a predictable fashion, and th results for $\nu(b)$ can be used to extrapolate the limiting behavior ($b \to \infty$ The results of the square lattice calculations for SAW are shown in Figu 16, where $1/\nu$ is plotted versus $1/\ell n\ b$. The best estimate for $1/\nu$ is th value of the intercept which gives the best fit to the data. From this procedure we find $1/\nu = D = 1.341 \pm 0.006$, i.e., $\nu = 0.746 \pm 0.004$. Thi result for ν agrees well with the presumably exact result [63], $\nu = 3/4$ d = 2.

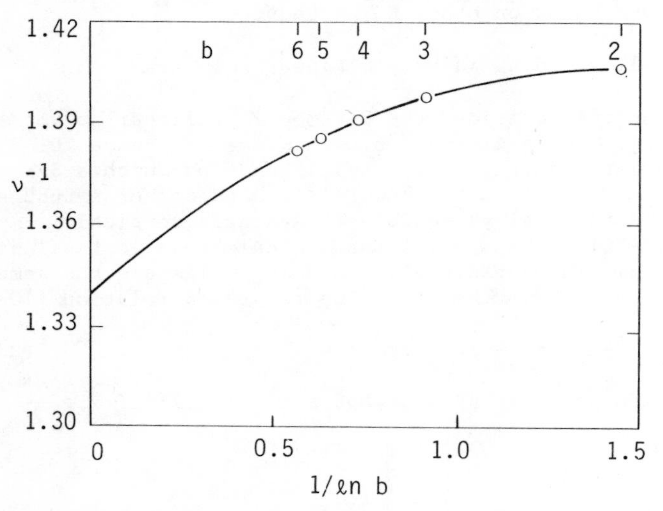

Figure 16: Plot of $1/\nu$ against $1/\ell n\ b$ for SAW on a squar lattice. Data fro reference 12.

The above procedure can be readily extended to any dimension, with corresponding increase in the labor of calculating the exact recursion relations. To study d = 3 [20], a simple cubic lattice is considered. For b = 2 cell, shown in Figure 17, the recursion relation obtained is [20]

$$K' = K^2 + 4K^3 + 8K^4 + 12K^5 + 14K^6 + 16K^7 + 10K^8 \qquad (25)$$

giving $K^* = 0.2973$ and $\nu = 0.5875$. The result for ν may be compared wit the best experimental result of 0.586 ± 0.004 [64] and the best theoretical estimate of 0.588 ± 0.0015 [65]. The result for K^* may be compared with the most reliable estimate $K_c = 1/\mu = 0.2135 \pm 0.0001$ [66] For b = 3, one finds [20] $K^* = 0.276$ and $\nu = 0.581$. Unfortunately the exact recursion relations for larger cells cannot be calculated within a reasonable amount of computer time. Therefore, other approaches are needed to extend the d = 3 results.

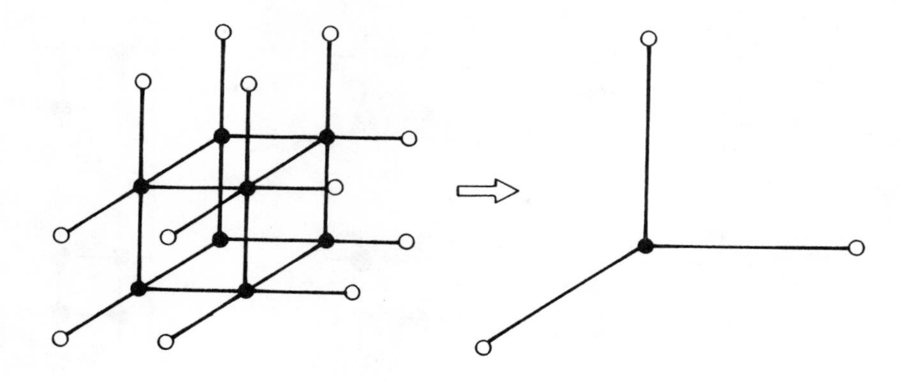

Figure 17: Example of a 2x2x2 cell on a simple cubic lattice used in d=3 calculations.

B. Lattice Animals and Branched Polymers

By introducing polyfunctional units, three or more monomers can branch out from a single site, and the resulting molecule becomes a branched polymer [5]. However, depending on the weight q assigned to the perimeters, branched configurations with different statistics will occur [19,58]. To study the dilute limit of branched polymers [11,18], which corresponds to the statistics of lattice animals, all polymers of the same size N must be weighted equally [19], i.e., $q = 1$ in (15).

Let us first consider branched polymers without loops. When loop formation is prohibited, the resulting configurations are tree-graphs self-avoidingly embedded on a lattice. The recursion relation for bond lattice animals without loops on the square lattice with $b = 2$ (Figure 13) is [11]

$$K' = K^2 + 4K^3 + 14K^4 + 22K^5 + 16K^6 + 4K^7 \qquad (26)$$

which has a fixed point at $K^* = 0.2725$, with $\nu = 0.5760$. Allowing the branch ends to join to other branch ends, or to polyfunctional units, to form loops, the number of possible configurations increases [11]. This set of all possible configurations corresponds to ordinary lattice animals or randomly branched polymers with loops. The recursion relation on the square lattice with $b = 2$ is [11]

$$K' = K^2 + 4K^3 + 14K^4 + 24K^5 + 21K^6 + 8K^7 + K^8 \qquad (27)$$

Now, the fixed point is at $K^* = 0.2702$ and $\nu = 0.5712$. These recursion relations have also been obtained for a cell with $b = 3$ [11]. A cell-to-cell extrapolation [11] of these results gives $\nu = 0.6273$ [11] for branched polymers without loops (tree-like animals) and $\nu = 0.6370$ [11] for branched polymers with loops (bond lattice animals).

Now let us consider site lattice animals on a square lattice. The simplest cell on a square lattice is a cell with 4 sites (See Figure 18). The 6 possible configurations of site animals which starting from the

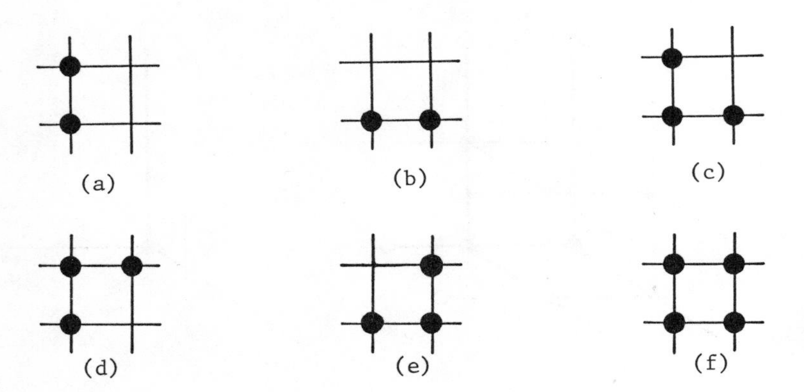

Figure 18: The 6 possible spanning configurations of site animals on a 2x2 cell on a square lattice.

origin span the b = 2 cell are shown in Figure 18. Note that all 6 configurations contribute to the recursion relation under rule r_0. The recursion relation

$$K' = 2K^2 + 3K^3 + K^4 \quad \text{(rule } r_0\text{)} \tag{28}$$

has a fixed point at $K^* = 0.3247$ with $\nu = 0.7976$ [23]. Under rule r_1 th recursion relation is

$$K' = K^2 + 3K^3 + K^4 \quad \text{(rule } r_1\text{)} \tag{29}$$

which has a fixed point at $K^* = 0.4142$ and $\nu = 0.7094$ [23]. For rule r_2 only configurations (c)-(f) are renormalized and the recursion relation

$$K' = 3K^3 + K^4 \quad \text{(rule } r_2\text{)}. \tag{30}$$

This recursion relation has a fixed point at $K^* = 0.5321$ and $\nu = 0.6040$ [23].

The exact recursion relations for cells sizes of up to b = 5 on a square lattice have been determined [23]. The results for ν for rules r r_1 and r_2 are summarized in Figure 19 where $1/\nu$ is plotted versus $1/\ell nb$. Using (24) the finite-b results can be extrapolated to the limit $b \to \infty$. We find [23] $1/\nu = D = 1.59$, 1.54 and 1.48 for rules r_0, r_1 and r_2, respectively. However, a much better procedure is to combine the three sets of data, because asymptotically they are expected to converge to th same result. As shown in Figure 19, we have determined $1/\nu$ by finding t value of the intercept which gives the best overall fit to the three set of data simultaneously. From this procedure we find $1/\nu = D = 1.54 \pm 0.02$, i.e., $\nu = 0.649 \pm 0.009$. This result is listed in Table 3 along with the estimates obtained by other techniques for site animals, bond animals, as well as bond animals without loops (i.e., branching trees), which are all expected to be in the same universality class.

In d = 3, the recursion relations have been determined [23] for sit animals using rules r_0, r_1, r_2 and r_3 on a cell of size b = 2 on the

Table 3

Comparison of the CR result for ν and D for site lattice animals in d=2 with the results obtained by other methods. I. Site animals, II. bond animals, and III. bond animals without loops. [Adapted from reference 23].

Method	ν	D
I. Site animals:		
Cluster renormalization	0.649 ± 0.009 [a]	1.54 ± 0.02
Monte-Carlo	0.660 ± 0.007 [b]	1.52 ± 0.02
	0.65 [c]	1.54
	0.65 ± 0.02 [d]	1.54 ± 0.05
Field Theory	0.61 [e]	1.64
Flory Theory	0.625 [f]	1.60
Finite size scaling	0.6408 ± 0.0003 [g]	1.5605 ± 0.0007
II. Bond animals:		
Cluster renormalization	0.6370 [h]	1.570
III. Bond animals without loops: [i]		
Cluster renormalization	0.6273 [h]	1.594
Monte-Carlo	0.615 [j]	1.63

a. [23]

b. [77,78]

c. [79]

d. [68]

e. [67]

f. [30,31]

g. [80]

h. [11]

i. Animals with and without loop are in the same universality class, [11,18,22].

j. [69]

simple cubic lattice. The values of ν are 0.71, 0.58, 0.49, and 0.46, respectively. It is not possible to obtain a closed-form recursion relation for larger cells within a reasonable computer time. However, it is possible to give a rough estimate of ν in d = 3 in the following way [23]. Let us define the ratio $f(b) = \nu/\overline{\nu}(b)$, where ν is the correct value of this exponent whereas $\overline{\nu}(b)$ is the average value of ν obtained from RG calculations with a cell of size b using rules r_0-r_d. From the data of Figure 19 one finds $f(2) \simeq 0.91$ in d = 2. Assuming that f(2) is independent of d and using the value $\overline{\nu}(2) \simeq 0.56$ one finds $\nu \simeq 0.51$ in d = 3. Although this simple procedure is not expected to be too reliable, its estimate is in reasonable agreement with the exact result $\nu = 1/2$ [67], and the Monte-Carlo result $\nu = 0.53 \pm 0.02$ [68] for site animals and $\nu \simeq 0.46$ [69] for bond animals without loops, which are in the same universality class as bond animals [22].

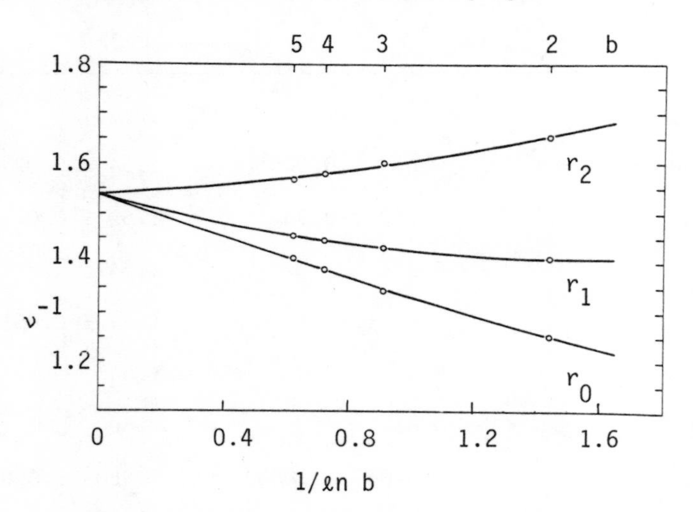

Figure 19: Plot of $1/\nu$ against $1/\ell n$ b for site animals on a square lattice. The curves through the points are the best quadratic fit to the data.

C. Percolation Clusters and Polymer Networks

Stauffer [49] and de Gennes [50] have suggested that the percolation model is a good model for the formation of a polymer network in the sol-gel transition. Recently the relation between gelation and percolation has been reviewed in detail [52].

The usual percolation generating function is obtained when K = p = 1-q in (6), (7). Percolation can be studied by two different but equivalent methods [21]. The first approach considers for each value of p all possible states of the lattice. Each such state may contain a number of clusters. In the alternate approach (see Section IV) one constructs only single clusters (lattice animals) and weights them according to their size N and perimeter size N_p. If one is interested in the statistics of clusters at the percolation threshold either approach can be used in an RG calculation [21]. Here we discuss the first approach for which the recursion relations for site percolation are well known [60].

As an example consider site percolation on a b = 2 cell on a square lattice. The recursion relation for rule r_2 (without restricting the number of clusters) is [60]

$$p' = p^4 + 4p^3q \tag{31}$$

which has a fixed point at $p^* = 0.768$ and the correlation length exponent $\nu_p = 1.395$ [60], where ν_p is defined by

$$\xi_p = |p_c - p|^{-\nu_p} \tag{32}$$

In order to calculate ν we associate a fugacity K to every occupied site and weight the empty sites by an independent parameter q. This recursion relation is [21]

$$K' = K^4 + 4K^3q \tag{33}$$

Using $q = q^* = 1-p^* = 0.232$ obtained from (31), one finds $K^* = p^* = 0.768$ (as expected) and $\nu = 0.559$ [21].

The exponent ν for site percolation on a square lattice has been further calculated [21] using the recursion relations obtained by Reynolds _et al_. [60] for cell sizes of up to b = 5. The results for rules r_0, r_1 and r_2 are shown in Figure 20. Fitting the data to (24) one finds [21] $\nu = 0.50$, 0.51 and 0.54 for rules r_0, r_1 and r_2, respectively. However, by determining the value of the intercept which gives the best overall fit to the 3 sets of data simultaneously one finds $1/\nu = D = 1.92 \pm 0.07$, i.e., $\nu = 0.52 \pm 0.02$. Using scaling arguments, Stauffer [34] has shown that the fractal dimension D is equivalent to the percolation magnetic scaling lower Y_H. The result for D obtained above is in agreement with the presumably exact value [70] of $91/48 \simeq 1.90$ for Y_H.

The results for bond percolation and for site and bond in d = 3 are obtained in a similar manner and are discussed in detail in reference [21].

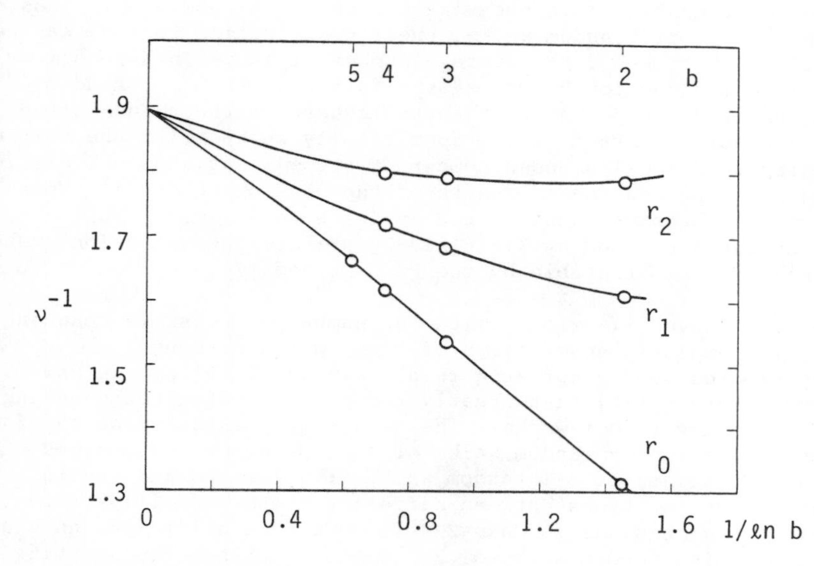

Figure 20: Plot of $1/\nu$ vs. $1/\ell n$ b for site percolation in d=2.

VII. APPLICATIONS: CROSSOVER PHENOMENA AND UNIVERSALITY

In this section we present applications of CR to crossover phenomena and determination of universality classes.

A. Excluded Volume Effects

The question of main interest in this section is the asymptotic behaviour of walks for which the strength of the excluded volume can be varied continuously. For such walks we investigate the crossover from the universality class described by unrestricted random walks to the universality class described by self-avoiding walks [71]. A lattice model which allows us to investigate the transition between these two universality classes has been proposed by Domb and Joyce [72]. In the Domb-Joyce model [72] each intersection of the polymer chain or walk with itself receives a weight (1-w). Hence for w=1 the model corresponds to a self-avoiding walk and for w=0 the model reduces to a standard random walk.

In order to develop a renormalization group method for the Domb-Joyce model, we must first develop a renormalization group approach for unrestricted random walks [71]. However the application of a PSRG method to the random walk problem is not straightforward. The difficulty is that the number of spanning random walks on a finite size cell is infinite. However, recently a new PSRG that eliminates this difficulty by consistently limiting the maximum number of steps allowed in a finite size cell has been developed [71]. We first review this approach [71] and then use it to study the Domb-Joyce model [72].

Note that at the critical fugacity K_c, which corresponds to the limit $N \to \infty$, only random walks of length $R^2 \propto N$ are important. This relation between R and N implies that although the recursion relation (15) contains the weights of many configurations having a much larger number of bonds or steps than the "critical" random walks, their contribution to (15) can be neglected. That is for $K=K_c$, the average number of bonds in a random walk on a cell of linear dimension b varies asymptotically as $b^{\frac{1}{2}}$; random walks with a number of steps larger than the square of their end-to-end displacement are not important [71]. Specifically we restrict the sum in (15) to spanning random walks whose number of steps N satisfy the relation $N \leq R^2$, where R is the end-to-end length of the walk on the cell. Note that this criterion becomes exact in the infinite cell limit. This procedure which was suggested by Klein (1982, private communication) makes the random walk problem tractable by the PSRG method [71].

Although the above criterion limits the number of possible spanning random walks, the explicit enumeration of these walks becomes prohibitively time-consuming for even relatively small cells. We now describe a novel method [71] that greatly reduces the effort of counting the number of spanning random walks. The method [71] is based on the fact that the number of spanning random walks of N steps can be determined given only the end-points of all random walks of N-1 steps and the fact that there are no correlations between different steps. We illustrate this method in two dimensions for a b=2 cell on a square lattice and count the number of spanning random walks of R = 4 and 5. Since the starting point of the random walk is the lower-left corner of the cell, the first step must be either up or to the right. We represent these two

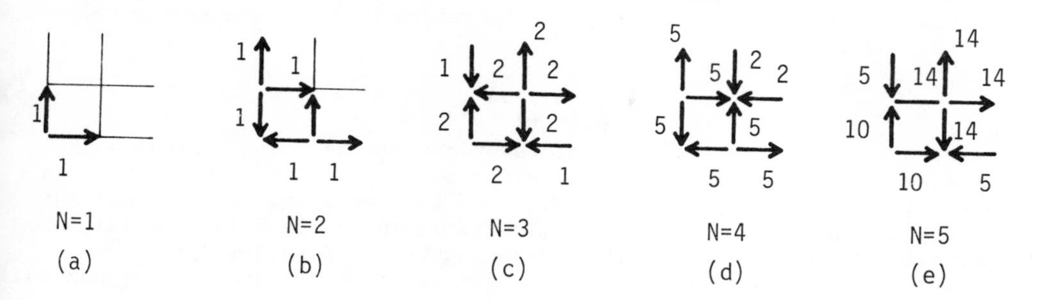

Figure 21: The number associated with each arrow is the number of possible walks with the same end-points.

possibilities by two arrows which represent the end-point of the walk (see Figure 21a). The number associated with each arrow denotes the number of walks with end-points at the corresponding site. In order to obtain the N = 2 walks, we note that each arrow in Figure 21a can move in three possible directions, e.g. the vertical arrow can continue upward, turn to the right or reverse direction. The end points of the 6 possible random walks of 2 steps are shown in Figure 21b. Since the spanning walks are those that terminate at one of the two vertical bonds, there is one spanning random walk of weight K^2. To obtain the walks of N = 3 steps we consider all the possible moves of the arrows in Figure 21b representing all the possible end-points of the N = 2 walks. For example we see that the two walks ending at site M in Figure 21 b can both go up and to the right as well as each reverse their direction. The result is shown in Figure 21c from which it can be seen that there are two spanning walks of weight K^3. We continue this iterative process and obtain the sequence of end-points shown in Figure 21d for N = 4 steps and in Figure 21 for N = 5 steps. A visual inspection of Figure 21 gives us the recursion relation [71]

$$K' = K^2 + 2K^3 + 5K^4 + 14K^5 \tag{34}$$

from which we obtain $K^*=0.3470$ and $\nu = 0.5853$ [71]. In comparison the corresponding recursion relation for the self-avoiding walk problem is (23) with $K^* = 0.4656$ and $\nu = 0.7152$.

Our counting procedure allows us to calculate by hand the recursion relations for b = 3 and 4 on the square lattice [71]. The recursion relations for b = 2-4 are given in reference [71]. The results for ν for b = 2,3 and 4 on the square lattice are 0.58, 0.56, and 0.54 [71], respectively. In comparison the exact value is $\nu = 1/2$.

Thus far we have only considered a one-parameter PSRG treatment of unrestricted random walks. In order to study the crossover from random walks to self-avoiding walks, we introduce a two-parameter PSRG in the context of the Domb-Joyce model [71]. In this model [72] an interaction $-w\delta_{ij}$ is introduced between every pair of points i and j of a configuration of the walk on the lattice. $\delta = 1$ if the ith and jth points of the walk occupy the same lattice site and is zero otherwise. The effect of this interaction is that each N step configuration of the chain has an associated factor of

$$\prod_{i=0,j=i+2}^{N-1\ N} \prod (1-w\delta_{ij}) = (1-w)^t \quad (35$$

where t is the number of self-intersections of the chain. Since each intersection of the chain with itself receives a weight (1-w), the case w=0 weights all random walks equally and the value w=1 excludes all but self-avoiding walk configurations. That is in the Domb-Joyce model, the parameter w can be continuously varied so that the random and self-avoiding walk correspond to the limits of no excluded volume and ful excluded volume respectively.

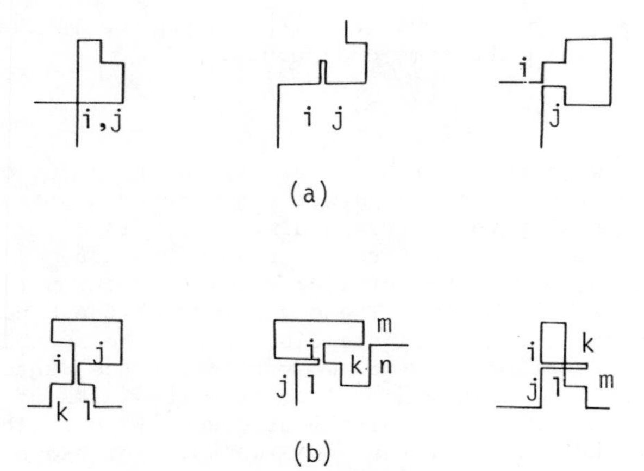

Figure 22: (a) Three types of single self-intersections. (b) Examples of 2, 3 and 4 self-intersections, respectively.

In Figure 22a we show the three types of single self-intersections which receive a weight (1-w). In Figure 22b we show examples of walks with two, three and four self-intersections which receive one, two and three factors of (1-w), respectively.

In order to study the Domb-Joyce model using a two-parameter PSRG, w must renormalize both the fugacity K for each step in the walk and the interaction parameter w [71]. The renormalization of the fugacity K follows the same procedure as before, except that in the K' recursion relation a walk of N steps and t self-intersections receives a weight $K^N(1-w)^t$. To renormalize w we must consider walks which rescale to an intersecting walk on the rescaled lattice. To obtain a self-intersecting walk on the renormalized lattice [73] a walk on the original lattice must first span across the cell and then return to self-intersect at the origi as shown in Figure 23. This renormalization procedure leads to a recursion relation of the form

$$K'^2(1-w') = \Sigma\, c_{Nt} K^N (1-w)^t \quad (36)$$

where c_{Nt} is the number of N-step walks which starting from the origin span the cell and then return to the origin by making t

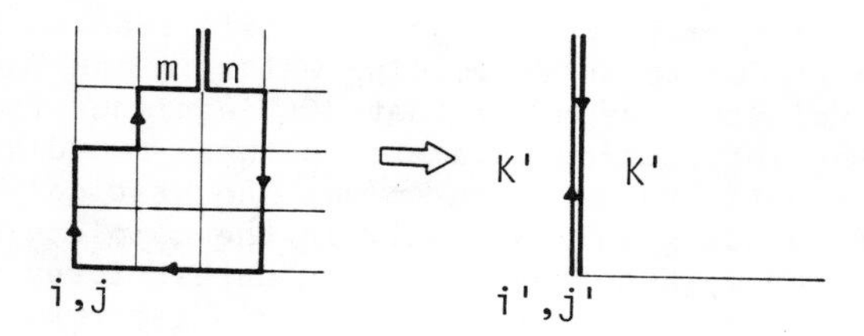

Figure 23: Examples of random walks which return to the origin and are renormalized to $K'^2(1-w')$.

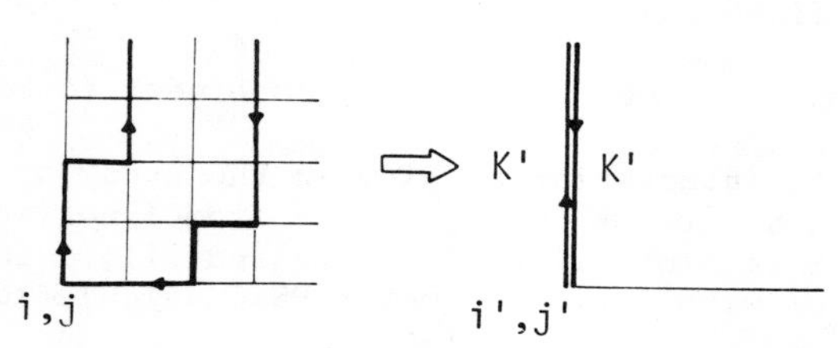

self-intersections. The rule for the maximum number of steps N to be considered in (36) is similar to that used in (34) with the provision that the restriction on the maximum number of steps apply separately to the part of the walk that first spans the cell and the part that returns to the origin.

For d = 1 the recursion relations for K' and w' are [71]

$$K'=K^2\{1+2K^2(1-w)^2\} \tag{37a}$$

$$K'^2(1-w')= K^4(1-w)^2\{1+4K^2(1-w)^2+4K^4(1-w)^4\} \tag{37b}$$

The direction of the renormalization group flows (see Figure 24) is from the random walk fixed point (w=0) to the self-avoiding walk fixed point (w=1), indicating that random walks and self-avoiding walks belong to two different universality classes. The direction of the flow implies that for any finite excluded volume (w>0), the flow is to the self-avoiding walk fixed point. Therefore for any finite w the critical behaviour is that of the self-avoiding walks.

For d = 2 the recursion relations for K' and w' for a 2x2 cell on the square [71] lattice are

$$K'=K^2+2K^3+K^4\{1+2(1-w)+2(1-w)^2\} +K^5\{8(1-w)+6(1-w)^2\} \tag{38a}$$

$$\begin{aligned}K'^2(1-w')=&K^4(1-w)^2+K^5\{2(1-w)+2(1-w)^2\}+K^6\{8(1-w)^2\\&+4(1-w)^3+2(1-w)^4\}+K^7\{4(1-w)^2+26(1-w)^3+18(1-w)^4\}\\&+K^8\{8(1-w)^3+42(1-w)^4+27(1-w)^5+4(1-w)^6\}+K^9\{28\\&(1-w)^4+94(1-w)^5+18(1-w)^6\}+K^{10}\{6(1-w)^4+62(1-w)^5\\&+110(1-w)^6+18(1-w)^7\}\end{aligned} \tag{38b}$$

The global flow diagram for these recursion relations is similar to that of d=1. There are only two unstable fixed points at $(K^*,w^*)=(0.3470,0)$ and (0.4656,1.0) corresponding to random walks (w=0) and self-avoiding walks (w=1) respectively.

In summary we have used a cell PSRG to study the crossover behavior from random to self-avoiding walks in the Domb-Joyce model. As expected the relevant parameter that distinguishes random from self-avoiding walks is the interaction parameter between two distinct sites along the chain. For any finite excluded volume the critical behavior is that of a self-avoiding walk and only in the complete absence of any excluded volume does the walk correspond to an unrestricted random walk. This conclusion is consistent with the Monte-Carlo renormalization-group results of Kremer et al. [74] who studied a hard sphere chain model in d=2-5 and found [74] that for d=2,3 there are two fixed points corresponding to random walks and self-avoiding walks, whereas for $d \geq 4$ there exists only the random walk fixed point.

B. Crossover from Linear to Branched Polymers

An interesting problem in the study of linear and branched polymers is the nature of the crossover from linear to branched polymers as the branching probability is increased [11]. In this section we study this problem with a two parameter PSRG calculation.

In order to treat linear polymers and branched polymers with the same PSRG transformation, consider a two parameter model [11]. K as before, is the monomer fugacity and f represents the probability that a randomly chosen site on a polymer is an f-functional unit [11]. When f is zero, only linear chains can occur; as f increases, branched polymers are produced. With this additional parameter, the recursion relations (23), and (27), can be combined [11] into one general expression.

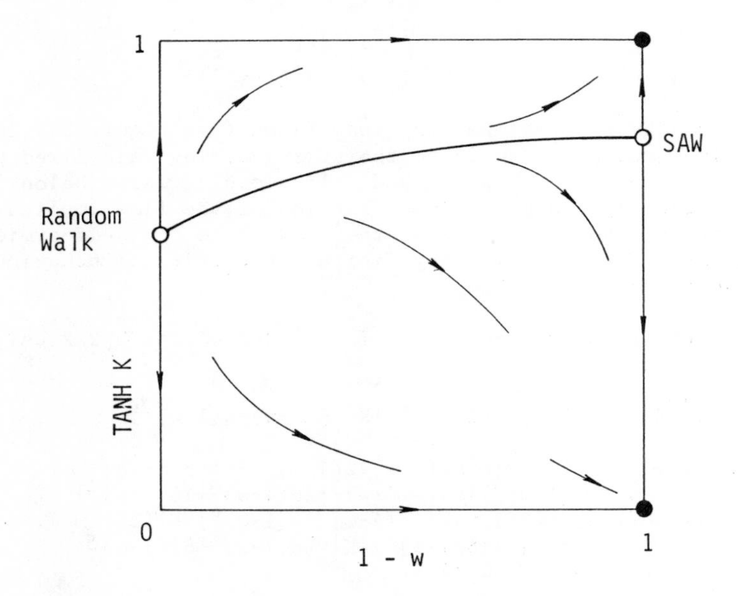

Figure 24: Crossover from random walk to self-avoiding walk. The flow shows that for all finite excluded volume chains have the statistics of SAW.

The result is [11],

$$K' = K^2 + K^3(2+2f) + K^4(1+12f+f^2) + K^5(10f+14f^2) + K^6(2f + 13f^2 + 6f^3) + K^7(2f^2 + 4f^3 + 2f^4) + K^8f^4 \quad (39)$$

Clearly (39) reduces to (23) if f is zero, while for f = 1, it reduces to (27).

The f-functional units may be renormalized as follows [11]: A configuration of polyfunctional units on the original cell renormalizes to a single branch point if the polymer configuration on the site part of the cell (the shaded region in Figure 25) spans the cell either horizontally

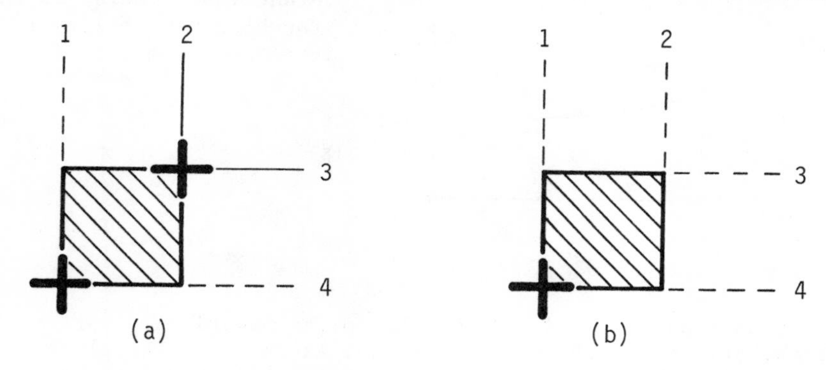

Figure 25: (a) Example of a configuration that renormalizes to a polyfunctional unit. (b) Example of a configuration that does not renormalize to a polyfunctional unit, becuase it cannot extend to either 1 or 2 or to 3 or 4.

or vertically. This requires that a polymer start at 0 and be able to extend to either 1 or 2 <u>or</u> to 3 or 4. For example, Figure 25a, shows a site configuration which renormalizes to a branch point whereas Figure 25b shows one which does not. These conditions define a recursion formula for f', the renormalized branching probability, as a function of f. When all sites on a polymer are occupied by polyfunctional units (f=1), all sites on the renormalized polymer must be occupied by branching units (f'=1). To preserve a probabilistic interpretation of f and f', we write f' as the sum of all branched polymers that renormalize according to the criterion we described above, divided by all possible configurations [11]. The denominator includes all diagrams of the type shown in Figure 25, whereas the numerator includes only the spanning configurations, such as the one shown in Figure 25a. In this way we find [11]

$$f' = N/D. \quad (40)$$

where,

$$N = f(1-f)^3(4K+7K^2+10K^3) + f^2(1-f)^2(2K+8K^2+18K^3+3K^4) + f^3(1-f)(3K^2+14K^3+3K^4) + f^4(4K^3+K^4)$$

$$D = 2K(1-f)^2+K^2(2-f-f^2)+2K^3(1+f)+K^4f.$$

The coupled recursion relations (39) and (40) together constitute a two parameter PSRG transformation. When $f \to 0$, $f' \to 0$ and (39) reduces to (23). When $f \to 1$, $f' \to 1$ and (39) reduces to that for the branched polymer.

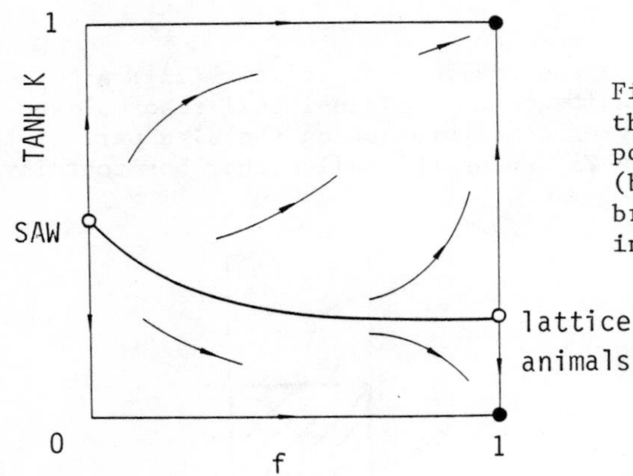

Figure 26: Flow diagram showing the crossover from SAW (linear polymers) to lattice animals (branched polymers) as the branching probability f is increased from 0 to 1.

Equations (39) and (40) may be solved numerically for the fixed points, critical surface, and critical exponents. The global flow is shown in Figure 26. There are two unstable fixed points at $(K^*,f^*) = (0.4656,0.0)$ and $(0.2725,1.0)$, corresponding to the linear polymer fixed point and the branched polymer fixed point, respectively [11]. The direction of the flow is from the line of zero branching probability (i.e., linear polymers) to the branched polymer fixed point, indicating that <u>linear polymers and branched polymers belong to two different universality classes</u> [11]. The branched polymer fixed point controls the flow away from the $f = 0$ axis. The critical surface is shown as the solid line in Figure 26. Along this line the critical behavior is that of the branched polymer.

Linearization of the recursion relations (39) and (40) near the linear polymer fixed point gives two eigenvalues along the K and f axes, respectively. Since both eigenvalues are greater than unity, the linear polymer fixed point is unstable in both directions. Linearizing the recursion relations near the branched polymer fixed point gives only one non-zero eigenvalue. Thus, only one eigenvalue is relevant and the two parameter group does not change the critical exponents of the branched polymer, but demonstrates that linear polymers and branched polymers belong to two different universality classes [11].

C. Effects of Loops in Polymers

Recently there has been considerable interest [22,46-48,74,75] in theoretical studies of polymers with loops in relation to the circular structure of natural macromolecules [75] (supercoiled DNA), and the fundamental question of the effects of topological constraints on conformation of polymers [46]. In general, polymers with loops and those without loops have considerably different topological structures and therefore it is also of fundamental interest to examine the effects of loops on conformation of both linear and branched polymers.

Rings: Let us first develop a single parameter CR approach for a model of ring polymers [22]. Consider a SAW that starts from the origin of a cell and extends across the cell in one particular direction. If the walk does not return to the cell to self-intersect, it is a SAW and is renormalized as discussed in section VI-A. However, if the walk extends across the cell in one direction but returns to the cell from another direction and self-intersects at the origin, then the configuration is a ring or a closed polygon (See Figure 27). To renormalize these configurations, we map them into two bonds with a fugacity K'^2; one bond in the direction of outgoing walk and one bond in the direction of incoming bond, as shown in Figure 27.

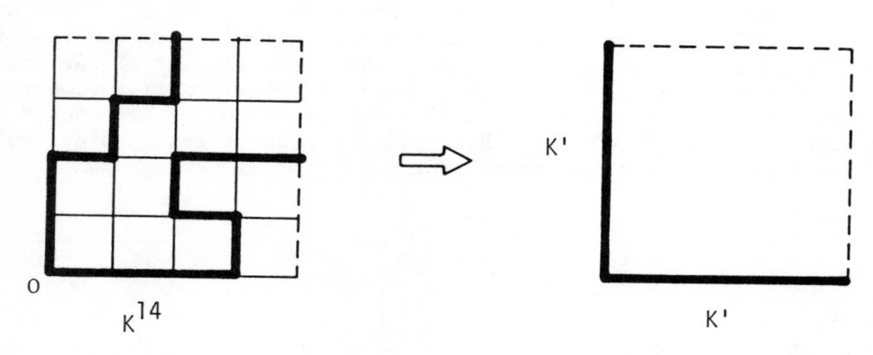

Figure 27: A SAW that leaves a cell, but returns to it and self-intersects at the origin is a ring polymer and is rescaled to two single steps under the CR.

The recursion relations for rings on cells of side b=2 and 3 on the square lattice are

$$K'^2 = K^4 + 2K^5 + K^6 \quad (b=2)$$

$$K'^2 = K^6 + 4K^7 + 12K^8 + 15K^9 + 12K^{10} + 7K^{11} \quad (b=3) \tag{41}$$

which give $\nu = 0.8392$ and $\nu = 0.8012$ for b=2 and 3, respectively. The cell-to-cell transformation [11] (from 3x3 → 2x2) gives $\nu = 0.7467$ [22] in remarkably close agreement with the (possibly) exact result $\nu = 0.75$ [63] for SAW in d=2. Even though the closeness of the numerical values of ν for SAW and rings is interesting, it is not sufficient to demonstrate that the two models belong to the same universality class.

In order to study the question of the universality of SAW and rings consider a two-parameter CR [22]. Let K be the fugacity for each step of a SAW that starts from the origin and gets across the cell in a particular direction, and let r be the fugacity for each step in the walk which returns to the cell and finally forms a ring by joining to the first step of the walk at the origin. Under this RG transformation [22] each configuration spanning in a particular direction contributes to K', and each configuration that spans in both directions renormalizes to K'r', where r' is the renormalized fugacity of a returning walk. On the 2x2 cell on a square lattice the two coupled recursion relations are

$$K' = K^2 + 2K^3 + K^4 + K^2r^2 + K^2r^3 + K^3r^2 \tag{42a}$$

and

$$K'r' = K^2r^2 + K^2r^3 + K^3r^2 \tag{42b}$$

The flow diagram for the above recursion relations is shown in Figure 28. Note that for all r the flow is to the SAW fixed point at r=0, K=K*, and therefore rings and SAW are described by the same critical fixed point and belong to the same universality class. Similar recursion relations are obtained in d=3 on a simple cubic lattice [22]. The flows, and therefore the conclusions, are the same as in d=2. The universality of SAW and rings is not entirely unexpected, because in the scaling limit, whether a linear animal has open ends or is a closed self-avoiding ring should not change its critical properties. However, the present result provides a direct confirmation of this expectation.

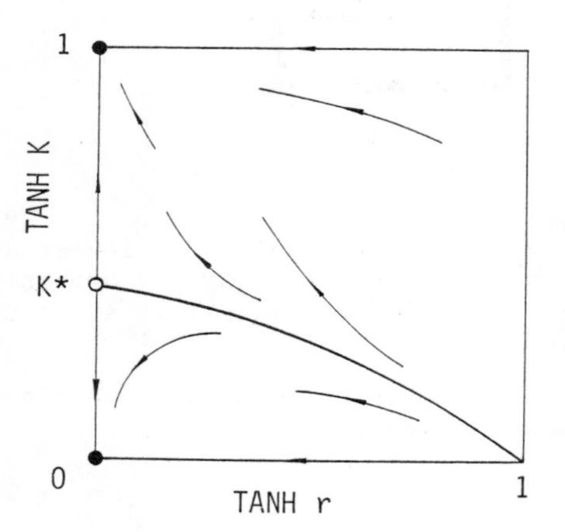

Figure 28: Flow diagram showing that ring polymers (r>0) are in the same universality class as linear polymers (r=0). Open circle denotes the linear polymer (SAW) fixed point. Solid dots are trivial fixed points.

The Trail Problem: A linear animal with loops—called a trail by Malakis [76]—may be viewed as the path of a random walk on a lattice with the constraint that no bond on the lattice can be visited more than once. A trail is analogous to a SAW, but— in contrast to SAW—a trail can visit a lattice site more than once. In one dimension and on lattices with coordination number 3 trails and SAW are equivalent. In general, a simple relationship does not exist between trails and SAW. Malakis [76] has enumerated all trails of up to 17 steps on a square lattice and from studies of their end-to-end length has concluded that the critical exponents for SAW and the trail problem are the same on the square lattice. Here we study the trail model using the CR method and explicitly show that within the RG scheme, SAW and trails are described by the same fixed point.

The single-parameter CR approach can be readily applied to trails

[22] on a square lattice by following the general approach of section VI-A for SAW, but with the added condition that spanning SAW with loops also contribute to K'. There are no trails with loops on a 2x2 cell on a square lattice, but for the 3x3, the recursion relation for trails is [22]

$$K' = K^3 + 3K^4 + 9K^5 + 5K^6 + 9K^7 + 2K^8 + 4K^9 + \ell(2K^8 + 4K^9 + 2K^{10}), \quad (43)$$

where ℓ is the fugacity for loops. Placing no restrictions on loops, i.e., setting $\ell = 1$ in (43), gives $K^* = 0.4450$ and $\nu = 0.7135$ [22], which can be compared with the corresponding value $\nu = 0.7178$ for SAW on the same cell [11] . The recursion relation for trails on a simple cubic lattice with b=2 is

$$K' = K^2+4K^3+8K^4+ 12K^5+14K^6+16K^7+10K^8+ \ell(2K^6+7K^7+17K^8+9K^9)+\ell^2K^{10}, \quad (44)$$

which has a fixed point at $K^* = 0.2955$ and gives $\nu = 0.5816$ when $\ell = 1$. Again the result for ν for trails is in close agreement with $\nu = 0.5875$ for SAW on the same cell [20].

To show that trails and SAW are in the same universality class, we construct a two-parameter RG in which the loop fugacity ℓ is also renormalized [22]. The renormalized fugacity ℓ' for loops contains only the weights of those configurations that span the cell and have a loop. For the 3x3 cell on the square lattice and the 2x2x2 cell on the simple cubic lattice, respectively, the recursion relations are

$$\ell' = \ell(2K^8 + 4K^9 + 2K^{10}) \quad (45)$$

and

$$\ell' = \ell(2K^6 + 7K^7 + 17K^8 + 9K^9) + \ell^2K^{10} \quad (46)$$

The global flows for (43) and (45) and for (44) and (46) are very similar qualitatively and the case of d=3 [i.e., (44) ad (46)] is shown in Figure 29. There is only one critical fixed point at $\ell = 0$ and for all ℓ the flow is to this fixed point [22]. Thus, loops are irrelevant and trails are in the same universality class as SAW.

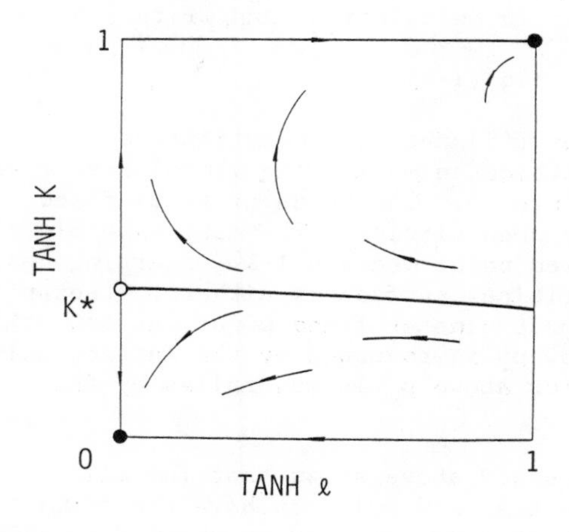

Figure 29: Flow diagram showing that polymers with loops are in the same universality class as polymers without loops. Solid dots are trivial fixed points.

Trees and Animals: A branched animal with no loops is a tree graph self-avoidingly embedded on a lattice; branched animals with loops correspond to ordinary random animals. The CR study of loops in random animals [22] follows the same reasoning as for trails: Each animal is weighted by $K^{N}\ell^{N_\ell}$ where N is the size and N_ℓ is the number of loops in the animal. The fugacity K' contains the weights of all spanning configurations [cf. section (VI-B)], but ℓ' contains the weights of only those spanning configurations having a loop.

The flow diagram obtained for K' and ℓ' is similar to the flows for the trail problem (Figure 29). Again, there is only a single critical fixed point at $\ell = 0$ and $K = K^*$, corresponding to clusters without loops. This fixed point controls the entire critical behavior for animals with loops. Thus, in random animals loop fugacity is a non-critical parameter and branched trees and random animals are in the same universality class.

D. Screening Effects in Branched Polymer Solutions

The effects of screening on the conformation of branched polymers can be studied using the generalized lattice animal model described in Section IV-A. In this model [19] each lattice animal is weighted by two parameters: A weight K for each element (site or bond) and a weight q for each perimeter. In the random animal limit ($K \to 0$, $q \to 1$), concentration of branched macromolecules is small and for large polymers the excluded volume effect is dominant; no screening is expected in this limit. However, as K approaches the percolation threshold p_c (i.e. $q \to q_c=1-p_c$) concentration of macromolecules becomes high and screening effects become important. This crossover from the excluded volume behavior (lattice animals) to the screened case (percolation) can be readily studied using a two-parameter CR [19].

For simplicity, consider generalized site animals on a triangular lattice [19]. The simplest cell has 3 sites, which the RG reduces to a single site, and $b=\sqrt{3}$. The recursion relation for K is simply

$$K'= K^2+3K^3q \tag{47}$$

where q is the weight for empty sites. To renormalize q treat it as a probability that a site is empty (as in percolation) and write [19]

$$q'= q^4+3q^3(1-q) \tag{48}$$

The coupled recursion relations (47) and (48) constitute a two-parameter CR for generalized lattice animals. The global flow diagram for these equations is shown in Figure 30. The three critical fixed points of interest are indicated by open circles. The most unstable fixed point occurs at the percolation fixed point where q=1-K. Starting near this fixed point the flow on the critical surface is either to the animal fixed point, at q=1, or to the compact cluster fixed point, at q=0. Note that all the critical behavior below p_c is governed by the lattice animal fixed point and all critical behavior above p_c is controlled by the compact cluster fixed point.

The fixed point structure discussed above shows that for all concentrations below p_c , very large branched polymers have the statistics of lattice animals, i.e. are dominaited by excluded volume effect. There is a discontinous change in the conformation of branched polymers at p_c. At this point branched polymers shrink in size and their fractal dimension discontinously decreases from that of lattice animals to the percolation clusters. Above p_c, there is a final collapse of branched polymers to a compact structure with fractal dimension equal to d. Thus, the percolation threshold p_c correspond to a special point at which different molecules in solution screen the excluded volume effect and lead to a discontinous change in the conformation of branched polymers.

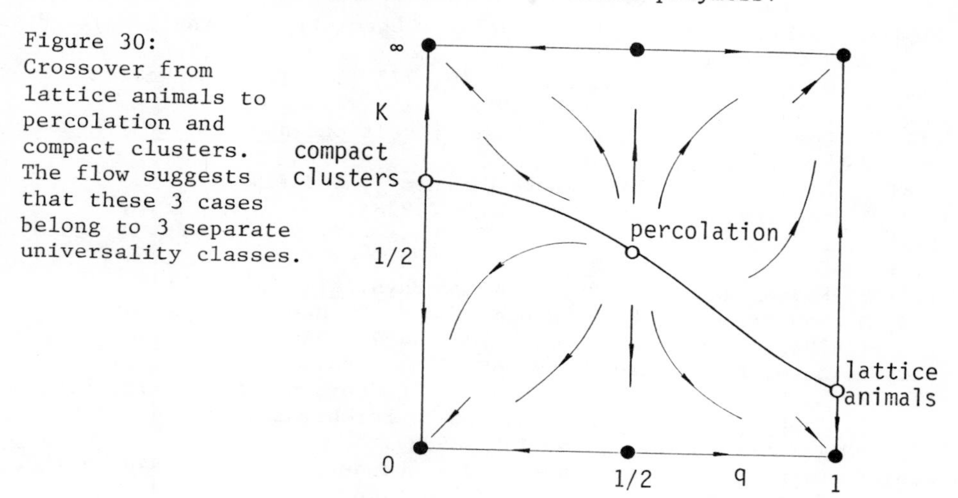

Figure 30: Crossover from lattice animals to percolation and compact clusters. The flow suggests that these 3 cases belong to 3 separate universality classes.

VIII. CONCLUDING REMARKS

This review centered around a particular position-space renormalization-group approach for polymers , which we called the Cluster Renormaliztion. The emphasis was mainly on the applications of CR to the study of crossover phenomena and universality classes in polymer models. We showed that this PSRG is well suited for calculations of phase diagrams and study of universality classes in polymer systems in two and three dimensions. The results indicate that due to the flexibility and simplicity of this type of renormalization, it can be applied to a wide range of problems in polymer physics.

Since applications of renormalization-group to the scaling properties of polymers is a rapidly growing field, it would be presumptous to claim completeness in this brief review. This review was primarily concentrated on topics close to the author's interests and the works carried out by him and his collaborators. The excellent monographs and reviews cited in the references discuss other topics and approaches and contain references to early literature.

ACKNOWLEDGEMENTS

Some of the works discussed here was carried out in collaboration with Antonio Coniglio, Harvey Gould, and Peter Reynolds. I would like to thank them and Mohamed Daoud, Hans Herrmann, William Klein, Hisao Nakanishi, Sid Redner, Gene Stanley, Dietrich Stauffer, and Chris Unger for valuable discussions.

REFERENCES

[1] P. G. de Gennes, Phys. Lett. 38A, 339 (1972).
[2] P. G. de Gennes, Scaling Concepts in Polymer Physics, (Cornell U. Press, Ithaca, NY) (1979).
[3] H. E. Stanley, Introduction to Phase Transitions and Critical Phenomena (Oxford U. Press, NY) (1971).
[4] C. Domb and M. S. Green, Phase Transitions and Critical Phenomena, (Academic Press, NY).
[5] P. J. Flory, Principles of Polymer Chemistry, (Cornell Univ. Press, Ithaca, NY) (1971).
[6] K. G. Wilson, Rev. Mod. Phys. 47, 773 (1975), K. G. Wilson, Sci. Am. 241, 158 (August 1979).
[7] C. Domb and M. S. Green, Phase Transitions and Critical Phenomena, Vol. 6 (Academic Press, NY).
[8] P. G. de Gennes, Riv. Nuovo Cimento 7, 363 (1977).
[9] B. Shapiro, J. Phys. C 11, 2829 (1978).
[10] S. L. A. de Queiroz, and C. M. Chaves, Z. Physik B40, 99 (1980).
[11] F. Family, J. Phys. A13, L325 (1980).
[12] S. Redner and P. J. Reynolds, J. Phys. A14, 2679 (1981).
[13] M. Napiorkowski, E. M. Hauge, and P. C. Hemmer, Phys. Lett. 72A, 193 (1979), J. Malakis, Physica 104A, 427 (1980).
[14] S. Muto, Prog. Theo. Phys. 65, 1081 (1981).
[15] H. E. Stanley, P. J. Reynolds, S. Redner, and F. Family, in Real-Space Renormalization, ed. T. Burkhardt and J. M. J. van Leeuwen (New York: Springer), p. 169 (1982).
[16] K. Kremer, A. Baumgartner, and K. Binder, Z. Physik B40, 331 (1981), A. Baumgartner, J. Phys. A13, L39 (1980).
[17] J. des Cloiseaux, J. de Physique, 36, 281 (1975).
[18] T. C. Lubensky and J. Isaacson, Phys. Rev. A20, 2130 (1979).
[19] F. Family and A. Coniglio, J. Phys. A13, L403 (1980).
[20] F. Family, J. de Physique 42, 189 (1981).
[21] F. Family and P. J. Reynolds, Z. Physik 45, 123 (1981).
[22] F. Family, Phys. Lett. 92A, 341 (1982).
[23] F. Family, J. Phys. A16, L97 (1983).
[24] F. Family in International Union of Pure and Applied Chemistry 28th Macromolecular Symposium, University of Massachusetts, Amherst, Massachusetts, 1982, edited by J. Chien, R. Lenz and O. Vogl (IUPAC, New York, 1982) p. 533.
[25] B. H. Zimm, and W. H. Stockmayer, J. Chem. Phys. 17, 1301 (1949).
[26] P. G. de Gennes, Biopolymers 6, 715 (1968).
[27] Reference 5, chapter XII.
[28] M. E. Fisher, J. Phys. Soc. Japan 26 (Suppl.), 44 (1969).
[29] P. G. de Gennes, C. R. Acad. Sci. Paris 291, 17 (1980).
[30] J. Isaacson and T. C. Lubensky, J. Physique-Lettres 41, L-469 (1980).
[31] M. Daoud and J. F. Joanny, J. Physique 42, 1359 (1981).
[32] B. B. Mandelbrot, Fractal Geometry of Nature (W. H. Freeman, San Francisco) (1982).
[33] H. E. Stanley, J. Phys. A10, L211 (1977)
[34] D. Stauffer, Phys. Rep. 54, 1 (1979).
[35] S. R. Forrest and T. A. Witten, J. Phys. A12, L109 (1979).
[36] T. Witten and L. M. Sander, Phys. Rev. Lett. 47, 1400 (1981) and Phys. Rev. B27, 5686 (1983).
[37] T. Vicsek and J. Kertesz, J. Phys. A14, L31 (1981).
[38] E. T. Gawlinski and H. E. Stanley, J. Phys. A14, L291 (1981).
[39] F. Harary, Graphical Theory and Theoretical Physics ed. F. Harary

(London: Academic Press) pp. 1-41.
[40] C. Domb, J. Phys. A 9, L141 (1976).
[41] W. S. Golomb, Polyominoes (New York: Scribner's) (1965), W. F. Lunnon, Computers in Number Theory, eds. A. O. L. Atkin and B. J. Birch, (London: Academic Press) pp. 347-372, D. H. Redelmeier, Discrete Math. 36, 191 (1981).
[42] C. Domb, Adv. Chem. Phys. 15, 229 (1969), D. S. McKenzie, Phys. Rep. 27C, 35 (1976).
[43] D. S. Gaunt, A. J. Guttman, and S. G. Whittington, J. Phys. A12, 75 (1979), D. S. Gaunt, J. L. Martin, G. Ord, G. M. Torrie, and S. G. Whittington, J. Phys. A13, 1791 (1980).
[44] P. G. Watson, J. Phys. C3, L28 (1970).
[45] J. L. Martin, M. F. Sykes, and F. T. Hioe, J. Chem. Phys. 46, 3478 (1976).
[46] J. des Cloizeaux and M. L. Mehta, J. de Physique 40, 665 (1979), J. des Cloizeaux, J. de Physique Lett. 42, L-433 (1981).
[47] A. Baumgartner, J. Chem. Phys. 76, 4275 (1982).
[48] Yi-der Chen, J. Chem. Phys. 74, 2034; 76, 2447, 5160 (1981).
[49] D. Stauffer, J. Chem. Soc. Faraday Trans. II 72, 1354 (1976).
[50] P. G. de Gennes, J. de Physique 37, L1 (1976).
[51] J. W. Essam, Rep. Prog. Phys. 43, 833 (1980).
[52] D. Stauffer, A. Coniglio and M. Adam, Adv. Polymer Sci. 44, 105 (1982).
[53] M. Sahimi, in Proceedings of the Workshop on the Mathematics and Physics of Disordered Media (Springer-Verlag) (1983).
[54] F. Family, J. Phys. A15, L583 (1982).
[55] D. S. Gaunt, M. F. Sykes and H. Ruskin, J. Phys. A9, 1899 (1976), M. F. Sykes and M. Glen, J. Phys. A9, 87 (1976), A. J. Guttmann, J. Phys. A15, 1987 (1982).
[56] D. S. Gaunt and H. Ruskin, J. Phys. A11, 1369 (1978), A. J. Guttmann, and D. S. Gaunt, J. Phys. A11, 949 (1978), D. S. Gaunt, J. Phys. A13, L97 (1980), M. F. Sykes, D. S. Gaunt and M. Glen, J. Phys. A14, 287 (1981).
[57] J. A. M. S. Duarte and H. J. Ruskin, J. de Physique 42, 1585 (1981) D. S. Gaunt, M. F. Sykes, G. M. Torric, and S. G. Whittington, J. Phys. A15, 3209 (1982).
[58] A. B. Harris, and T. C. Lubensky, Phys, Rev. B24, 2656 (1981).
[59] T. W. Burkhardt and J. M. J. van Leeuwen, ed., Real-Space Renormalization, (Springer-Verlag) (1982).
[60] P. J. Reynolds, H. E. Stanley and W. Klein, Phys. Rev. B21, 1223 (1980).
[61] H. Gould, F. Family and H. E. Stanley, Phys. Rev. Lett. 50, 686 (1983).
[62] P. D. Eschbach, D. Stauffer and H. J. Herrmann, Phys. Rev. B23, 422 (1981).
[63] B. Nienhuis, Phys. Rev. Lett. 49, 1062 (1982).
[64] J. P. Cotton, J. Physique-Lett. 41, L-231 (1980).
[65] J. C. le Guillou, and J. Zinn-Justin, Phys. Rev. B32, 3976 (1980).
[66] M. G. Watts, J. Phys. A8, 61 (1975).
[67] G. Parisi, and N. Sourlas, Phys. Rev. Lett. 46, 871 (1981).
[68] H. Gould and K. Holl, J. Phys. A14, L443 (1981).
[69] W. A. Seitz and D. J. Klein, J. Chem. Phys. 75, 5190 (1981).
[70] R. B. Pearson, Phys. Rev. B22, 2579 (1980), B. Nienhuis, E. K. Riedel, and M. Schick, J. Phys. A13, L189 (1980).
[71] F. Family and H. Gould, Submitted to J. Phys. A (1983).
[72] C. Domb and G. S. Joyce, J. Phys. C5, 956 (1972).

[73] J. A. Marqusee and J. M. Deutch, J. Chem. Phys. 75, 5179 (1981).
[74] J. J. Prentis, J. Chem. Phys. 76, 1574 (1982).
[75] W. R. Bauer, F. H. C. Crick, and J. H. White, Sci. Am. 243, 118 (July 1980).
[76] A. Malakis, J. Phys. A9, 1283 (1976).
[77] D. Stauffer, Phys. Rev. Lett. 41, 1333 (1978).
[78] H. J. Herrmann, Z. Phys. B 32, 335 (1979).
[79] H. P. Peters, D. Stauffer and K. Loewenich, Z. Phys. B 34, 399 (1979).
[80] B. Derrida and L. de Seze, J. Physique 43, 475 (1982).

Random-Walk Model of Chain Polymer Adsorption Behavior at Critical Energy and Relation to a Reflecting Boundary Condition

Robert J. Rubin
National Bureau of Standards
Washington, D.C. 20234

ABSTRACT

It is found that there is a subtle difference between the set of energy-weighted random walks generated in the discrete random-walk model of polymer chain adsorption at the critical energy and the corresponding set of random walks generated in the presence of a reflecting boundary. This difference is lost in the continuous random flight model of adsorption.

INTRODUCTION

The first investigation of a simple random walk model of the adsorption of an isolated polymer chain at a solution surface was reported thirty years ago.[1] There now exists an extensive literature which has been summarized in a number of recent reviews.[2,3] The purpose of this paper is to describe some subtle differences between the set of energy-weighted random walks generated in a discrete random walk model of polymer chain adsorption at the critical energy and the corresponding set of random walks generated in the presence of a reflecting boundary. These subtle differences have their analog in the lattice model of self-avoiding walks at the theta point where volume exclusion of monomers is exactly compensated by monomer-monomer attraction.

The discrete random walk model and some of its already-known properties are described in the second section. Next, in the third section, behavior of the random walk model at the critical energy of adsorption is examined in detail and compared with the analogous model when the solution surface acts as a reflecting surface. There are both differences and similarities in properties of the two models. For example, for random walks which start at the surface in the two models, the probability of returning to the surface at the Nth step, if $N >> 1$, is the same. However the average number of visits to the surface in the two models, while proportional to $N^{1/2}$ for $N >> 1$, differ by a factor $\exp\theta_c$ where $\theta_c = \epsilon_c/kT$ is the reduced critical energy of adsorption. The average number of visits in the adsorption model is the smaller. This difference is not accounted for in de Gennes' continuum limit random flight model of weak adsorption.

DISCRETE RANDOM WALK MODEL

In our lattice model of a polymer chain in a solution bounded by a plane adsorbing surface,[4] the solution phase is represented by an array of layers of sites parallel to the bounding surface layer. The surface layer of sites is counted as the $k = 0$ layer, and successive layers of sites in the solution phase are labeled by $k = 1, 2, 3, \cdots$. Each distinct random walk configuration of N steps corresponds to a distinct polymer chain configuration consisting of $N + 1$ monomers. In the model, each random walk configuration of N steps which starts in layer k_0 with η steps lying in the surface layer has the same *a priori* probability; but relative to an N-step configuration with $\eta' < \eta$ steps lying in the surface layer, the probability of the former is larger by the Boltzmann weighting factor $\exp[(\eta - \eta')\theta]$, where $\theta = \epsilon/kT$ and $\epsilon > 0$ is the energy gained for each monomer unit of a configuration in the surface layer.

0094-243X/84/1090073-11 $3.00 1984 American Institute of Physics

In order to mimic the presence of the boundary surface layer in the generation of random walk configurations, we can use either one of two equivalent methods in a nearest-neighbor random walk model. First, in a semi-infinite set of layers, make $k = -1$ an absorbing layer.[5] Then, any random walk of N steps starting at $k_0 \geqslant 0$ and ending in, or to the right of, layer $k = 0$ at step N could not have crossed layer $k = 0$ (visited the absorbing layer $k = -1$) at an intermediate step. In addition, Rubin[4] introduced a counting variable to keep track of the number of visits to the surface layer $k = 0$ during the course of the random walk. It is then a straightforward matter to calculate the energy-weighted probability of the surviving N-step walks which visit $k = 0$ η times. The second method for generating random walk configurations in the presence of a boundary is based on a formal procedure developed by Rubin and Weiss[6] for counting visits of a random walk to a set of points. In the present application, the set consists of two points, $k = -1$ and $k = 0$, and the random walk, which starts at $k_0 \geqslant 0$, is performed on the infinite line, $-\infty < k < \infty$. It is again a simple matter to calculate the weighted probability of N-step walks which visit $k = 0$ η times and $k = -1$ not at all.

For the first of the above methods, we summarize the recurrence relations between the energy-weighted probabilities, $P(k, N + 1)$ and $P(k', N)$, where $P(k', N)$ is the probability that the walker is in layer k' at the Nth step. The relations, if $k = -1$ is an absorbing site, are

$$P(k, N+1) = (a/2)P(k+1, N) + (1 - a)P(k, N) + (a/2)P(k-1, N), \; k \geqslant 1 \qquad (1)$$

and

$$P(0, N+1) = e^{\theta}\left[(a/2)P(1, N) + (1 - a)P(0, N)\right], \qquad (2)$$

where $1 - a = Z'/Z$ is the probability that a random walk step is taken within a layer [Z is the nearest-neighbor coordination number and Z′ is the nearest-neighbor coordination number within a single layer.] and $a/2$ is the probability that a random walk step is taken from one layer to another. In Equation (2) for $P(0, N+1)$, the transition probabilities are increased by the factor $\exp\theta$ to account for the greater weight of those configurations of $N + 1$ steps where the $N + 1$ st step is made in the surface layer $k = 0$. Equations (1) and (2) can be written in matrix form as

$$\mathscr{P}(N + 1) = \mathscr{W}\mathscr{P}(N) \qquad (3)$$

or

$$\mathscr{P}(N + 1) = \mathscr{W}^{N+1}\mathscr{P}(0) \qquad (4)$$

where the vector $\mathscr{P}(N)$ is

$$\mathscr{P}(N) = \begin{pmatrix} P(0, N) \\ P(1, N) \\ P(2, N) \\ \vdots \end{pmatrix} \qquad (5)$$

and the weighted transition probability matrix $\mathscr{W}$ is

$$\mathscr{W} = \begin{pmatrix} (1 - a)e^{\theta} & (a/2)e^{\theta} & 0 & 0 & \cdots \\ a/2 & 1 - a & a/2 & 0 & \cdots \\ 0 & a/2 & 1 - a & a/2 & \cdots \\ 0 & 0 & a/2 & 1 - a & \\ \vdots & \vdots & \vdots & & \end{pmatrix}. \qquad (6)$$

The starting vector, $\mathscr{P}(0)$, is

$$\mathscr{P}(0) = \begin{bmatrix} 0 \\ \vdots \\ 0 \\ 1 \\ 0 \\ \vdots \end{bmatrix} \tag{7}$$

where the only nonzero component of $\mathscr{P}(0)$ is $P(k_0, 0) = 1$. In case the random walk starts in the surface layer $k = 0$, the correct count of visits of the random walk to the surface layer is achieved most conveniently by setting $P(0,0)$, the only nonzero component of $\mathscr{P}(0)$, equal to e^{θ}

$$\mathscr{P}(0) = \begin{bmatrix} e^{\theta} \\ 0 \\ \vdots \end{bmatrix} \tag{7a}$$

It is clear from the explicit expression for the components of $\mathscr{P}(N + 1)$ in Equation (4) that the sum of the components of $\mathscr{P}(N)$ is proportional to the weighted probability that the random walk survives at least to the Nth step. Furthermore this sum has the structure

$$\sum_{k=0}^{\infty} P(k, N) = \sum_{\eta=0} A_{\eta}(N)e^{\eta\theta} \tag{8}$$

where $A_{\eta}(N)$ is the fraction of all N-step walks which start at k_0, never visit $k = -1$, and visit $k = 0$ η times. Rubin[4] obtained explicit expressions for the weighted probabilities $P(k, N)$ of Equations (1) and (2) corresponding to the starting condition (7a) by using the method of generating functions. Using the same method, the generating function solution of Equations (1) and (2) corresponding to the starting condition (7) is

$$\Gamma_{k_0}(\varnothing, z) = \left[1 - z(1 - a + a\cos\varnothing)\right]^{-1} \left\{ e^{ik_0\varnothing} + \frac{[1 - e^{-\theta} - \tfrac{1}{2}aze^{-i\varnothing}]\, \mathscr{G}(k_0, z)}{1 + \tfrac{1}{2}az\mathscr{G}(1, z) - (1 - e^{-\theta})\mathscr{G}(0, z)} \right\} \tag{9}$$

where the generating function, $\zeta_{k_0}(k, z)$, for $P(k, N)$ is

$$\zeta_{k_0}(k, z) = \sum_{z=0}^{\infty} P(k, N)\, z^N \tag{10}$$

and

$$\Gamma_{k_0}(\varnothing, z) = \sum_{k=0}^{\infty} \zeta_{k_0}(k, z)\, e^{ik\varnothing} \quad . \tag{11}$$

The explicit expression for $\zeta_{k_0}(k, z)$ obtained from Equation (9) is

$$\zeta_{k_0}(k, z) = \frac{1}{2\pi}\int_{-\pi}^{\pi} d\varnothing\, e^{-ik\varnothing}\Gamma(\varnothing, z)$$

$$= \mathscr{G}(k_0 - k, z) + \frac{1 - e^{-\theta} - \tfrac{1}{2}az\mathscr{G}(k_0, z)\mathscr{G}(k + 1, z)}{1 + \tfrac{1}{2}az\mathscr{G}(1, z) - (1 - e^{-\theta})\mathscr{G}(0, z)} \tag{12}$$

where

$$\mathscr{G}(k, z) = \left\{(1 - z)[1 - (1 - 2a)z]\right\}^{-1/2} \left\{\frac{1 - (1 - a)z - \{(1 - z)[1 - (1 - 2a)z]\}^{1/2}}{az}\right\}^{|k|} \quad (13)$$

is the perfect, infinite lattice random walk generating function or propagator. That is, the generating function solution of the random walk recurrence equation on the infinite lattice

$$\hat{P}(k, N+1) = (a/2)\hat{P}(k+1, N) + (1 - a)\hat{P}(k, N) + (a/2)\hat{P}(k - 1, N), -\infty < k < \infty \quad (14)$$

with the starting condition $\hat{P}(k, 0) = \delta_{k,0}$ is[3]

$$\hat{\Gamma}(\varnothing, z) = [1 - z(1 - a + a\cos\varnothing)]^{-1}$$

$$= \sum_{k=-\infty}^{\infty} \mathscr{G}(k, z)e^{ik\varnothing} \quad (15)$$

where

$$\mathscr{G}(k, z) = \sum_{N=0}^{\infty} \hat{P}(k, N)z^N \ .$$

Thus, $\zeta_{k_0}(k, z)$ in Equation (12), the generating function for the solution of the modified random walk recurrence Equations (1) and (2) for the initial condition (7), is expressed in terms of the generating function for the solution of the translationally invariant recurrence Equation (14). Such relations are familiar in solutions of locally modified recurrence equation.[4,7]

Having obtained the generating function solution (9)—(11) of the recurrence Equations (1) and (2), we have the generating function of the sum of the weighted probabilities that the random walk survives up to the Nth step

$$\Gamma_{k_0}(O,z) = \left(\frac{1}{1 - z}\right)\left\{1 - \frac{\left[1 - (1 - \tfrac{1}{2}az)e^{\theta}\right] \mathscr{G}(k_0, z)}{\mathscr{G}(0, z) - \left[\mathscr{G}(0, z) - 1 - \tfrac{1}{2}az\ \mathscr{G}(1, z)\right]e^{\theta}}\right\} \quad (16)$$

$$= \sum_{N=0}^{\infty}\left\{\sum_{k=0}^{\infty} P(k, N)\right\} z^N \ . \quad (17)$$

Equation (16) can be written as a simple power series in e^{θ}

$$\Gamma_{k_0}(0, z) = \left(\frac{1}{1 - z}\right)\left\{1 - \frac{\mathscr{G}(k_0, z)}{\mathscr{G}(0, z)} + \frac{\mathscr{G}(k_0, z)}{\mathscr{G}(0, z)}\left[\frac{1}{\mathscr{G}(0, z)} + \frac{az}{2}\left(\frac{\mathscr{G}(1, z)}{\mathscr{G}(0, z)} - 1\right)\right] \times \right.$$

$$\left. \sum_{\eta=1}^{\infty}\left[1 - \frac{1}{\mathscr{G}(0, z)} - \frac{az}{2}\frac{\mathscr{G}(1, z)}{\mathscr{G}(0, z)}\right]^{\eta-1} e^{\eta\theta}\right\} \ .$$

$$= \sum_{\eta=0}^{\infty} \mathscr{H}_{\eta}(z)e^{\eta\theta} \quad (18)$$

The coefficient of $e^{\eta\theta}$ in Equation (18) is the generating function of $A_\eta(N)$ in Equation (8), the fraction of N-step random walks which start at k_0, never visit $k = -1$ and visit $k = 0$ η times is,

$$\mathcal{H}_\eta(z) = \left(\frac{1}{1-z}\right) \frac{\mathcal{G}(k_0, z)}{\mathcal{G}(0, z)} \left[\frac{1}{\mathcal{G}(0, z)} + \frac{az}{2}\left(\frac{\mathcal{G}(1, z)}{\mathcal{G}(0, z)} - 1\right)\right] \times$$

$$\left[1 - \frac{1}{\mathcal{G}(0, z)} - \frac{az}{2}\frac{\mathcal{G}(1, z)}{\mathcal{G}(0, z)}\right]^{\eta-1}, \eta \geqslant 1 . \tag{19}$$

The generating function of $A_0(N)$, which is also the probability of never visiting $k = 0$ in an N-step walk, is

$$\mathcal{H}_0(z) = \left(\frac{1}{1-z}\right)\left(1 - \frac{\mathcal{G}(k_0, z)}{\mathcal{G}(0, z)}\right) . \tag{20}$$

In the special case where the random walk starts in the surface layer, and the initial starting location is counted as a visit to the surface layer, see Equation (7a), Equation (18) remains valid if k_0 is set equal to zero. However, Equation (16) may be written more simply as

$$\Gamma_0(0, z) = \left(\frac{e^\theta}{1-z}\right) \frac{1 + \tfrac{1}{2}az[\mathcal{G}(1, z) - \mathcal{G}(0, z)]}{\mathcal{G}(0, z) - [\mathcal{G}(0, z) - 1 - \tfrac{1}{2}az\,\mathcal{G}(1, z)]e^\theta} ; \tag{21}$$

and the generating function for $A_\eta(N)$, $\eta \geqslant 1$, is

$$\mathcal{H}_\eta(z) = \left(\frac{1}{1-z}\right)\left[\frac{1}{\mathcal{G}(0,z)} + \frac{az}{z}\left(\frac{\mathcal{G}(1,z)}{\mathcal{G}(0,z)} - 1\right)\right] \times$$

$$\left[1 - \frac{1}{\mathcal{G}(0,z)} - \frac{az}{2}\frac{\mathcal{G}(1,z)}{\mathcal{G}(0,z)}\right]^{\eta-1} . \tag{22}$$

$\mathcal{H}_1(z)$ is the generating function for the probability of not revisiting layer 0 up to step N, $A_1(N)$.

Our principle concern in this paper is to compare some properties of the discrete random walk model of polymer chain adsorption at a solution surface and at the adsorption-desorption transition with the analogous random walk model of polymer chain configurations near a reflecting surface. This investigation grew out of our attempt to study in detail the relation between the discrete random walk adsorption model and de Genne's continuous random flight adsorption model.[8,9] In de Genne's model, a mixed boundary condition

$$\left(\frac{\partial P}{\partial \chi} + KP\right)_{\chi = 0} = 0$$

of the random flight equation

$$\frac{\partial P}{\partial N} = \frac{1}{\sigma}\frac{\partial^2 P}{\partial \chi^2}$$

is imposed at the solution boundary surface $\chi = 0$. In this model of adsorption, the adsorbed state corresponds to $K > 0$. The case of negligible adsorption energy corresponds to the absorbing boundary condition $K = -\infty$ [$P(0, N) = 0$], and the adsorption-desorption transition corresponds to the reflecting boundary condition $K = 0 \left[\left(\frac{\partial P}{\partial \chi} \right)_{\chi = 0} = 0 \right]$.

We will first summarize results obtained in the discrete model.[4] Rubin[4] treated the special case where the random walk starts in the solution surface layer, $k_0 = 0$, and found that for $N >> 1$ the total weighted probability of all surviving N-step walks is

$$\sum_{k=0}^{\infty} P(k, N) \cong \begin{cases} (2e^{-\theta} - 2 + a)^{-1} (2a/N\pi)^{1/2}, \ \theta < \theta_c & \text{(23a)} \\ 1 \quad , \theta = \theta_c & \text{(23b)} \\ z_+^{-(N+1)} f(z_+) , \theta > \theta_c & \text{(23c)} \end{cases}$$

where z_+ is the root of the denominator in Equation (16) or (21),

$$\mathcal{G}(0, z) - [\mathcal{G}(0, z) - 1 - \tfrac{1}{2}az\, \mathcal{G}(1, z)]e^{\theta} = 0$$

which lies between $z = 0$ and $z = 1$, and where $\exp\theta_c = 2/[2 - a]$ is the critical value at which $z_+ = 1$.

Rubin also evaluated the quantity

$$<\eta> = \frac{d}{d\theta} \ln \left[\sum_{k=0}^{\infty} P(k, N) \right] = \left[\sum_{\eta=1}^{\infty} \eta A_\eta(N) e^{\eta\theta} \right] \Big/ \left[\sum_{\eta=0}^{\infty} A_\eta(N) e^{\eta\theta} \right]$$

which is the weighted average value of the number of visits of the random walk in the surface layer. The result for $N >> 1$ is[4]

$$<\eta> \cong \begin{cases} \left[1 - \left(\frac{2-a}{2} \right) e^{\theta} \right]^{-1}, \quad \theta < \theta_c & \text{(24a)} \\ (2-a)(2N/a\pi)^{1/2}, \quad \theta = \theta_c & \text{(24b)} \\ N\left\{ 1 - \frac{1/2}{1 - e^{-\theta}} \left[1 - \left(1 + \left(\frac{a}{1-a} \right)^2 \left(\frac{e^{-\theta}}{1 - e^{-\theta}} \right) \right)^{-1/2} \right] \right\}, \quad \theta > \theta_c & \text{(24c)} \end{cases}$$

Three different forms of the asymptotic dependence of $\sum_{k=0}^{\infty} P(k, N)$ and of $<\eta>$ on N for $N >> 1$ are obtained, depending upon whether $\theta < \theta_c$, $\theta = \theta_c$ or $\theta > \theta_c$. These different asymptotic forms have a simple physical interpretation. Consider first the simplest or unweighted case, $\theta = 0$, where the probabilities of all surviving N-step configurations have the same weight regardless of the number of visits of any configuration to layer $k = 0$. In this case, the relation

$$\sum_{k=0}^{\infty} P(k, N) \cong (2/Na\pi)^{1/2}, \quad \theta = 0 \tag{25}$$

expresses the known result for a one-dimensional random walk that the walk will ultimately end in the absorbing point, $k = -1$, i.e., $\sum_{k=0}^{\infty} P(k, N) \to 0$ as $N \to \infty$. As the magnitude of the weighting factor, e^{θ}, per visit to layer 0 increases, $\sum_{k=0}^{\infty} P(k, N)$ for surviving random walk configurations increases until $\sum_{k=0}^{\infty} P(k, N) \cong 1$ at $e^{\theta} = 2(2 - a)^{-1}$. For still larger values of e^{θ}, $\sum_{k=0}^{\infty} P(k, N)$ grows exponentially with N.

The associated average number of visits of N-step configurations to surface layer 0 in the unweighted case $\theta = 0$, is, according to Equation (24a)

$$<\eta> \cong 2/a\,. \tag{26}$$

Thus, members of the surviving set of unweighted N-step random walk configurations, in the limit $N \rightarrow \infty$, visit the surface layer an average of 2/a times. As the weighting factor e^{θ} increases, the relative contribution to $<\eta>$ or $\sum_{k=0}^{\infty} P(k, N)$ of random walk configurations in this same set which visit k = 0 η times is increased by the factor $e^{\eta\theta}$. According to Equations (24a)—(24c), the value of $<\eta>$ increases with θ in the range $0 \leqslant \theta < \theta_c$, but its asymptotic value is independent of N until at $\theta = \theta_c$, $<\eta> \sim N^{1/2}$. Finally, for $\theta > \theta_c$, the value of $<\eta>$ is proportional to N, i.e., a nonzero average fraction of all steps lie in the surface layer. In this energy range, the typical random walk configuration corresponds to a polymer chain in an adsorbed state.

MODEL BEHAVIOR AT CRITICAL ENERGY

In this section we compare two properties of the discrete random walk models which characterize behavior at the boundary surface. For random walk configurations which start in the surface layer, we calculate both the probability of returning to the surface layer at the Nth step, P(0, N), and the average number of visits to the surface layer in an N step walk. Each of these quantities is calculated for energy-weighted random walk configurations at the critical energy of adsorption and for random walk configurations originating next to a reflecting surface. The random walk adsorption model has already been defined in the preceding section. The transition probability matrix in this case, Equation (6), reduces to the following form at $e^{\theta_c} = 2/(2 - a)$

$$\mathcal{W}_c = \begin{pmatrix} \frac{2(1-a)}{2-a} & \frac{a}{2-a} & 0 & 0 & \cdots \\ a/2 & 1-a & a/2 & 0 & \cdots \\ 0 & a/2 & 1-a & a/2 & \cdots \\ 0 & 0 & a/2 & 1-a & \\ \vdots & \vdots & \vdots & & \end{pmatrix} \tag{27}$$

The matrix $\mathcal{W}_c$ is stochastic, i.e., all row sums are equal to unity. This property can be used to account for the asymptotic normalization of $\sum_{k=0}^{\infty} P(k, N)$ at e^{θ_c}, Equation (23b). The transition probability matrix in the case of a reflecting boundary has the form[3]

$$\mathcal{W}_r = \begin{pmatrix} 1-a/2 & a/2 & 0 & 0 & \cdots \\ a/2 & 1-a & a/2 & 0 & \cdots \\ 0 & a/2 & 1-a & a/2 & \cdots \\ 0 & 0 & a/2 & 1-a & \\ \vdots & \vdots & \vdots & & \end{pmatrix} \tag{28}$$

The matrix $\mathcal{W}_r$ is not only stochastic but it is readily verified by summing the associated set of recurrence equations for the component P(k, N)'s that the total probability is conserved at each step,

$$\sum_{k=0}^{\infty} P(k, N+1) = \sum_{k=0}^{\infty} P(k, N) . \tag{29}$$

i.) The probabilities $P_c(0, N)$ and $P_r(0, N)$

In the adsorption model, the generating function for $P_c(0, N)$, the probability of returning to the surface layer at step N at the critical energy for random walks which start in layer $k = 0$ is obtained from Equation (12) by substituting therein $k_0 = 0$, $e^{\theta_c} = 2/(2 - a)$, and the expression for $\mathscr{G}(k, z)$ given in Equation (13). After some simplification, one obtains

$$\zeta_c(0, z) = \frac{2}{a(2 - a + az)} \left(\left[\frac{1 - (1 - 2a)z}{1 - z} \right]^{1/2} - 1 + a \right) . \tag{30}$$

$P_c(0, N)$ is the coefficient of z^N in the expansion of $\zeta_c(0, z)$ in ascending powers of z. If one follows the method used by Rubin[4] to calculate the asymptotic value of $P_c(0, N)$ from its generating function, then it follows that the principal contribution to $P_c(0, N)$ comes from the most singular term in the expansion of $\zeta_c(0, z)$ in the neighborhood of $z = 1$. Substituting $z = 1 + y$ in Equation (30), the most singular term is

$$\zeta_c(0, y) = \left(\frac{2}{-ay} \right)^{1/2} + \cdots \tag{31}$$

The associated asymptotic value of $P_c(0, N)$ is then (see Rubin,[4] Appendix)

$$P_c(0, N) \cong \left(\frac{2}{a\pi N} \right)^{1/2} . \tag{32}$$

In calculating the analogous probability $P_r(0, N)$ for return to the starting layer which is adjacent to a reflecting layer, we can make use of the isomorphism between the reflecting-wall random walk on the layers $k = 0, 1, 2, ...$ and the random walk on the infinite line, Equation (14), with the starting probability distribution

$$\hat{P}(k, 0) = \delta_{k,0} + \delta_{k,-1} . \tag{33}$$

According to Equation (15), the generating function solution of Equation (14) for $\hat{P}(0, N)$ corresponding to the starting distribution (33) is

$$\mathscr{G}(0, z) + \mathscr{G}(-1, z) = \frac{1}{az} \left(\left[\frac{1 - (1 - 2a)z}{1 - z} \right]^{1/2} - 1 \right) \tag{34}$$

where we have used the definition of $\mathscr{G}(k, z)$, Equation (13), in obtaining the final form in (34). This generating function expression in turn is the generating function, $\zeta_r(0, z)$, for $P_r(0, N)$, i.e.,

$$\zeta_r(0, z) = \frac{1}{az} \left(\left[\frac{1 - (1 - 2a)z}{1 - z} \right]^{1/2} - 1 \right) . \tag{35}$$

It is clear in comparing the expressions for $\zeta_c(0, z)$ and $\zeta_r(0, z)$ in Equations (30) and (35) that these generating functions are unequal. Nevertheless, the most singular term in the expansion of $\zeta_r(0, z)$ in the neighborhood of $z = 1$, obtained by substituting $z = 1 + y$ in Equation (35) and expanding in powers of y is identical with that obtained for $\zeta_c(0, y)$ in Equation (31)

$$\zeta_r(0, y) = \left(\frac{2}{ay}\right)^{1/2} + \cdots \tag{36}$$

Thus, in the limit N >> 1 the two probabilities are asymptotically equal, namely,

$$P_c(0, N) \cong P_r(0, N)\ . \tag{37}$$

At the same time, if one calculates $P_c(0, 1)$ and $P_c(1, 1)$ and $P_r(0, 1)$ and $P_r(1, 1)$ using the respective starting conditions $P_c(k, 0) = \frac{2}{2 - a}\delta_{k,0}$ and $P_r(k, 0) = \delta_{k,0}$ and the transition probability matrices $\mathscr{W}_c$ and $\mathscr{W}_r$, Equations (28) and (29), one obtains

$$P_c(0, 1) = (1 - a)\left(\frac{2}{2 - a}\right)^2$$

$$P_c(1, 1) = \frac{a}{2 - a}$$

$$P_r(0, 1) = 1 - a/2$$

and

$$P_r(1, 1) = a/2\ .$$

The values of $P_c(0, 1)$ and $P_r(0, 1)$ can also be obtained as the coefficients of z in the expansions of the generating functions, $\zeta_c(0, z)$ and $\zeta_r(0, z)$ in Equations (30) and (35). Thus after one step, the normalized weighted probabilities are different

$$\frac{P_c(0, 1)}{P_c(0, 1) + P_c(1, 1)} = 1 + \frac{2a(1 - a)}{(2 - a)^2}$$

$$\frac{P_r(0, 1)}{P_r(0, 1) + P_r(1, 1)} = 1 - a/2\ .$$

ii.) The average number of visits to layer 0, $<\eta>_c$ and $<\eta>_r$

The average value, $<\eta>_c$, has been calculated by Rubin[4] and appears in the preceding section, Equation (24b).

The average value, $<\eta>_r$, can be calculated using the formalism outlined recently by Rubin and Weiss[6] for counting the number of visits to a set of points. In the present application, the set consists of one point, k = 0. Again we take advantage of the isomorphism between the reflecting-wall random walk on the nonnegative integer lattice starting at k = 0 and the random walk on the infinite line with the starting distribution, Equation (33),

$$P(k, 0) = [\chi\delta_{k,0} + \delta_{k,-1}]\big|_{\chi = 1}\ .$$

If, for each visit to k = 0, random walk configurations are weighted by a factor χ, then the generating function for the probability of visiting k = 0 η times in an N-step walk, is the coefficient of χ^η in (see Equations (13), (21) and (23) in ref. 6)

$$\Gamma(0, \chi, z) = (1 - z)^{-1} \left\{ 1 + \frac{(\chi - 1)\, \mathcal{G}(-1, z)}{\chi + (1 - \chi)\, \mathcal{G}(0, z)} \right.$$

$$\left. + 1 + \frac{(\chi - 1)\, \mathcal{G}(0, z)}{\chi + (1 - \chi)\, \mathcal{G}(0, z)} \right\} \quad (38)$$

The generating function for, $<\eta>_r$, the average number of visits to k = 0 in an N-step walk is obtained by differentiating $\Gamma(0, \chi, z)$ in Equation (38) with respect to χ and then setting $\chi = 1$. The result is

$$\frac{\partial}{\partial \chi} \Gamma(0, \chi, z)\Big|_{\chi = 1} = (1 - z)^{-1} \left[\mathcal{G}(-1, z) + \mathcal{G}(0, z) \right]$$

$$= \frac{1}{az} \left\{ \frac{[1 - (1 - 2a)z]^{1/2}}{(1 - z)^{\frac{3}{2}}} - \frac{1}{1 - z} \right\} \quad . \quad (39)$$

The coefficient of z^N in the expansion of the right-hand side of Equation(39) is $<\eta(N)>_r$ the average number of visits to k = 0 in an N-step walk which starts at k = 0 in the case where there is a reflecting boundary next to k = 0. The principal contribution to $<\eta(N)>_r$ in the limit of N >> 1 comes from the most singular term in the expansion of the generating function (39) about the point z = 1. Substituting z = 1 + y in (39) and expanding in ascending powers of y, the most singular term is

$$\frac{\partial}{\partial \chi} \Gamma(0, \chi, y)\Big|_{\chi = 1} = (2/a)^{1/2} (-y)^{-\frac{3}{2}} + \cdots \quad (40)$$

The corresponding expression for the leading contribution to $<\eta(N)>_c$ in case N >> 1 comes from the most singular term in the expansion of the integrand of Equation (51) in ref. 4 about the point z = 1 [note that in the notation used in this paper z here is the same as y in Equation (51)]. The most singular term contributing to $<\eta(N)>_c$ is

$$(2 - a)(2a)^{-1/2} (-y)^{-\frac{3}{2}} \quad . \quad (41)$$

It follows from Equations (40) and (41) and the result in (25b) obtained by Rubin[4] that

$$<\eta(N)>_r \cong 2(2 - a)^{-1} <\eta(N)>_c$$

$$\cong e^{\theta_c} <\eta(N)>_c \quad (42)$$

or

$$<\eta(N)>_r \cong 2\,(2N/a\pi)^{1/2} \quad (43)$$

SUMMARY

In this paper we have compared two properties of the energy-weighted random walk model of chain polymer adsorption with the corresponding properties of a random walk with a reflecting boundary. In the adsorption model there is an absorbing boundary, but for each visit of a random walk configuration to the point adjacent to the absorbing point the probability of the configuration is increased by a factor e^{θ}. The probability that the unweighted random walk survives until step N decreases as $N^{-1/2}$, and as e^{θ} increases, there is a critical value e^{θ_c} at which the probability that the energy-weighted random walker survives until step N is asymtotically equal to unity. The probability of returning to the starting point in the two models is asymptotically the same at the critical energy for walks which start next to the boundary

$$P_c(0, N) \cong P_r(0, N) , \quad N >> 1 .$$

However, the average number of visits to the starting point is different

$$<\eta(N)>_r / <\eta(N)>_c \cong e^{\theta_c} , \quad N >> 1 \quad .$$

REFERENCES

1. H.L. Frisch, R. Simha, and F.R. Eirich, J. Chem. Phys. **21**, 365 (1953).
2. E. Dickenson and M. Lal, Adv. Mol. Relaxation Interactions **17**, 1 (1980).
3. G.H. Weiss and R.J. Rubin, Adv. Chemical Physics **52**, 363 (1983).
4. R.J. Rubin, J. Chem. Phys. **43**, 2392 (1965).
5. J. Pouchlý, Coll. Czech. Chem. Comm. **28**, 1804 (1963).
   E.A. DiMarzio, J. Chem. Phys. **42**, 2101 (1965).
6. R.J. Rubin and G.H. Weiss, J. Math. Phys. **23**, 250 (1982).
7. R.J. Rubin, J. Math. Phys. **1**, 309 (1960).
8. P.G. de Gennes, Rep. Prog. Phys. **32**, 187 (1969).
9. H. Lépine and A. Caillé, Can. J. Phys. **56**, 403 (1978).

WEIERSTRASSIAN AND LEVY RANDOM WALKS, AND THE SPHERICAL MODEL OF FERROMAGNETISM WITH LONG-RANGE INTERACTIONS

F. T. Hioe
Department of Physics
St. John Fisher College
Rochester, New York 14618

and

Department of Physics and Astronomy
University of Rochester
Rochester, New York 14627

ABSTRACT

A review of the Weierstrassian and Levy random walks and the spherical model of ferromagnetism with long-range interactions, is given. The three problems share the common features of having an infinite mean square correlation length, and of having a singular structure function at the origin. Questions concerning self-similar clustering, transience, persistence, phase transition and long-range correlation are discussed.

I. WEIERSTRASSIAN RANDOM WALKS

Hughes, Shlesinger and Montroll[1] recently constructed a random walk on a lattice having a hierarchy of self-similar clusters built into the distribution function of allowed jumps. The probability density function for a step x is assumed to be (in 1 D)

$$p(x) = \frac{a-1}{2a} \sum_{n=0}^{\infty} a^{-n}\{\delta(x-\Delta b^n) + \delta(x+\Delta b^n)\} . \tag{1.1}$$

If b is an integer greater than one, then the walk takes place on a lattice of spacing Δ, and the probability of a displacement of ℓ sites at a given step is

$$p(\ell) = \frac{a-1}{2a} \sum_{n=0}^{\infty} a^{-n}\{\delta_{\ell,-b^n} + \delta_{\ell,b^n}\} . \tag{1.2}$$

The structure function,[2-4] or the Fourier transform of the jump distribution function, for this lattice walk is

$$\lambda(k) = \frac{a-1}{a} \sum_{n=0}^{\infty} a^{-n} \cos(b^n k) \tag{1.3}$$

which is of the form of the celebrated function of Weierstrass[5]

$$W(x) = \sum_{n=0}^{\infty} A^n \cos(B^n \pi x) , \tag{1.4}$$

0094-243X/84/1090085-25 $3.00

which is continuous but nowhere differentiable[6]. The expansion of $\lambda(k)$ about $k=0$ is found[1] to be

$$\lambda(k) = 1 + k^{\mu}Q(k) + \frac{a-1}{a}\sum_{m=1}^{\infty}\frac{(-1)^m}{(2m)!}\frac{k^{2m}}{(1 - b^{2m}/a)} \tag{1.5}$$

where

$$\mu = \frac{\ell na}{\ell nb} < 2 \tag{1.6}$$

and

$$Q(k) = \frac{a-1}{a\ell nb}\sum_{n=-\infty}^{\infty}\Gamma(s_n)\cos(\tfrac{1}{2}\pi s_n)\exp(-\frac{2n\pi i\ell nk}{\ell nb}) ,$$
$$s_n = -\mu + \frac{2n\pi i}{\ell nb} . \tag{1.7}$$

The case $b^2 < a$ or $\mu > 2$ leads to the usual diffusive or Gaussian behavior.

The generalization to two-dimensions is

$$p(\underline{\ell}) = \frac{a-1}{4a}\sum_{n=0}^{\infty} a^{-n}\{(\delta_{\ell_1,-b^n} + \delta_{\ell_1,b^n})\delta_{\ell_2,0} + (\delta_{\ell_2,-b^n} + \delta_{\ell_2,b^n})\delta_{\ell_1,0}\} \tag{1.8}$$

and the structure function becomes

$$\lambda(\underline{k}) = \frac{a-1}{2a}\sum_{n=0}^{\infty} a^{-n}\{\cos(k_1 b^n) + \cos(k_2 b^n)\}$$
$$= 1 + \tfrac{1}{2}\{|k_1|^{\mu}Q(|k_1|)+|k_2|^{\mu}Q(|k_2|)\} + \frac{a-1}{2a}\sum_{m=1}^{\infty}\frac{(-1)^m(k_1^{2m}+k_2^{2m})}{(2m)!(1-b^{2m}/a)} . \tag{1.9}$$

The extension to d-dimensions is obvious.

By studying the probability $1/u$ that the walker never returns to his starting point, where

$$u = \frac{1}{(2\pi)^d}\int_{-\pi}^{\pi}\cdots\int \{1 - \lambda(\underline{k})\}^{-1}d^d\underline{k} \tag{1.10}$$

Hughes et al[1] suggested that the walk

$$\text{is persistent if } 1 \lesssim \mu < 2,$$
$$\text{is transient if } 0 < \mu < 1, \tag{1.11}$$

where persistent means certain to return infinitely often to any neighborhood of the starting point.

The probability density function for the walk (1.2) can be cast into an inverse power law

$$p(x) = \frac{a-1}{a}\frac{1}{x^{\mu}} \tag{1.12}$$

where

$$x = b^n, \; n = 0,1,2,3, \ldots \quad . \tag{1.13}$$

Equation (1.12) was derived by first letting

$$x = b^n \tag{1.14}$$

and noting from Eq. (1.6) that

$$a = b^{\mu}. \tag{1.15}$$

While (1.12) can be regarded as a discrete analog of the probability density function of a Levy random walk on a continuum line with probability density

$$p(x) \simeq \text{const.}\ |x|^{-1-\sigma}, \qquad (0 < \sigma < 2) \;, \tag{1.16}$$

there are important differences. For example, $p(x)$ given by (1.16) is normalizable only if $0 < \sigma < 2$, whereas $p(x)$ given by (1.12) is normalizable, i.e.

$$\sum_x p(x) = \text{finite} \tag{1.17}$$

for any value of $\mu > 0$. The latter case is due to the fact that in (1.12) x takes values only those given by Eq. (1.13) rather than all possible lattice points.

The anslysis of Hughes et al establishes a close relationship between random walks characterized by (1.2) or (1.12) and the Weierstrassian function. The Weierstrassian function, as was beautifully demonstrated in the computer plots of Berry and Lewis[7], is self-similar in that it is oscillatory on <u>every</u> length scale (and is thus nowhere differentiable). A self-similar clustering property of the set of points visited by Levy random flights and the associated "fractal" dimension for the problem have been discussed and illustrated earlier by Mandelbrot[8].

An important feature of Levy random walks (1.16) is the infinite mean-square displacement per step, as

$$\int_{-\infty}^{\infty} x^2 p(x)\, dx = \infty \;. \tag{1.18}$$

The structure function $\lambda(k)$, which is the Fourier transform of the jump distribution function $p(x)$, generally exhibits a singular behavior at k=0. The Weierstrassian random walk considered by Hughes et al suggests this behavior as indicated in Eq. (1.5), although the small k behavior of $Q(k)$ in (1.5) has not been explicitly presented. It should also be noted that in two and higher dimensions, the structure function $\lambda(\underline{k})$ of the Weierstrassian walks is not spherically symmetrical, as can be seen from Eq. (1.9). Spherical symmetry was assumed, however, by Hughes et al in discussing the persistent or transient nature of the Weierstrassian walks.

In the following section we will discuss another type of discrete Levy random walk on a general d-dimensional lattice (d=1,2,3,..) in which the probability of each jump is given by

$$p(\underline{\ell}) \propto |\underline{\ell}|^{-d-\sigma}, \tag{1.19}$$

where $\underline{\ell}$ can be any lattice site. The singular behavior of the structure function $\lambda(\underline{k})$ at $\underline{k}=0$ will be clearly demonstrated, and the problem will be shown to be closely connected to the spherical model of ferromagnetism with long-range interactions. The question of whether the walk is transient or persistent is related to the question of whether a phase transition does or does not occur at finite temperature in the spherical model.

II. LEVY RANDOM WALK ON A LATTICE

Much of the work concerning random walks on a general d-dimensional lattice with Levy's probability density function

$$p(\underline{\ell}) \propto |\underline{\ell}|^{-d-\sigma}, \tag{2.1}$$

where $\underline{\ell}$ = lattice site, $|\underline{\ell}| = \sqrt{\ell_1^2 + \ell_2^2 + \ldots + \ell_d^2}$, has been given by Joyce[9,10] in his extensive work on the spherical model with long-range interactions. Unlike the Weierstrassian random walk, the symmetric random walk with $p(\underline{\ell}-\underline{\ell}')$ given by (2.1) is a true discrete analogy of a Levy random walk on a continuum space with

$$p(x) \simeq \text{const.} |x|^{-d-\sigma}. \tag{2.2}$$

We will describe the results in this section, and will state them in the forms which will make their connection to the corresponding results for the spherical model more transparent.

Consider a random walk on a d-dimensional lattice in which the probability of going from a lattice site $\underline{\ell}$ to a site $\underline{\ell}'$ ($\neq\underline{\ell}$) in a single step is

$$p(\underline{\ell}-\underline{\ell}') = N_{d,\sigma}^{-1} \, |\underline{\ell}-\underline{\ell}'|^{-d-\sigma} \tag{2.3}$$

where $N_{d,\sigma}$ is a normalization constant given by

$$N_{d,\sigma} = \sum_{\ell}{}' \, |\underline{\ell}|^{-d-\sigma} . \tag{2.4}$$

(The prime denotes the exclusion of the term $\underline{\ell}=0$.) Montroll and Weiss[4] have shown that many of the properties of this random walk can be described in terms of the lattice Green's function

$$P(\underline{\ell}, z) = \frac{v_o}{(2\pi)^d} \int \frac{\cos \underline{\ell}\cdot\underline{k}}{1 - z\lambda(\underline{k})} \, d\underline{k} \tag{2.5}$$

where $\lambda(\underline{k})$ is the characteristic function given by

$$\lambda(\underline{k}) = \sum_{\underline{\ell}} p(\underline{\ell}) e^{i\underline{k}\cdot\underline{\ell}} \tag{2.6}$$

and where v_o is the volume of a basic cell. For the following common two- and three-dimensional lattices with nearest neighbor vectors of length a, v_o is given by

$$
\begin{aligned}
d=2 &\begin{cases} \text{sq} & v_o = a^2 \\ \text{triangular} & v_o = (\sqrt{3}/2)a^2 \end{cases} \\
d=3 &\begin{cases} \text{sc} & v_o = a^3 \\ \text{bcc} & v_o = (4/3^{3/2})a^3 \\ \text{fcc} & v_o = 2^{-1/2}a^3 \\ \text{diamond} & v_o = (8/3^{3/2})a^3 \ . \end{cases}
\end{aligned}
\tag{2.7}
$$

Generally, for a d-dimensional Bravais lattice with primitive lattice vectors $\underline{a}_1, \underline{a}_2, \ldots, \underline{a}_d$, where $\underline{a}_j = (a_{j1}, a_{j2}, \ldots, a_{jd})$ in the Cartesian coordinates, the volume of the basic cell is given by

$$
v_o = \begin{vmatrix} a_{11} & a_{12} & \cdots & a_{1d} \\ a_{21} & a_{22} & \cdots & a_{2d} \\ \vdots & & & \\ a_{d1} & \cdots & \cdots & a_{dd} \end{vmatrix} .
$$

The characteristic function $\lambda(\underline{k})$ for $p(\underline{\ell})$ given by (2.3) cannot, in general, be evaluated in closed form, but Joyce, using the method developed by Nijboer and DeWette[11], was able to obtain the following expansion about $\underline{k} = 0$: (see Appendix A)

$$
\lambda(\underline{k}) = 1 - (I_{d,\sigma}/v_o N_{d,\sigma})|\underline{k}|^{\sigma} + O(|\underline{k}|^{\xi}) \tag{2.8}
$$

where

$$
I_{d,\sigma} = [2^{1-\sigma}\pi^{\frac{1}{2}d}\Gamma(1 - \tfrac{1}{2}\sigma)]/[\sigma\Gamma(\tfrac{1}{2}d + \tfrac{1}{2}\sigma)], \tag{2.9}
$$

and

$$
0 < \sigma < 2 \ , \qquad \xi > \sigma \ . \tag{2.10}
$$

For the particular case $\sigma = 2$, the expansion becomes

$$
\lambda(\underline{k}) = 1 + \frac{\pi^{\frac{1}{2}d}[\Gamma(1+\frac{1}{2}d)]^{-1}}{4v_o N_{d,\sigma}} |\underline{k}|^2 \ell n|\underline{k}|^2 + O(|\underline{k}|^2) \ , \tag{2.11}
$$

while for $\sigma > 2$ the analytic term $O(|\underline{k}|^2)$ dominates the non-analytic term and the analysis of the usual random walk to nearest neighbor sites applies.

For the one-dimensional lattice with unit spacing, Joyce obtained the following complete expansion of $\lambda(k)$ about $k=0$:

$$\lambda(k) = 1-[I_{1,\sigma}/2\zeta(1+\sigma)]|k|^{\sigma} + \sum_{m=1}^{\infty} (-1)^m \frac{\zeta(1+\sigma-2m)}{\zeta(1+\sigma)} \frac{k^{2m}}{(2m)!} ,$$

where $\sigma > 0$ and $\sigma \neq 2,4,6,\ldots$. It is interesting to compare the above expression with Eq. (1.5) for the Weierstrassian random walk.

The normalization constant (2.4) can be expressed, in one-dimension, in terms of the Riemann zeta function $\zeta(s) = \sum_{n=1}^{\infty} n^{-s}$ as

$$N_{1,\sigma} = 2\zeta(1 + \sigma) .$$

In two-dimensions, we can make use of Hardy's result[12]

$$\sum_{m,n=1}^{\infty} (m^2 + n^2)^{-s} = \zeta(s)\beta(s) - \zeta(2s)$$

where

$$\beta(s) = \sum_{n=0}^{\infty} (-1)^n(2n + 1)^{-s}$$

to write

$$N_{2,\sigma} = 4\zeta(1 + \tfrac{\sigma}{2})\beta(1 + \tfrac{\sigma}{2}) - 4\zeta(2 + \sigma) .$$

In three and higher dimensions, we know of no closed-form expressions for $N_{d,\sigma}$.

As is well known, the lattice Green's function $P(\underline{\ell},z)$ is the generating function for $P_n(\underline{\ell})$, the probability that a random walker, starting at the origin $\underline{\ell} = 0$ will reach the site $\underline{\ell}$, not necessarily for the first time, after a walk of n steps, i.e.

$$P(\underline{\ell},z) = \sum_{n=0}^{\infty} P_n(\underline{\ell})z^n, \quad |z| < 1 \tag{2.12}$$

where

$$P_n(\underline{\ell}) = \frac{v_o}{(2\pi)^d} \int \lambda(\underline{k})^n \cos \underline{\ell}\cdot\underline{k}\; d\underline{k} . \tag{2.13}$$

We will introduce three generating functions whose significance will become clear later. The three generating functions will be called respectively the correlation function $\Gamma(\underline{\ell},z)$, the correlation length $\Lambda(z)$ ($= q(z)^{-1}$, $q(z)$ will be called the inverse correlation length) and the "susceptibility" $\chi(z)$.

The correlation function $\Gamma(\underline{\ell},z)$ is defined by

$$\Gamma(\underline{\ell},z) = \frac{P(\underline{\ell},z)}{P(\underline{0},z)} . \tag{2.14}$$

It will be recognized that for the special value of $z = 1$,

$$\Gamma(\underline{\ell},1) = \frac{P(\underline{\ell},1)}{P(\underline{0},1)} \tag{2.15}$$

gives the probability that a random walker who starts at $\underline{\ell}=0$ and continues to walk indefinitely will eventually reach the lattice site $\underline{\ell}$ where $\underline{\ell}\neq 0$. Of importance is the behavior of $\Gamma(\underline{\ell},z)$ as $z \to 1-$. From Eqs. (2.5) and (2.8), we expand $1 - z\lambda(\underline{k})$ in the form

$$1 - z\lambda(\underline{k}) = 1-z + E\,z\,k^{\sigma} + O(k^{\xi}) \qquad 0 < \sigma < 2, \quad \xi > \sigma \;, \tag{2.16}$$

where

$$E = \frac{I_{d,\sigma}}{v_o N_{d,\sigma}} \;.$$

It is convenient to write (2.16) in the form

$$1 - z\lambda(\underline{k}) = (Ez)[q^{\sigma} + k^{\sigma} + O(k^{\xi})] \tag{2.17}$$

where the quantity $q \equiv q(z)$ defined by

$$q = (Ez)^{-1/\sigma}(1-z)^{1/\sigma} \tag{2.18}$$

turns out to be a very useful quantity which we call the inverse correlation length. The asymptotic behavior of $\Gamma(\underline{\ell},z)$ as $z \to 1-$ can be investigated by substituting (2.17) into (2.5) and transforming to the hyperspherical polar coordinates. It was found that when $q \gtrsim 0$, the decay of the correlation function as $r = |\underline{\ell}| \to \infty$ is given by (see Appendix B)

$$\Gamma(\underline{\ell},z) \sim D_{\infty}\left(\frac{a}{r}\right)^{d-\sigma} (qr)^{-2\sigma}, \qquad (\sigma < 2) \tag{2.19}$$

with q small and <u>fixed</u>. On the other hand, if q is set equal to 0 first in (2.17) so that z=1, then the decay of correlation functions as $r = |\underline{\ell}| \to \infty$ is given by

$$\Gamma(\underline{\ell},1) \sim D_o\left(\frac{a}{r}\right)^{d-\sigma} \;. \tag{2.20}$$

Equation (2.18) expresses how the inverse correlation length vanishes (i.e. the correlation length becomes infinite) as $z \to 1$, the "critical point." The "critical" exponent associated with this behavior, normally denoted by ν, is thus given by

$$\nu = 1/\sigma \;. \tag{2.21}$$

Equation (2.20) expresses how the correlation function decays at the "critical" point. If we write Eq. (2.20) into the "normal" form

$$\Gamma(\underline{\ell},1) \sim D_o\left(\frac{a}{r}\right)^{d-2+\eta} \tag{2.22}$$

then the corresponding "critical" exponent η is given by

$$\eta = 2 - \sigma \quad . \tag{2.23}$$

The third useful generating function which we will introduce for our problem is the "susceptibility" $\chi(z)$ defined by

$$\chi(z) = \sum_{\underline{\ell}} \Gamma(\underline{\ell},z) = P(\underline{0},z)^{-1} \sum_{\underline{\ell}} P(\underline{\ell},z) \quad . \tag{2.24}$$

It is easy to show that

$$\sum_{\underline{\ell}} P(\underline{\ell},z) = (1-z)^{-1} \tag{2.25}$$

so we have

$$\chi(z) = P(\underline{0},z)^{-1}(1-z)^{-1} \quad . \tag{2.26}$$

The "critical" exponent γ for $\chi(z) \sim (1-z)^{-\gamma}$ is thus

$$\gamma = 1 \quad . \tag{2.27}$$

It will be noted from (2.21), (2.23) and (2.27) that the scaling relation

$$\nu = \gamma(2-\eta)^{-1} \tag{2.28}$$

is satisfied.

The fractional power of correlation length $\Lambda(z)^{\sigma'}$ can be expressed as

$$\Lambda(z)^{\sigma'} = \frac{\sum_{\underline{\ell}} \ell^{\sigma'} P(\underline{\ell},z)}{\sum_{\underline{\ell}} P(\underline{\ell},z)} \tag{2.29}$$

where

$$\sigma' < \sigma \quad . \tag{2.30}$$

The condition (2.30) is necessary so that the numerator on the right-hand side of (2.29) would converge for $z < 1$. Since $\Lambda(z)^{\sigma'} \simeq (1-z)^{-\sigma'/\sigma}$ as $z \to 1-$ from Eq. (2.18), and $\sum_{\underline{\ell}} P(\underline{\ell},z) = (1-z)^{-1}$ from Eq. (2.25), we find from Eq. (2.29) that

$$R(z) \equiv \sum_{\underline{\ell}} \ell^{\sigma'} P(\underline{\ell},z) \simeq (1-z)^{-\frac{\sigma'}{\sigma}-1} \quad \text{as } z \to 1- \quad . \tag{2.31}$$

If we consider R(z) to be the generating function for the mean σ'th power of the end-to-end distance of the walk:

$$R(z) = \sum_{n=0}^{\infty} \langle R_n^{\sigma'} \rangle z^n ,$$

then from Eq. (2.31), the asymptotic behavior of the mean σ'th power of the end-to-end distance $\langle R_n^{\sigma'} \rangle$ after n steps of the walk for large n is (see Appendix C)

$$\langle R_n^{\sigma'} \rangle \simeq A\, n^{\sigma'/\sigma} , \qquad (2.32)$$

where A depends on the lattice considered. Equation (2.32) is an analog of the more familiar $\langle R_n^2 \rangle \simeq An^{2\nu}$ for the case when the mean square displacement of n steps is finite if we note that $1/\sigma = \nu$. Here the mean square end-to-end displacement is infinite but the fractional moments exist so long as $\sigma' < \sigma$.

The behavior of $P(z) \equiv P(\underline{0},z)$ as $z \to 1-$ can be studied by substituting (2.17) into (2.5) and transforming into the hyperspherical polar coordinates. It was found that to leading order (see Appendix D)

$$\begin{aligned} P(z) &= P(1) - c_o(1-z)^{(d-\sigma)/\sigma}, && (1 < d/\sigma < 2) \\ &= P(1) + a_o(1-z)\ln(1-z), && (d/\sigma = 2) \\ &= P(1) - b_o(1-z) && (d/\sigma > 2) \ . \end{aligned} \qquad (2.33)$$

Thus the asymptotic formula for $P_n(\underline{0})$, the probability of returning to the origin after n steps (not necessarily for the first time) for large n is

$$P_n(0) \sim a\, n^{-d/\sigma} \qquad (0 < \sigma < 2) \qquad (2.34)$$

where

$$a = \frac{\Gamma(d/\sigma)}{\sigma 2^{d-1}\pi^{\frac{1}{2}d}\,\Gamma(\frac{1}{2}d)} \left[v_o^{(d+\sigma)/d} N_{d,\sigma}/I_{d,\sigma} \right]^{d/\sigma} . \qquad (2.35)$$

The quantity $P(1)^{-1}$ gives the probability of not returning to the origin (i.e. the probability of escape) and

$$1 - P(1)^{-1} \qquad (2.36)$$

gives the probability of eventual return to the origin. Numerical values for $P(1)^{-1}$ for one- and two-dimensional lattices have been given by Joyce[9]. Unlike the (finite-range) nearest neighbor random walk, P(1) can be finite, and thus the walk can be transient in one- and two-dimensions for Levy (long-range) random walk providing

$$0 < \sigma < \min\{2,d\} \ . \tag{2.37}$$

Other properties of Levy walk such as the expected number of distinct sites visited in a walk have been studied by Gillis and Weiss[13] for walks in one- and two-dimensions.

For completeness, some corresponding properties of the (finite range) nearest neighbor random walk are given in Appendix E.

The principal results of this section, Eqs. (2.18), (2.22) and (2.26) have close correspondences in the spherical model of ferromagnetism, which we will discuss in the following section.

III. SPHERICAL MODEL OF FERROMAGNETISM

In the spherical model of Berlin and Kac[14], we consider a d-dimensional lattice of N interacting spins which has a magnetic Hamiltonian

$$\mathcal{H} = - \sum_{<\underline{\ell},\underline{\ell}'>} J_{\underline{\ell},\underline{\ell}'}\sigma_{\underline{\ell}}\sigma_{\underline{\ell}'} - mH \sum_{\underline{\ell}} \sigma_{\underline{\ell}} \tag{3.1}$$

where the spin variables $\sigma_{\underline{\ell}}$ are scalar quantities which are allowed to take all values $-\infty < \sigma_{\underline{\ell}} < \infty$ subject to the constraint

$$\sum_{\underline{\ell}} \sigma_{\underline{\ell}}^2 = N \ . \tag{3.2}$$

The first summation in (3.1) is taken over all distinct pairs of lattice sites, and $J_{\underline{\ell},\underline{\ell}}$ is the exchange integral between spins at lattice sites $\underline{\ell}$ and $\underline{\ell}'$. It is assumed that $J_{\underline{\ell},\underline{\ell}'}$ depends only on $\underline{\ell}-\underline{\ell}'$. Stanley[15] established a precise correspondence between the spherical model and the n-component spins (Heisenberg) model by considering a d-dimensional lattice of N classical spins of n components which has a magnetic Hamiltonian

$$\mathcal{H}^{(n)} = - \sum_{<\underline{\ell},\underline{\ell}'>} J_{\underline{\ell},\underline{\ell}'} \ \underline{S}_{\underline{\ell}}^{(n)} \cdot \underline{S}_{\underline{\ell}'}^{(n)} \tag{3.3}$$

where $\underline{S}_{\underline{\ell}}^{(n)}$ is an n-dimensional vector of length $n^{\frac{1}{2}}$. He argued that in the limit $N,n \to \infty$ the free energy of this model is identical to that of the spherical model. His argument was later made rigorous by Kac and Thompson[16]. Thus in the modern language of renormalization group,[17] the spherical model corresponds to $n = \infty$.

The logarithm of the partition function $Z(\beta,H)$ of the system from which all the thermodynamic functions can be derived was shown to be given by

$$\ell n\, Z(\beta,H) = \lim_{N\to\infty} \frac{1}{N} \ell n\, Z_N(\beta,H)$$

$$= \frac{1}{2} K\xi_s + \frac{1}{2} \ell n 2\pi - \frac{1}{2} \ell n K + \frac{(\beta m H)^2}{2K(\xi_s - 1)} - \frac{v_o}{2(2\pi)^d} \int \ell n[\xi_s - \lambda(\underline{k})] d\underline{k} \tag{3.4}$$

where

$$K = \beta \sum_{\underline{\ell}} J(\underline{\ell}), \qquad \beta = \frac{1}{kT} \tag{3.5}$$

$$\lambda(\underline{k}) = \sum_{\underline{\ell}} J(\underline{\ell}) \cos \underline{k}\cdot\underline{\ell} / \sum_{\underline{\ell}} J(\underline{\ell}) \tag{3.6}$$

and where the saddle point ξ_s is determined as a function of β and H from the implicit saddle-point equation

$$K = \frac{(\beta m H)^2}{K(\xi_s - 1)} + \frac{v_o}{(2\pi)^d} \int \frac{d\underline{k}}{\xi_s - \lambda(\underline{k})} \quad . \tag{3.7}$$

If the integral which occurs in (3.7) is denoted by $R(\xi_s)$

$$R(\xi_s) \equiv \frac{v_o}{(2\pi)^d} \int \frac{d\underline{k}}{\xi_s - \lambda(\underline{k})} \tag{3.8}$$

the connection between the spherical model and the theory of random walks is provided by the relations

$$R(\xi_s) = \xi_s^{-1} P(\xi_s^{-1}) \tag{3.9}$$

and

$$p(\underline{\ell}-\underline{\ell}') = J(\underline{\ell}-\underline{\ell}') / \sum_{\underline{\ell}} J(\underline{\ell}) \quad . \tag{3.10}$$

The integral (3.8) has a branch point singularity at $\xi_s = 1$. When the saddle point coincides with the branch point at $\xi = 1$, a phase transition will occur in zero magnetic field at a Curie temperature given by

$$K_c = R(1) \tag{3.11}$$

providing the integral R(1) is finite. If $H \gtrless 0$, the dominant saddle-point does not reach the branch point $\xi_s = 1$ for any non-zero temperature, and no phase transition occurs. It follows that the spherical model with ferromagnetic interactions $J(\underline{\ell}-\underline{\ell}')$ has a phase transition ($K_c^{-1} > o$) if and only if the corresponding random walk is transient. Thus the spherical model with ferromagnetic interactions $J(\underline{\ell}-\underline{\ell}')$ does not exhibit a phase transition if

$$M_1 = \sum_{\underline{\ell}} |\underline{\ell}| J(\underline{\ell}) < \infty , \qquad d = 1 \tag{3.12}$$

$$M_2 = \sum_{\underline{\ell}} |\underline{\ell}|^2 J(\underline{\ell}) < \infty, \qquad d = 2 . \tag{3.13}$$

For the case $J(\underline{\ell}) \sim A|\underline{\ell}|^{-d-\sigma}$ as $|\underline{\ell}| \to \infty$ ($\sigma > 0$, $A > 0$), there is phase transition in d=1 and d=2 if $\sigma < d$ in which case $M_d = \infty$, but there is no phase transition for $\sigma = d$, even though $M_d = \infty$. Generally, there is phase transition, and the corresponding Levy walk is transient providing Eq. (2.37) is satisfied.

We shall set the magnetic field H equal to zero, and consider the behavior of various thermodynamic functions as $T \to T_c$ or as $\xi_s \to 1+$. It will be seen that this amounts to studying the behavior of various generating functions given in Section II as $z \to 1-$.

The reduced spin-spin correlation function of the spherical model is defined by

$$\Gamma_N(\underline{\ell},T,H) = \langle\sigma_o\sigma_{\ell}\rangle_N - \langle\sigma_o\rangle_N^2 . \tag{3.14}$$

In the thermodynamic limit $N \to \infty$, it was shown that

$$\Gamma(\underline{\ell},T,H) = \lim_{N \to \infty} \Gamma_N(\underline{\ell},T,H) = \frac{1}{K\xi_s} P(\underline{\ell}, \frac{1}{\xi_s}) \tag{3.15}$$

where $P(\underline{\ell},z)$ is the random walk generating function (2.5). Hence,

$$\Gamma(\underline{\ell},T,0) = K^{-1}P(\underline{\ell},z) = \frac{P(\underline{\ell},z)}{P(\underline{0},z)} \tag{3.16}$$

and at the critical point,

$$\Gamma(\underline{\ell},T_c,0) = \frac{P(\underline{\ell},1)}{P(\underline{0},1)} . \tag{3.17}$$

Equations (3.16) and (3.17) are seen to be exactly identical to Eqs. (2.14) and (2.15), respectively. The inverse correlation length q is defined in the same way as in Eqs. (2.17) and (2.18) by

$$\xi_s - \lambda(\underline{k}) = E[q^\sigma + k^\sigma + O(k^\zeta)] , \quad \zeta > \sigma , \tag{3.18}$$

where

$$q = E^{-1/\sigma}(\xi_s - 1)^{1/\sigma} . \tag{3.19}$$

The isothermal susceptibility χ_T of the system is defined by

$$\chi_{T,H} = m\left(\frac{\partial M}{\partial H}\right)_T \tag{3.20}$$

where $M(T,H)$ is the reduced magnetization per spin. It was shown that $\chi_{T,H} = \sum_{\underline{\ell}} \Gamma(\underline{\ell},T,H)$, (Fluctuation theorem), or

$$\chi_{T,H} = (m^2\beta)K^{-1}(\xi_s-1)^{-1}\left[1 - H\left(\frac{\partial\xi_s}{\partial H}\right)_T(\xi_s - 1)^{-1}\right] . \qquad (3.21)$$

Thus in zero field,

$$\chi \equiv \frac{kT}{m^2}\chi_{T,0} = K^{-1}(\xi_s - 1)^{-1} \qquad (3.22)$$

which parallels Eq. (2.26).

For the correlation function, the result follows immediately from Eqs. (2.19), (2.22) and (2.23):

$$\Gamma(\underline{\ell},T,H) \sim D_\infty\left(\frac{a}{r}\right)^{d-\sigma}(qr)^{-2\sigma} \qquad \begin{array}{l} q \text{ small and fixed} \\ r \to \infty \\ \sigma < 2 \end{array} \qquad (3.23)$$

and

$$\Gamma(\underline{\ell},T_c,0) \sim D_o\left(\frac{a}{r}\right)^{d-2+\eta} \qquad \text{as } r \to \infty \qquad (3.24)$$

where the critical exponent

$$\eta = 2 - \sigma \qquad (0 < \sigma < \min\{2,d\}) . \qquad (3.25)$$

For the inverse correlation length (3.19) and susceptibility (3.22), we must now express the right-hand sides in terms of the temperatures $T\text{-}T_c$. This is done first through Eq. (3.7) which we write (for H=o)

$$K = R(\xi_s) \qquad (3.26)$$

and then from Eqs. (3.9), (3.11) and (2.33), we write

$$\begin{aligned} K &= K_c - c(\xi_s - 1)^{(d-\sigma)/\sigma} , && (1 < \tfrac{d}{\sigma} < 2) \\ &= K_c - a(\xi_s - 1)\ell n(\xi_s - 1), && (\tfrac{d}{\sigma} = 2) \qquad (3.27) \\ &= K_c - b(\xi_s - 1) , && (\tfrac{d}{\sigma} > 2) . \end{aligned}$$

Therefore,

$$\begin{aligned} \xi_s - 1 &= C(K_c - K)^{\sigma/(d-\sigma)} && (1 < \tfrac{d}{\sigma} < 2) \\ &= A(K_c - K)\,|\ell n(K_c - K)|^{-1} && (\tfrac{d}{\sigma} = 2) \qquad (3.28) \\ &= B(K_c - K) && (\tfrac{d}{\sigma} > 2) . \end{aligned}$$

Substituting (3.28) into (3.19), we find

$$q \sim F(T - T_c)^{1/(d-\sigma)} \qquad (1 < \frac{d}{\sigma} < 2)$$
$$\sim F(T - T_c)^{1/\sigma} |\ell n(T - T_c)|^{-1/\sigma} \qquad (\frac{d}{\sigma} = 2) \qquad (3.29)$$
$$\sim F(T - T_c)^{1/\sigma} \qquad (\frac{d}{\sigma} > 2)$$

i.e. the critical behavior of the inverse correlation length can be expressed in the usual form ($t \equiv T - T_c$)

$$q \sim Ft^{\nu} \qquad (t \to 0+,\ H = 0) \qquad (3.30)$$

with

$$\nu = \frac{1}{d-\sigma} \qquad 1 < \frac{d}{\sigma} < 2$$
$$= \frac{1}{\sigma} \qquad \frac{d}{\sigma} > 2\ . \qquad (3.31)$$

Similarly, substituting (3.28) into (3.22) gives

$$\chi \sim C\,t^{-\sigma/(d-\sigma)} \qquad (1 < \frac{d}{\sigma} < 2)$$
$$\sim C\,t^{-1} |\ell nt| \qquad (\frac{d}{\sigma} = 2) \qquad (3.32)$$
$$\sim C\,t^{-1} \qquad (\frac{d}{\sigma} > 2)$$

as $t \to 0+$, $0 < \sigma < \min\{2,d\}$. Thus

$$\chi \sim C\,t^{-\gamma} \qquad (3.33)$$

with

$$\gamma = \sigma/(d-\sigma) \qquad (1 < \frac{d}{\sigma} < 2)$$
$$= 1 \qquad (\frac{d}{\sigma} > 2)\ . \qquad (3.34)$$

It will be noted from Eqs. (3.25), (3.31) and (3.34) that the scaling relation (2.28) is satisfied.

The following table summarizes the critical exponents for the Levy random walk and the spherical model with long-range interactions.

TABLE

	Levy random walk $(0 < \sigma < 2)$	Spherical model (long-range interactions)
Correlation function	$\eta = 2 - \sigma$	$\eta = 2 - \sigma$
Correlation length	$\nu = \frac{1}{\sigma}$	$\nu = \begin{cases} \frac{1}{d-\sigma} & (1 < \frac{d}{\sigma} < 2) \\ \frac{1}{\sigma} & (\frac{d}{\sigma} > 2) \end{cases}$
Susceptibility	$\gamma = 1$	$\gamma = \begin{cases} \frac{\sigma}{d-\sigma} & (1 < \frac{d}{\sigma} < 2) \\ 1 & (\frac{d}{\sigma} > 2) \end{cases}$
	Scaling relation $\nu = \gamma(2-\eta)^{-1}$	

The work reported here was supported in part by the U.S. Department of Energy.

ACKNOWLEDGMENTS

I am very grateful to Professor Elliott Montroll and Dr. Michael Shlesinger for introducing to me the problem of Levy random walks. I have benefited greatly from many discussions with Dr. Shlesinger, in particular for illuminating me with the concept of fractal dimensions. Part of the work reported here was done during a visit to the Institute for Physical Science and Technology (IPST), University of Maryland in the month of July 1982. The hospitality and financial support of IPST is gratefully acknowledged.

APPENDIX A

Nijboer and DeWette[11] obtained a useful formula for converting slowly converging lattice sums into expressions with good convergence.

Consider a three-dimensional lattice and let the unit cell of the lattice be specified by the basic vectors $\underline{a}_1, \underline{a}_2, \underline{a}_3$. The volume of the unit cell is

$$v_o = \underline{a}_1 \cdot (\underline{a}_2 \times \underline{a}_3) \; . \tag{A1}$$

The position vector of any lattice point is

$$\underline{\ell} = \ell_1 \underline{a}_1 + \ell_2 \underline{a}_2 + \ell_3 \underline{a}_3 \; , \tag{A2}$$

where ℓ_1, ℓ_2, ℓ_3 are positive and negative integers. The unit cell of the reciprocal lattice is specified by the vectors $\underline{b}_1, \underline{b}_2, \underline{b}_3$ defined by the relations

$$\underline{a}_i \cdot \underline{b}_j = \delta_{ij} \tag{A3}$$

or

$$\underline{b}_1 = v_o^{-1}(\underline{a}_2 \times \underline{a}_3), \; \underline{b}_2 = v_o^{-1}(\underline{a}_3 \times \underline{a}_1), \; \underline{b}_3 = v_o^{-1}(\underline{a}_1 \times \underline{a}_2) \; .$$

The reciprocal lattice vector is written as

$$\underline{h} = h_1 \underline{b}_1 + h_2 \underline{b}_2 + h_3 \underline{b}_3 \; , \tag{A4}$$

where h_1, h_2, h_3 are positive and negative integers. The lattice sum to be considered is

$$S = S(\underline{R}, \underline{k}, \alpha) = \sum_{\underline{\ell}} \frac{e^{2\pi i \underline{k} \cdot \underline{\ell}}}{|\underline{\ell} - \underline{R}|^{\alpha}} \tag{A5}$$

where $\underline{R}$ is any space vector

$$\underline{R} = R_1 \underline{a}_1 + R_2 \underline{a}_2 + R_3 \underline{a}_3 \; , \tag{A6}$$

and Σ' means that the term $\underline{\ell} = 0$ is excluded from the summation.

The formula of Nijboer and DeWette gives, for d-dimensions,

$$S = \frac{1}{\Gamma\left(\frac{\alpha}{2}\right)} \left[\sum_{\underline{\ell}}{}' \frac{\Gamma\left(\frac{\alpha}{2}, \pi|\underline{\ell}-\underline{R}|^2\right)}{|\underline{\ell}-\underline{R}|^{\alpha}} e^{2\pi i \underline{k} \cdot \underline{\ell}} - \frac{\gamma\left(\frac{\alpha}{2}, \pi R^2\right)}{R^{\alpha}} + \frac{\pi^{\alpha - \frac{d}{2}}}{v_o} \sum_{\underline{h}} |\underline{h}-\underline{k}|^{\alpha - d} \Gamma\left(-\frac{\alpha}{2} + \frac{d}{2}, \pi |\underline{h}-\underline{k}|^2\right) e^{2\pi i (\underline{h}-\underline{k}) \cdot \underline{R}} \right] \tag{A7}$$

where $\Gamma(n,x)$ is the incomplete Gamma function defined by

$$\Gamma(n,x) = \int_x^\infty e^{-t} t^{n-1} dt \tag{A8}$$

and

$$\gamma(n,x) = \Gamma(n) - \Gamma(n,x) = \int_o^x e^{-t} t^{n-1} dt \,, \tag{A9}$$

and where α is any real number for which S exists. The expression holds for $\underline{R} \neq 0$ and $\underline{k} \neq 0$.

For $\underline{R} = 0$, the second term should be replaced by its limit value

$$\frac{\gamma(\frac{\alpha}{2}\,,\ \pi R^2)}{R^\alpha} \to \frac{\pi^{\alpha/2}}{\alpha/2} \ . \tag{A10}$$

For $\underline{k} = 0$ the second summation should carry a prime and the term $\pi^{\alpha/2}/v_o(\frac{\alpha}{2} - \frac{d}{2})$ should be added. In that case the sum only exists for $\alpha > d$. (A11)

Thus for the following sum for the d-dimensional lattice

$$S_{d,\sigma}(\underline{k}) \equiv \sum_{\underline{\ell}}{}' \ |\underline{\ell}|^{-(d+\sigma)} \cos(\underline{\ell}\cdot\underline{k})$$

$$= \frac{1}{\Gamma(\frac{d+\sigma}{2})} \left[\sum_{\underline{\ell}}{}' \ \frac{\Gamma\{\frac{d+\sigma}{2}\,, \pi c^2 |\ell|^2\}}{|\ell|^{d+\sigma}} \prod_{j=1}^{d} \cos \ell_j k_j - \left(\frac{2}{d+\sigma}\right) (\pi c^2)^{(d+\sigma)/2} \right.$$

$$\left. + \frac{\pi^{\frac{d}{2}+\sigma}}{v_o} \sum_{\underline{h}} \left|\underline{h} - \frac{\underline{k}}{2\pi}\right|^\sigma \Gamma\left\{-\frac{\sigma}{2}\,, \frac{\pi}{c^2}\left|\underline{h} - \frac{\underline{k}}{2\pi}\right|^2\right\} \right] \tag{A12}$$

$$S_{d,\sigma} \equiv S_{d,\sigma}(\underline{o}) \equiv N_{d,\sigma} \equiv \sum_{\underline{\ell}}{}' \ |\ell|^{-(d+\sigma)}$$

$$= \frac{1}{\Gamma(\frac{d+\sigma}{2})} \left[\sum_{\underline{\ell}}{}' \frac{\Gamma\{\frac{d+\sigma}{2}\,,\ \pi c^2 |\underline{\ell}|^2\}}{|\underline{\ell}|^{d+\sigma}} - \left(\frac{2}{d+\sigma}\right) (\pi c^2)^{(d+\sigma)/2} \right.$$

$$\left. + \frac{2\pi^{(d+\sigma)/2}}{v_o \sigma} + \frac{\pi^{\frac{d}{2}+\sigma}}{v_o} \sum_{\underline{h}}{}' \ |\underline{h}|^\sigma \ \Gamma\left\{-\frac{\sigma}{2}\,,\ \frac{\pi}{c^2} |\underline{h}|^2\right\} \right] \tag{A13}$$

where the parameter c can be chosen to speed up the convergence.

For the expansion about $\underline{k} = 0$ of $\lambda(\underline{k}) = S_{d,\sigma}(\underline{k})/S_{d,\sigma}(\underline{o})$ given by Eq. (2.8), the first term is clearly 1, and the term in $|k|^\sigma$ is obtained from (A12) as

$$N_{d,\sigma}^{-1} \frac{1}{\Gamma(\frac{d+\sigma}{2})} \frac{\pi^{\frac{d}{2}+\sigma}}{v_o} |\frac{k}{2\pi}|^\sigma \Gamma(-\frac{\sigma}{2}) .$$

APPENDIX B

Consider

$$P(\underline{\ell},z) = \frac{v_o}{(2\pi)^d} \int \frac{e^{i\underline{k}\cdot\underline{\ell}}}{1-z\lambda(\underline{k})} d^d k \tag{B.1}$$

$$\simeq \frac{v_o}{(2\pi)^d} \frac{1}{Ez} \int \frac{e^{i\underline{k}\cdot\underline{\ell}}}{q^\sigma + k^\sigma + O(k^\xi)} d^d k, \quad 0<\sigma<2, \; \xi>\sigma \tag{B2}$$

where

$$E = I_{d,\sigma}/v_o N_{d,\sigma}(L) , \quad q = (Ez)^{-1/\sigma}(1-z)^{1/\sigma} . \tag{B3}$$

Use the hyperspherical coordinates

$$dS = (\sin\theta_1)^{d-2}(\sin\theta_2)^{d-3} \ldots (\sin\theta_{d-2})d\theta_1 \ldots d\theta_{d-2}d\phi \tag{B4}$$

$$e^{i\underline{k}\cdot\underline{\ell}} = e^{ikr\cos\theta_1} \quad \text{where } k = |\underline{k}|, \quad r = |\underline{\ell}| \tag{B5}$$

and use the relation

$$\int_0^\pi e^{ikr\cos\theta_1}(\sin\theta_1)^{d-2}d\theta_1 = \int_{-1}^1 e^{ikrx}(1-x^2)^{\frac{d}{2}-1} dx$$

$$= \sqrt{\pi} \left(\frac{2}{kr}\right)^{\frac{d}{2}-1} \Gamma\left(\frac{d-1}{2}\right) J_{\frac{d}{2}-1}(kr) , \tag{B6}$$

$$d^d k = k^{d-1} dk \, dS , \tag{B7}$$

where the surface area of a unit hypersphere

$$S = \frac{2\pi^{d/2}}{\Gamma(d/2)} , \tag{B8}$$

and

$$\int_{o}^{\pi} (\sin\theta_1)^{d-2} d\theta_1 = \frac{2^{d-2}[\Gamma(\frac{d-1}{2})]^2}{\Gamma(d-1)} \quad . \tag{B9}$$

We find

$$\int \frac{e^{i\underline{k}\cdot\underline{\ell}}}{q^\sigma + k^\sigma + O(k^\xi)} d^d\mathbf{k}$$

$$= \frac{S}{\int_{o}^{\pi} (\sin\theta_1)^{d-2} d\theta_1} \sqrt{\pi}\, \Gamma\left(\frac{d-1}{2}\right)\left(\frac{2}{r}\right)^{\frac{d}{2}-1} \int_{o}^{\pi} \frac{k^{d/2} J_{d/2-1}(kr)}{q^\sigma + k^\sigma} dk \tag{B10}$$

$$= \frac{2\pi^{d/2}}{\Gamma(d/2)} \frac{\Gamma(d-1)}{2^{d-2}[\Gamma(\frac{d-1}{2})]^2} \sqrt{\pi}\, \Gamma(\frac{d-1}{2})\, 2^{\frac{d}{2}-1} r^{1-\frac{d}{2}d} \int_{o}^{\pi} \frac{k^{d/2} J_{d/2-1}(kr)}{q^\sigma + k^\sigma} dk$$

$$= A\, r^{1-\frac{1}{2}d} \int_{o}^{\pi} \frac{k^{d/2} J_{d/2-1}(kr)}{q^\sigma + k^\sigma} dk \tag{B11}$$

where
$$A = \frac{\pi^{(d+1)/2}\Gamma(d-1)}{2^{\frac{d}{2}-2}\, \Gamma(\frac{d}{2})\Gamma(\frac{d-1}{2})} \quad .$$

Let $kr = z$

$$r^{1-\frac{1}{2}d} \int_{o}^{\pi r} \frac{(kr)^{d/2}\, r^{-d/2}\, J_{d/2-1}(kr)}{(qr)^\sigma + (kr)^\sigma} r^{\sigma-1} d(kr)$$

$$= \left(\frac{1}{r}\right)^{d-\sigma} \int_{o}^{\infty} \frac{z^{\frac{1}{2}d}\, J_{\frac{1}{2}d-1}(z)}{x^\sigma + z^\sigma} dz \quad , \qquad x = qr. \tag{B12}$$

Since $|\underline{\ell}|$ should strictly be r/a, we have the basic asymptotic formula

$$P(\underline{\ell},z) \sim A(\frac{a}{r})^{d-\sigma}\, D(qr) \tag{B13}$$

where

$$D(x) = \int_0^{\pi} \frac{z^{\frac{1}{2}d} J_{\frac{1}{2}d-1}(z)}{x^{\sigma} + z^{\sigma}} \, dz \, . \tag{B14}$$

The limiting behavior of D(x) for large and small values of x has been analyzed in detail by Theumann[18]. His main results below

$$D(x) \sim B_{\infty} x^{-2\sigma} \qquad x \to \infty \tag{B15}$$

and

$$D(x) = B_o + B_1 x^{\sigma} + \ldots \quad \text{as } x \to 0+ \tag{B16}$$

lead to Eqs. (2.19) and (2.20).

APPENDIX C

The theorems of Darboux[19] and Hamy[20] are useful for deducing the asymptotic form of the nth coefficient, f_n say, of a function F(w) if the behavior of the function near its singularities w_i is known. The theorems were studied by Ninham[21] and can be summarized as follows:

Let F(w) be a function of w which is analytic with the exception of a finite number of singularities w_i. Suppose that near each singularity w_i, F(w) can be represented by

$$F(w) = (w-w_i)^{-\alpha_i} G_i(w) + H_i(w) \ , \tag{C1}$$

where $G_i(w)$ and $H_i(w)$ are regular at w_i. Then the nth Taylor coefficient, f_n of F(w) in

$$F(w) = \sum_{n=o}^{\infty} f_n w^n \tag{C2}$$

can be represented asymptotically in the form

$$f_n \sim \sum_i \frac{\phi_i(n)}{w_i^n} \ , \tag{C3}$$

where

$$\phi_i(n) = A_{oi}\binom{\alpha_i + n-1}{n} + A_{1i}\binom{\alpha_i + n-2}{n} + A_{2i}\binom{\alpha_i + n-3}{n} + \ldots \tag{C4}$$

The smallest $w_i = w_c$ is the dominant singularity and its index α is the critical exponent. We shall call such functions as of Darboux type.

The above consideration can be applied to problems of the

following type: Let F(T) be a thermodynamic property and $\phi(N)$ the appropriate configurational property such that

$$F(T) = \sum_{N=0}^{\infty} \phi(N) e^{-\beta NJ} = \sum_{N=0}^{\infty} \phi(N) z^N , \tag{C5}$$

where

$$z = e^{-\beta J}, \quad \beta = 1/kT . \tag{C6}$$

Singular behavior in F(T) arises at a temperature T_c corresponding to the radius of convergence of this power series. If as $T \to T_c$, the behavior of F(T) is given by

$$F(T) \simeq A(T-T_c)^{-h-1} , \tag{C7}$$

then the asymptotic behavior of $\phi(N)$ as $N \to \infty$ is

$$\phi(N) \sim a\, N^h , \tag{C8}$$

where

$$a = A/\Gamma(h+1) . \tag{C9}$$

In arriving at (C8) and (C9) from (C3) and (C4), we have made use of the Stirling's approximation

$$p! \sim p^p\, e^{-p}\sqrt{2\pi p} \tag{C10}$$

to write

$$\begin{aligned}\frac{(n-\alpha-1)!}{(-\alpha-1)!\,n!} &\simeq \frac{(n-\alpha-1)^{n-\alpha-1} e^{-(n-\alpha-1)}\sqrt{n-\alpha-1}}{(-\alpha-1)!\; n^n e^{-n}\sqrt{n}} \\ &\simeq \frac{(n-\alpha-1)^{-\alpha-1}(n-\alpha-1)^n e^{\alpha+1}}{(-\alpha-1)!\; n^n} \\ &\simeq \frac{n^{-\alpha-1}}{(-\alpha-1)!} .\end{aligned} \tag{C11}$$

Equations (C7) and (C8) are the results used to obtain Eq. (2.32) from Eq. (2.31).

Conversely, suppose that the coefficients f_n of the function

$$F(z) = \sum_{n=o}^{\infty} f_n z^n \tag{C12}$$

behave asymptotically as

$$f_n \sim \frac{1}{n^{1+\theta}} \sum_{r=o}^{M} \frac{a_r}{n^r} \quad \text{as } n \to \infty \quad . \tag{C13}$$

Assuming the singularity of F(z) at z=1 is of the Darboux type, then[10] if $\theta \neq 1,2,3,\ldots,$

$$F(z) = b_o + b_1(1-z) + b_2(1-z)^2 + \ldots$$
$$+ (1-z)^{\theta}[c_o + c_1(1-z) + c_2(1-z)^2 + \ldots\ ] \tag{C14}$$

where

$$c_o = a_o\Gamma(-\theta) \quad . \tag{C15}$$

When $\theta = 1,2,3,\ldots,$ then

$$F(z) = b'_o + b'_1(1-z) + b'_2(1-z)^2 + \ldots$$
$$+ (1-z)^{\theta}\ \ell n(1-z)[c'_o + c'_1(1-z) + c'_2(1-z)^2 + \ldots\ ]\ . \tag{C16}$$

APPENDIX D

Consider

$$P(z) \equiv P(\underline{0},z) = \frac{v_o}{(2\pi)^d}\int \frac{1}{1-z\lambda(\underline{k})}\, d^dk \simeq \frac{v_oS}{(2\pi)^d}\int_o^{\pi} \frac{k^{d-1}dk}{\zeta + Ezk^{\sigma}} \tag{D1}$$

where

$$\zeta = 1-z, \quad E = I_{d,\sigma}/v_oN_{d,\sigma}(L), \quad S = 2\pi^{d/2}/\Gamma(d/2)\ . \tag{D2}$$

$$P(z) - P(1) = \frac{v_oS}{(2\pi)^dE}\int_o^{\pi}\left\{\frac{k^{d-1}}{\zeta(E^{-1} + \zeta^{-1}k^{\sigma})} - k^{d-1-\sigma}\right\}dk$$

$$= \frac{v_oS}{(2\pi)^dE}\int_o^{\pi}\frac{-\zeta E^{-1}k^{d-\sigma-1}}{\zeta(E^{-1} + \zeta^{-1}k^{\sigma})}\, dk$$

$$= -\frac{v_oS}{(2\pi)^dE}\int_o^{\pi}\frac{k^{d-\sigma-1}}{1 + E\zeta^{-1}k^{\sigma}}\, dk \quad . \tag{D3}$$

Let $x = E\zeta^{-1}k^{\sigma}$, we find

$$P(1) - P(z) \simeq \frac{v_o S \zeta^{(d-\sigma)/\sigma}}{(2\pi)^d E^{d/\sigma}} \int_o^{E\pi^\sigma/\zeta} \frac{x^{(d-2\sigma)/\sigma}}{1+x} dx \quad . \tag{D4}$$

Three possibilities arise as $z \to 1-$:

(a) If $-1 < \frac{d-2\sigma}{\sigma} < 0$, or $\frac{1}{2} d < \sigma < d$, then

$$P(1) - P(z) \sim \zeta^{(d-\sigma)/\sigma} \quad . \tag{D5a}$$

(b) If $\frac{d-2\sigma}{\sigma} = 0$, or $\sigma = \frac{1}{2} d$, then

$$P(1) - P(z) \sim -\zeta \ln \zeta \; . \tag{D5b}$$

(c) If $\frac{d-2\sigma}{\sigma} > 0$, or $0 < \sigma < \frac{1}{2} d$, then

$$P(1) - P(z) \sim \zeta \; . \tag{D5c}$$

These are the results given by Joyce[9] and presented in Eqs. (2.33).

APPENDIX E

For the (finite-range) nearest neighbor random walk whose characteristic function is given by

$$\lambda(\underline{k}) = 1 - C k^2 + O(k^4) \tag{E1}$$

many of its properties can be derived from those of the Levy walk by simply setting $\sigma = 2$. However, since the finite-range random walk is transient only when $d \geq 3$, this restriction has to be implied.

For example, setting $\sigma = 2$ in Eqs. (2.33), (2.34), (2.18), (2.21), (2.23), (2.32) and (2.20), we find

$$\begin{aligned} P(z) &= P(1) - c(1-z)^{d/2-1} && (d=3) \\ &= P(1) - a(1-z)\ln(1-z) && (d=4) \\ &= P(1) - b(1-z) && (d \gtrsim 5) \end{aligned} \tag{E2}$$

$$P_n(\underline{0}) \sim a\, n^{-d/2} \qquad (\text{any } d) \tag{E3}$$

$$q \sim (1-z)^{\frac{1}{2}} \qquad (d \gtrsim 3) \tag{E4}$$

$$\nu = 1/2 \qquad (\text{any } d) \tag{E5}$$

$$\eta = 0 \qquad (d \gtrsim 3) \tag{E6}$$

$$\langle R_n^2 \rangle \sim A\, n^{\frac{1}{2}} \qquad (\text{any } d) \tag{E7}$$

$$\Gamma(\underline{\ell},1) \sim D_o \left(\frac{a}{r}\right)^{d-2} \qquad (d \gtrsim 3) \; . \tag{E8}$$

A distinct difference, however, was found for the behavior of $\Gamma(\underline{\ell},z)$ (as $|\underline{\ell}| \to \infty$) which, instead of decaying like an inverse power law as given by Eq. (2.19), decays exponentially for finite range random walk[10]:

$$\Gamma(\underline{\ell},z) \sim D_\infty (qa)^{\frac{1}{2}(d-3)} \left(\frac{a}{r}\right)^{\frac{1}{2}(d-1)} e^{-qr} \qquad \text{as } r \to \infty,\ q \text{ small and fixed},\ d \geq 3, \tag{E9}$$

where

$$q = C(1-z)^{\frac{1}{2}} . \tag{E10}$$

The function which plays the part of D(qr) in (C13), instead of that given by (C14), is, for the finite range random walk,

$$\mathcal{D}(x) = K_{\frac{1}{2}d-1}(x) \qquad d \geq 3 \tag{E11}$$

where $K_{\frac{1}{2}d-1}(x)$ is the modified Bessel function of the second kind, and we have

$$\mathcal{D}(x) \sim D_\infty x^{\frac{1}{2}(d-3)} e^{-x} \qquad \text{as } x \to \infty \tag{E12}$$

and

$$\lim_{x \to 0+} \mathcal{D}(x) = D_o . \tag{E13}$$

REFERENCES

1. B. Hughes, M. Shlesinger and E. W. Montroll, Proc. Nat. Acad. Sci. 78, 3287 (1981).
2. E. W. Montroll, J. Soc. Indust. Appl. Math. 4, 241 (1956).
3. E. W. Montroll, SIAM 16, 193 (1964).
4. E. W. Montroll and G. H. Weiss, J. Math. Phys. 6, 167 (1965).
5. Weierstrass, Abhandlungen aus der Functionenlehre p. 97 (1872).
6. G. H. Hardy, Trans. Amer. Math. Soc. 17, 301 (1916).
7. M. V. Berry and Z. V. Lewis, Proc. Roy. Soc. (London) A 370, 459 (1980).
8. B. B. Mandelbrot, Fractals: Form, Chance, and Dimension, Freeman, San Francisco 1977.
9. G. S. Joyce, Phys. Rev. 146, 349 (1966).
10. G. S. Joyce, in Phase Transitions and Critical Phenomena, Vol. 2. Edited by C. Domb and M. S. Green, Academic Press, New York, 1973.
11. B. R. A. Nijboer and F. W. DeWette, Phsica 23, 309 (1957).
12. G. H. Hardy, Messenger of Mathematics 49, 85 (1919).
13. J. E. Gillis and G. H. Weiss, J. Math. Phys. 11, 1307 (1970).
14. T. H. Berlin and M. Kac, Phys. Rev. 86, 821 (1952).
15. H. E. Stanley, Phys. Rev. 176, 718 (1968).
16. M. Kac and C. J. Thompson, Phys. Norveg. 5, 163 (1971).

17. See for example, C. Domb and M. S. Green (Editors) "Phase Transitions and Critical Phenomena" Vol. 6, Academic Press, New York 1976.
18. W. K. Theumann, Phys. Rev. B2, 1396 (1970).
19. M. G. Darboux, J. Math. 3, 377 (1878).
20. M. Hamy, J. Math. 4, 203 (1908).
21. B. W. Ninham, J. Math. Phys. 4, 679 (1963).

Stochastic Model for Exciton Lineshapes at Finite Temperatures

Bruce J. West* and Katja Lindenberg+
University of California at San Diego, La Jolla, CA 92093
and
Center for Studies of Nonlinear Dynamics
La Jolla Institute, La Jolla, CA 92037

ABSTRACT

Stochastic Hamiltonian models are often used to describe the behavior of excitons in molecular aggregates. In such descriptions, the exciton-phonon interactions is modeled as a fluctuating potential. In this paper we show that the equations of motion obtained from stochastic Hamiltonian models lack dissipative terms that must be present if the excitons are to achieve thermal equilibrium. We consider a particular fully dynamical model whose results guide us in the choice of the appropriate dissipative term. We thus show explicitly that the stochastic Hamiltonian models correspond to infinite temperature, and we construct stochastic equations of motion that are applicable at finite temperatures. As a particular application of the model we consider optical lineshapes and compare results in the presence and absence of the dissipative contribution.

INTRODUCTION

One of the most useful theories of energy transfer in molecular crystals is originally due to Haken and Strobl[1,2] and has since been used and extended by other authors.[3-6] In this theory, the exciton-phonon coupling is treated phenomenologically and semiclassically via the introduction of a stochastic term in the Hamiltonian and is at long times equivalent to a random walk process on a discrete lattice. The Hamiltonian is assumed to have the form

$$H_s = \sum_k E_k a_k^+ a_k + \sum_{n,m} V_{nm}(t)\, a_n^+ a_m \quad . \tag{1.1}$$

Here a_k^+ and a_k are creation and annihilation operators of an exciton with momentum k and energy E_k. The operators a_n^+ and a_n create and destroy an exciton localized at the nth site of the lattice and are discrete Fourier transforms of the operators a_k^+ and a_k. The fluctuating functions $V_{nm}(t)$ model the fluctuations in the site energy (n=m) and in the interactions (n≠m) of localized excitons due to the scattering of excitons by phonons. The fluctuations are assumed to be Gaussian with first and second moments given by

$$\langle V_{nm}(t)\rangle = 0 \tag{1.2}$$

+Permanent address: Chemistry Department, B-014, UCSD.

*Permanent address: 8950 Villa La Jolla Drive Suite 2150
La Jolla, CA 92037

and

$$\langle V_{nm}(t)\, V_{n'm'}(t')\rangle = \Gamma_{|n-m|}\,\delta(t-t')[\delta_{nn'}\delta_{mm'} + \delta_{nm'}\delta_{n'm}(1-\delta_{nn'})]. \tag{1.3}$$

The brackets < > denote an ensemble average. This model is formulated for rapidly relaxing fluctuations and is hence appropriate only for a small exciton bandwidth relative to the phonon bandwidth.

The Haken-Strobl (H-S) model was originally[1,2] used to obtain evolution equations for the exciton density matrix elements $\rho_{nn'}$. From this equation the exciton diffusion coefficient was obtained via a calculation of the second moment of the diagonal elements of the density matrix. The results obtained from this approach are well known to be valid only in the infinite temperature limit. Thus the temperature dependence of transport coefficients can not be determined by this procedure. It also follows that the stochastic Hamiltonian leads to a density matrix whose diagonal elements ρ_{kk} in k-space approach a value independent of k as $t \to \infty$. On the other hand, at finite temperatures one expects a canonical equilibrium distribution:

$$\rho_{kk}(t\to\infty) \to \text{const.} \qquad \text{(Haken-Strobl)} \quad , \tag{1.4a}$$

$$Z_{ex}^{-1}(\beta)\, e^{-\beta\varepsilon_k} \qquad \text{(correct thermodynamic behavior)} \; . \tag{1.4b}$$

Here $Z_{ex}(\beta)$ is the exciton partition function and ε_k is the energy of the exciton with wave vector k in the presence of the phonon bath.

The Haken-Strobl model has also been used to calculate exciton lineshapes.[3] With the restrictions imposed by (1.3) on the temporal correlations of the fluctuations, the lineshapes are always Lorentzian with peak shifts and widths determined by the spectral strengths $\Gamma_{|n-m|}$. Just as for the transport coefficients, the spectra obtained in this model are of course temperature independent.

The Haken-Strobl model has been generalized in a number of useful ways. In particular, Sumi[4] and Silbey and his coworkers[5,6] have considered the effects of fluctuations with finite correlation times. Sumi[4] and Blumen and Silbey[5] considered non-delta-correlated site energy fluctuations, with

$$\langle V_{nn}(t)\, V_{mm}(t')\rangle = D^2 \exp(-\gamma|t-t'|)\delta_{nm} \quad . \tag{1.5}$$

Non-delta-correlated transfer integrals have been dealt with by Jackson and Silbey.[6] The parameter γ in (1.5) is a measure of the phonon bandwidth, and D is the average amplitude of the potential fluctuations at each lattice site. We note that this generalization (and in fact any generalization of the stochastic model to date) does not per se introduce a temperature.

The problem of exciton dynamics at finite temperatures has been approached using fully dynamical models.[3,7-14] The two most fully developed theories are those of the MIT group (Silbey et al.)[7-10] and of the Rochester group (Kenkre et al.).[3,11-14] Although formally elegant, the first theory has been limited in application due to the technical difficulties in calculating physical observables such as lineshapes. The latter theory, although perhaps somewhat more readily applied, can only be used to obtain observables associated with the diagonal elements of the density matrix. In particular, it cannot be used to calculate lineshapes. As a result, to our knowledge there exists no calculation that predicts the temperature dependence of exciton lineshapes. Further, the temperature dependence of exciton transport coefficients has only recently been investigated for a dynamical model.[10]

In spite of the restriction to infinite temperatures of the Haken-Strobl model and its subsequent generalizations, their tractability has engendered their widespread use. We note that these models are assumed to be capable of representing a gamut of exciton-phonon interactions from weak to strong and from local to nonlocal. The variety of possible interactions is manifest in different statistical properties chosen for $V_{nm}(t)$. It would therefore be highly desirable to have available a phenomenological stochastic model built on premises similar to those adopted by Haken and Strobl but applicable at finite temperatures. Such a model would enable one to calculate temperature dependences of transport and spectral properties. In this paper we propose such a model. We believe ours to be the first stochastic model of transport of excitons in a molecular medium valid at finite temperatures. We note that essentially the same formalism is applicable to the transport of other excitations (e.g. polarons and electrons).

In Section 2 we review the H-S-type models and some of the lineshape results obtained previously for these models. In Section 3 we present our model and discuss the physical basis for its differences from the H-S-type models. In this discussion we emphasize the importance of the fluctuation-dissipation relation (absent in the H-S-type models) to ensure the proper thermodynamic behavior at finite temperatures. We calculate lineshapes for our model and discuss the differences between our results and those obtained earlier by other authors.[4,5] We end with a discussion in Section 4.

2. REVIEW OF HAKEN-STROBL-TYPE MODELS

In this section we review some of the results of the H-S-type models. We present the results in terms of the equations of motion of the exciton operators in order to make explicit comparisons with the new results presented in the next section.

The exciton operators are assumed to obey Bose-Einstein commutation relations, an approximation which is valid at low exciton density. The equations of motion of the exciton operators in the Heisenberg representation using the Hamiltonian (1.1) are then

$$\dot{a}_k(t) = i[H, a_k]$$
$$= -iE_k a_k(t) - i \sum_{k_1} F_{kk_1}(t)\, a_{k_1}(t) \qquad (2.1)$$

where the fluctuating coefficient $F_{kk_1}(t)$ is related to the Hamiltonian potential fluctuations in (1.1) by

$$F_{kk_1}(t) = \frac{1}{N} \sum_{n,m} e^{-ikn}\, V_{nm}(t)\, e^{ik_1 m} \quad . \qquad (2.2)$$

N is the total number of lattice sites. We note that the fluctuating coefficient in (2.1) is a c-number and not an operator.

The dynamical properties of the exciton are determined by the solution of Eq. (2.1). We define the exciton operators in the interaction representation by

$$\hat{a}_k(t) \equiv e^{iE_k t}\, a_k(t) \quad . \qquad (2.3)$$

In terms of these, the equation of motion (2.1) becomes

$$\dot{\hat{a}}_k(t) = - i \sum_{k_1} M_{kk_1}(t)\, \hat{a}_{k_1}(t) \quad , \qquad (2.4)$$

where the elements of the matrix $\underline{\underline{M}}$ are given by

$$M_{kk_1}(t) = F_{kk_1}(t)\, e^{i(E_k - E_{k_1})t} \quad . \qquad (2.5)$$

If we define a vector $\hat{\underset{\sim}{a}}(t)$ with components $\{\hat{a}_k(t)\}$, then the solution of (2.4) can be compactly written as

$$\hat{\underset{\sim}{a}}(t) = \left[e^{-i \int_0^t \underline{\underline{M}}(\tau) d\tau} \right]_T \underset{\sim}{a}(0) \qquad (2.6)$$

where the subscript T denotes time ordering, and $a_k(0)$ is the inital value of $a_k(t)$.

Expression (2.6) can be used to obtain the density matrix and the corresponding transport coefficients.[5,6] We leave the calculation of transport properties to another paper.[15] The result (2.6) can also be used to calculate lineshapes associated with exciton transport, and this will be the focus of the calculations presented herein.

The lineshape $I(\omega)$ is given by[4-6]

$$I(\omega) = \frac{1}{\pi} \text{ Re} \int_0^\infty dt\, e^{i\omega t} \langle 0| \langle \hat{\mu}(t) \rangle\, \hat{\mu}(0) |0\rangle \quad . \qquad (2.7)$$

Here it is assumed that the exciton energies are much larger than $k_B T$ and therefore only the exciton ground state $|0\rangle$ enters in the quantum mechanical average. The dipole moment operator is defined by

$$\mu = \sum_k \mu_{-k} (a_k^+ + a_{-k}) \tag{2.8}$$

where μ_k is the dipolar strength of an exciton in state k and is non-zero only for k near zero. $\hat{\mu}(t)$ in (2.7) is the dipole operator in the Heisenberg representation. In view of (2.8) we can write

$$I(\omega) = \sum_k |\mu_k|^2 I_k(\omega) \tag{2.9}$$

where

$$I_k(\omega) = \text{Re}\,\frac{1}{\pi}\int_0^\infty dte^{i\omega t} \langle 0|\langle a_k(t)\rangle\, a_k^+(0)|0\rangle\ . \tag{2.10}$$

The analysis of the integrand of Eq. (2.10) requires that we ensemble average the solution (2.6):

$$\langle \hat{\underset{\sim}{a}}(t)\rangle = \left\langle \left[e^{-i\int_0^t \underline{\underline{M}}(\tau)d\tau} \right]_T \right\rangle \underset{\sim}{a}(0) \quad . \tag{2.11}$$

The average on the right side can be expressed in terms of the cumulants of $\underline{\underline{M}}(\tau)$.[16-18] We retain only the second cumulant (the first cumulant vanishes because of (1.2)), thus obtaining a result that is exact if the fluctuations are delta-correlated and Gaussian, and approximate otherwise.[6,18,19] The higher cumulants are of higher order in the correlation time than the second one for Gaussian fluctuations. Thus we write

$$\langle \hat{a}(t)\rangle = \left[\exp \underline{\underline{K}}_2(t)\right] \underset{\sim}{a}(0) \tag{2.12}$$

where

$$\underline{\underline{K}}_2(t) = \int_0^t dt_1 \int_0^{t_1} dt_2 \left\langle \underline{\underline{M}}(t_1)\, \underline{\underline{M}}(t_2) \right\rangle \quad . \tag{2.13}$$

The matrix $\underline{\underline{K}}_2(t)$ depends on the statistics of the fluctuations $V_{nm}(t)$ in (1.1). If we choose the fluctuations to be local and site diagonal, then using (2.2) and (2.5) we obtain

$$\left\langle M_{kk_1}(t)\, M_{k_1k'}(t') \right\rangle = \frac{1}{N}\,\phi(t-t')\; e^{i\left(E_k - E_{k_1}\right)(t-t')}\, \delta_{k,k'} \tag{2.14}$$

where

$$\phi(t-t') \equiv \langle V_{nn}(t)\, V_{nn}(t')\rangle \tag{2.15}$$

has been assumed independent of site n. The analysis can easily be generalized to other fluctuations, including site dependent ones, but this will be left to subsequent work.[15] The cumulant $\underline{\underline{K}}_2$ in this approximation is thus diagonal, with elements

$$-D(k,t) \equiv \left[\underline{\underline{K}}_2(t)\right]_{kk}$$

$$= -\int_0^t dt_1 \int_0^{t_1} dt_2\ \phi(t_1-t_2) g_k(t_1-t_2) \tag{2.16}$$

where

$$g_k(\tau) \equiv \frac{1}{N}\sum_{k_1} e^{i\left(E_k-E_{k_1}\right)\tau}$$

$$= e^{iE_k\tau}\int d\omega g_{ex}(\omega)\ e^{-i\omega\tau} \tag{2.17}$$

and where $g_{ex}(\omega)$ is the normalized exciton density of states. We shall refer to $g_k(\tau)$ as the exciton spectral correlation function.

The phonon correlation function (2.15) is usually taken to be a simple exponential,[4-6]

$$\phi(t) = \frac{\mathcal{D}^2}{\gamma}\ e^{-\gamma|t|} \quad . \tag{2.18}$$

In the limit $\gamma\to\infty$, the fluctuations become delta-correlated (H-S limit):

$$\phi(t) = \mathcal{D}^2\delta(t) \quad . \tag{2.19}$$

We note that in the work of Sumi[4] and of Silbey et al.,[5,6] $\mathcal{D}^2/\gamma$ is D^2. The exciton density of states is most frequently taken to be semicircular,

$$g_{ex}(\omega) = \frac{2}{\pi B^2}\sqrt{B^2-(\omega-E_M)^2} \qquad \text{for } E_M-B\leq\omega\leq E_M+B$$

$$= 0 \qquad \text{for } \begin{cases}\omega>E_M+B\\ \omega<E_M-B\end{cases} \tag{2.20}$$

where E_M is the mean exciton energy and the exciton bandwidth is 2B. Other analytically tractable densities of states that have been considered include a Lorentzian and a Gaussian.[5] With (2.20) and (2.18) in (2.16) it is now possible to calculate the lineshape (2.10):[5]

$$I_k(\omega) = \frac{1}{\pi} \text{Re} \int_0^\infty d\tau\, e^{i(\omega-E_k)\tau} \exp\left[- \frac{\mathcal{D}^2}{\gamma} \int_0^\tau d\tau_1 \int_0^{\tau_1} d\tau_2\, e^{-\gamma\tau_2}\, g_k(\tau_2) \right] . \tag{2.21}$$

The lineshape (2.21) is determined by three characteristic energy parameters. $D \equiv \mathcal{D}/\gamma^{1/2}$ is the root mean squared level of the potential fluctuations, γ is a measure of the phonon bandwidth, and B measures the exciton bandwidth. Sumi[4] divides the parameter space into three regions, depending on the relative values of these parameters. When $D^2 >> B^2 + \gamma^2$, the fluctuations at each lattice site are strong and broaden the absorption line. The line is then Gaussian (except perhaps for high frequencies ($\omega \gtrsim D$)) of width proportional to D:[4,5]

$$I_k(\omega \lesssim D) = \frac{1}{\sqrt{2\pi}\, D} \exp\left[-(\omega-E_k)^2/2D^2 \right] \quad . \tag{2.22}$$

The full width at half maximum (FWHM) for (2.22) is $2(2 \ln 2)^{1/2}D$. This result is essentially independent of the specific choice of the exciton density of states. The second parameter regime corresponds to $D<<B$ and $B>>\gamma$. The fluctuations are now weak and the exciton bandwidth is larger than the phonon bandwidth. Since the exciton bandwidth is proportional to the exciton transfer rate among lattice sites, large B implies rapid exciton transfer. Thus the absorption line is motionally narrowed in this regime. The line is Lorentzian near the center of the exciton band, it is centered at E_k, and has a FWHM much smaller than in the Gaussian case:[4,5]

$$I_k(\omega) = \frac{\Gamma_1/\pi}{(\omega-E_k-\Gamma_2)^2 + \Gamma_1^2} \quad . \tag{2.23}$$

where Γ_2 is of order D^2E_k/B^2 and $\Gamma_1 \simeq 2D^2/B$ for the semicircular density of states ($\Gamma_1 \simeq D^2/B$ for a Lorentzian $g_{ex}(\omega)$). The FWHM is then much smaller than for the Gaussian lineshape and also much smaller than the exciton bandwidth B. The third parameter regime corresponds to weak fluctuations and a phonon bandwidth that is greater than that of the exciton, i.e. $D<<\gamma$ and $\gamma>>B$. The lineshape is again motionally narrowed but now due to the high phonon transfer rate. The line is again Lorentzian, i.e. of the form (2.23), but now Γ_2 is of order D^2E_k/γ^2 and $\Gamma_1 \simeq 2D^2/\gamma$ near the center of the exciton band. Exhaustive discussions and calculations for all these cases can be found in Sumi.[4]

We stress that in the above discussion the fourth important energy parameter of the problem, namely the thermodynamic temperature k_BT, is absent. This is of course due to the underlying assumption, not always explicitly recognized, that in the H-S-type models the temperature is infinite. The temperature has usually been introduced *parametrically* by means of the ad hoc assumption that the strength, $\mathcal{D}^2$, of the potential fluctuations grows, usually linearly, with temperature. Using this assumption, Sumi[4] has carried out some explicit comparisons of the above theory with temperature-dependent FWHM measurements in crystalline anthracene by Morris and Sceats.[20] We note that in these systems the energies $k_B T$, γ, and B can be of the same order of magnitude. Sumi's procedure for determining the temperature dependence of the lineshape is incomplete. The theory we develop in the next section properly introduces $k_B T$ into the description of exciton transport.

3. TEMPERATURE-DEPENDENT MODEL

The model with which we propose to replace the H-S-type models is described by the equations of motion (with $\{[A][B]\}_s \equiv AB + BA$)

$$\dot{a}_k(t) = - i\, E_k a_k(t) - i \sum_{k_1} F_{kk_1}(t) a_{k_1}(t) - \sum_{k_1} \sum_{k_2} \sum_{k_3} \left(E_{k_3} - E_{k_2}\right) \int_0^t d\tau\, K^{kk_1}_{k_2 k_3}(t-\tau) \cdot \left\{ \left[a^+_{k_2}(\tau) a_{k_3}(\tau) \right] \left[a_{k_1}(t) \right] \right\}_s \tag{3.1}$$

and its hermetian conjugate for each k. In our model, equation (3.1) replaces (2.1) for the H-S-type models. The fluctuating coefficient $F_{kk_1}(t)$ is the same as in (2.1) and is therefore again assumed to be capable of representing a gamut of interactions. The new contribution appearing in our model is the last term in (3.1). We identify this as a *dissipative* contribution which has been absent in all prior stochastic formulations of the exciton transport problem. The crux of the model lies in the choice of the dissipative kernel $K^{kk_1}_{k_2 k_3}(t)$, which must satisfy the physical constraints described below.

To motivate our generalization we have chosen a particular dynamical Hamiltonian in the Appendix and derived Eq. (3.1). The particular Hamiltonian that we used (linear coupling model) is only important insofar as it establishes the *connection* between the fluctuating and dissipative portions of the dynamics. It is this connection that is sufficient to ensure thermal equilibration of the excitons, regardless of the underlying dynamics. Therefore, the form (3.1) is a valid finite-temperature description even for more complicated interactions. The differences between models appear in the statistics of $F_{kk_1}(t)$ and in the choice of the corresponding kernel $K^{kk_1}_{k_2 k_3}(t)$. Our result is in the spirit of the

Einstein relation between the equilibrium distribution of fluctuations of a system and its entropy, a relation that is "independent of all special assumptions regarding the laws which may govern the elementary processes."[21]

The choice of the kernel is dictated by the fact that the last term in (3.1) results from the average interaction of the exciton with the heat bath. Contrary to the usual statements,[4-6] this average interaction *cannot* be included in E_k because it is *irreversible* and can therefore not be incorporated into a purely excitonic Hamiltonian. The combined exciton-phonon system is thermodynamically isolated, and the exciton system is thermodynamically closed. It then follows that the kernel must be completely specified by a *fluctuation-dissipation relation* between it and the fluctuations $F_{kk_1}(t)$. This relation depends on temperature and on the nature of the interactions between the excitons and the phonons, and can generally be expressed as

$$K^{kk_1}_{k_2k_3}(t) = \int_0^\infty d\tau f(\beta,t,\tau)\, \Phi^{kk_1}_{k_2k_3}(\tau) \tag{3.2}$$

where Φ is the symmetrized[22] correlation function

$$\Phi^{kk_1}_{k_2k_3}(\tau) = \tfrac{1}{2}\left\langle F_{kk_1}(t)\, F_{k_2k_3}(t+\tau) + F_{kk_1}(t+\tau)\, F_{k_2k_3}(t) \right\rangle \tag{3.3}$$

and $f(\beta,t,\tau)$ is a scalar function of the temperature. We note that (3.2) is unusual in that the dissipative response to the bath is not instantaneous. The dissipative "response time" is temperature dependent. As an example we mention the result (A23) in the Appendix, which gives the explicit relation (3.2) for the linear coupling model and motivates its generalization. We thus contend that Eq. (3.1) is generic, i.e. transcends the linear coupling model considered in the Appendix.

Due to limitations of space we leave the complete analysis of $f(\beta,t,\tau)$ to another paper.[15] Here we restrict our considerations to the temperature regime where k_BT is greater than the phonon bandwidth (yet finite). In this regime the dissipative response to the bath is instantaneous so that

$$f(\beta,t,\tau) \simeq f_o(\beta)\delta(t-\tau) \quad . \tag{3.4}$$

Our conclusion is again based on the analysis in the Appendix. The more general validity of this relation for models other than the particular one considered in the Appendix will be established elsewhere.[15] We note that (3.4) does *not* imply delta-correlated fluctuations, but rather an instantaneous response to the fluctuations. We then obtain the simpler fluctuation-dissipation relation

$$K^{kk_1}_{k_2k_3}(t) = f_o(\beta)\, \Phi^{kk_1}_{k_2k_3}(t) \quad . \tag{3.5}$$

To compare and contrast our results to those of the preceding section we again choose the fluctuations to be local and site diagonal (cf. (2.15)) with a simple exponential correlation function (cf. (2.18)). In this approximation (3.5) reduces to

$$K^{kk_1}_{k_2k_3}(t) = f_o(\beta)\, \frac{\mathcal{D}^2}{N\gamma}\, e^{-\gamma|t|}\, \delta_{k-k_1+k_2-k_3,0} \quad . \tag{3.6}$$

It is generally assumed that to leading order dissipative coefficients are temperature independent. (This is indeed the case for the linear coupling model.) We also recall from the preceding section that a temperature dependence is often introduced in $\mathcal{D}$ (or D) in the H-S-type models. The two dependences must then compensate so that

$$f_o(\beta) D^2 = \text{constant} \equiv \varepsilon \tag{3.7}$$

independent of temperature. In the linear coupling model (cf. (A24)) one obtains $f_o(\beta) = \beta/2$ and $D^2 \sim k_B T$, as usually assumed in the H-S-type models. Then

$$K^{kk_1}_{k_2k_3}(t) = \frac{\varepsilon}{N}\, e^{-\gamma|t|}\, \delta_{k-k_1+k_2-k_3,0} \quad . \tag{3.8}$$

The coefficient ε has dimensions of energy and constitutes an *additional parameter*, previously unrecognized in stochastic formulations, that plays an important role in the dynamics of the exciton. The linear coupling model provides us with an immediate interpretation of this parameter as the characteristic strength of the exciton-phonon coupling:

$$\varepsilon \sim \frac{\Gamma^2}{\Omega_{ph}} \quad . \tag{3.9}$$

Here Γ is a typical exciton-phonon coupling energy (cf. (A1)) and Ω_{ph} is a typical phonon energy.

We can now insert the particular kernel (3.8) into the equation of motion(3.1):

$$\dot{a}_k(t) = - i\, E_k a_k(t) - i \sum_{k_1} F_{kk_1}(t)\, a_{k_1}(t)$$
$$- \frac{\varepsilon}{N} \sum_{k_2} \sum_{k_3} \left(E_{k_3} - E_{k_2}\right) \int_0^t d\tau\, e^{-\gamma(t-\tau)} \left\{ \left[a^+_{k_2}(\tau) a_{k_3}(\tau)\right] \left[a_{k+k_2-k_3}(t)\right] \right\}_s \tag{3.10}$$

Equation (3.10) allows for the first time an analysis of exciton dynamics at finite temperaturesbased on a stochastic model. As described in the previous section, Sumi[4] and Silbey et al.[5,6,23] have emphasized the importance of considering the exciton bandwith (B), the phonon bandwidth (γ) and the fluctuation level (D) in the transport and spectral behavior of the excitons. In the spirit of their discussion we are now able to introduce the two additional energy parameters that affect exciton dynamics, namely, the temperature $k_B T$ and the exciton-phonon coupling strength ε. Heretofore the former has been assumed to be infinite (i.e. much larger than γ and B) and consequently the latter has been assumed to vanish. Let us now consider what happens when $k_B T$ is finite, albeit still larger than γ, and ε is nonvanishing. Since Eq. (3.10) is nonlinear, its detailed analysis is quite difficult. We therefore restrict our investigation here to a set of approximations that are often made (in other mode coupling contexts)[24] without more justification than their physical plausibility. From this analysis we can determine at least the qualitative, and perhaps semiquantitative, effects of the dissipative contribution in (3.10).

The primary dissipative contribution is the term with $k_3 = k$. We retain only this term in the following analysis. Furthermore, we replace $a^+_{k_2}(\tau)$ by $a^+_{k_2}(t)\exp\left[i\,E_{k_2}(\tau-t)\right]$, and similarly for $a_{k_3}(\tau)$. This replacement approximates the time evolution of $a^+_{k_2}$ and a_{k_3} by their evolution in the absence of the heat bath. With this replacement the integral in (3.10) can be done explicitly with the result

$$\dot{a}_k(t) = -\,i\,E_k a_k(t) - i\sum_{k_1} F_{kk_1}(t)\,a_{k_1}(t)$$

$$-\,\frac{\varepsilon}{N}\sum_{k_1}\frac{\left(E_k - E_{k_1}\right)}{i\left(E_k - E_{k_1} + i\gamma\right)}\left[e^{\,i\left(E_k - E_{k_1} + i\gamma\right)t} - 1\right]\left\{\left[a^+_{k_1}(t)\right]\left[a_{k_1}(t)\right]\right\}_s a_k(t). \tag{3.11}$$

Our final approximation is the replacement of the symmetrized product in (3.11) by its average value at thermal equilibrium:

$$\left\langle\left\{\left[a^+_{k_1}(t)\right]\left[a_{k_1}(t)\right]\right\}_s\right\rangle = 1 + \frac{2}{Z_{ex}(\beta)}\,e^{-\beta E_{k_1}}$$

$$\equiv 1 + 2\,N_{k_1}(\beta)\quad . \tag{3.12}$$

We note that this assumption is only plausible because of the existence of the fluctuation-dissipation relation built into (3.11), which ensures the eventual equilibration of the system. We then rewrite (3.11) as

$$\dot{a}_k(t) = -i\, E_k a_k(t) - i \sum_{k_1} F_{kk_1}(t)\, a_{k_1}(t) - \lambda_k(t) a_k(t) \quad . \tag{3.13}$$

In terms of the exciton density of states (cf. (2.17)), the coefficient $\lambda_k(t)$ is given by

$$\lambda_k(t) = \varepsilon \int d\omega g_{ex}(\omega)\,[1+2N(\beta,\omega)]\, \frac{(E_k-\omega)\,[\gamma+i(E_k-\omega)]}{(E_k-\omega)^2+\gamma^2} \left[1-e^{i(E_k-\omega+i\gamma)t}\right] \tag{3.14}$$

where $N(\beta,\omega) \equiv e^{-\beta\omega}/Z_{ex}(\beta)$.

Equation (3.13) is once again linear and can be integrated following the procedure of Section 2. In the interaction representation we now have

$$\hat{a}_k(t) \equiv e^{i\, E_k t + \Lambda_k(t)}\, a_k(t) \tag{3.15}$$

where

$$\Lambda_k(t) \equiv \int_0^t \lambda_k(t')dt' \quad . \tag{3.16}$$

The equation of motion (3.13) then transforms to

$$\dot{\hat{a}}_k(t) = -i \sum_{k_1} M_{kk_1}(t)\, \hat{a}_{k_1}(t) \tag{3.17}$$

where in place of (2.5) we have

$$M_{kk_1}(t) = F_{kk_1}(t)\; e^{i\left(E_k-E_{k_1}\right)t}\; e^{\Lambda_k(t)-\Lambda_{k_1}(t)} \quad . \tag{3.18}$$

The solution to (3.17) is then the analog of (2.6) with average value

$$\langle \hat{\underset{\sim}{a}}(t) \rangle = \exp[-D(k,t)]\; \underset{\sim}{a}(0) \tag{3.19}$$

where in place of (2.16) we now have

$$D(k,t) = \int_0^t dt_1 \int_0^{t_1} dt_2\; \phi(t_1-t_2)\; G_k(t_1,t_2) \quad . \tag{3.20}$$

The modified correlation function $G_k(t_1,t_2)$ replaces the exciton spectral correlation function g_k in (2.16) and is given by

$$G_k(t_1,t_2) = e^{i E_k(t_1-t_2)} e^{[\Lambda_k(t_1)-\Lambda_k(t_2)]} \int d\omega g_{ex}(\omega) e^{-i\omega(t_1-t_2)} \cdot e^{-[\Lambda(\omega,t_1)-\Lambda(\omega,t_2)]} \tag{3.21}$$

where $\Lambda(\omega,t)$ is $\Lambda_k(t)$ with E_k replaced by ω:

$$\Lambda(\omega,t) = \varepsilon \int d\omega' g_{ex}(\omega') [1+2N(\beta,\omega')] \frac{(\omega-\omega')[\gamma+i(\omega-\omega')]}{(\omega-\omega')^2+\gamma^2} \cdot \left[t - \frac{e^{i(\omega-\omega'+i\gamma)t}-1}{i(\omega-\omega')-\gamma} \right] . \tag{3.22}$$

Finally, the lineshape (2.10) is given by

$$I_k(\omega) = \frac{1}{\pi} \mathrm{Re} \int_0^\infty d\tau e^{i(\omega-E_k)\tau-\Lambda_k(\tau)} \exp\left[-\int_0^\tau d\tau_1 \int_0^{\tau_1} d\tau_2\, \phi(\tau_1-\tau_2) G_k(\tau_1,\tau_2) \right] \tag{3.23}$$

The lineshape (3.23) has the same "formal" structure as that for the H-S-type models (cf. (2.21)) but with the important inclusion of the new dissipative term $\Lambda_k(t)$ appearing in both the τ-integrand and in the modified exciton spectral correlation function $G_k(\tau_1,\tau_2)$. To compare (3.23) with (2.21) we examine the lineshapes in the different parametric regimes *accessible by the approximations that led to (3.23)*. The parameter space has now been increased by the addition of the exciton-phonon coupling strength ε. We note that the center of the absorption line is shifted from E_k by terms of order ε that we are systematically neglecting in this discussion.

When $D^2 >> B^2+\gamma^2$ and $D >> \varepsilon$, the absorption line is again broadened at each lattice site. For moderate frequencies one obtains a Gaussian of the form (2.22), i.e.

$$I_k(\omega \underset{\sim}{<} D') = \frac{1}{\sqrt{2\pi}\, D'} \exp[-(\omega-E_k)^2/2D'^2] \tag{3.24}$$

but now with width proportional to

$$D' = [D^2+c_1 B^2 \varepsilon/k_B T]^{\frac{1}{2}} \tag{3.25}$$

where c_1 is a real constant of order unity. Thus the FWHM is no longer linear in temperature but contains corrections dependent on the exciton bandwidth B, the coupling strength ε and the *inverse* of the temperature.[20] The correction can be appreciable for singlet exciton lineshapes in a number of molecular crystals at room temperature. If $D<\varepsilon$ but still $D^2>>B^2+\gamma^2$, then the lineshape is still Gaussian but the correction becomes negligible and we recover (2.22).

If $D<<B$, $B>\gamma$ and $\varepsilon<B$ then the lineshape is approximately Lorentzian. This regime corresponds to Eq. (2.23) in the previous section but with the width Γ_1 now given by

$$\Gamma_1 = 2\frac{D^2}{B} + c_2\frac{\gamma\varepsilon}{k_BT} \tag{3.26}$$

where c_2 is again a real constant of order unity. The size of the correction is unrestricted in this regime. For instance, if $k_BT<B$ then using (3.7) we can see that in fact $\gamma\varepsilon/k_BT>D^2/B$. Thus the temperature dependence of the lineshape in this regime can be quite different from that of (2.23).

The final regime that we can consider is $D<<\gamma$, $\gamma>B$ and $\varepsilon<\gamma$. The lineshape, as before (cf. (2.23), is approximately Lorentzian but again with a new width

$$\Gamma_1 = 2\frac{D^2}{\gamma} + c_3\frac{B^2\varepsilon}{\gamma\, k_BT} \tag{3.27}$$

where c_3 is real and of O(1). In this regime the correction is of relative order $B^2/(k_BT)^2$ and hence small.

We end this section by making the following points. First, we note that regardless of the particular form of the lineshape, the dissipative contribution of our model induces a temperature dependent broadening in addition to that obtained in the H-S-type models. Second, the additional broadening is inversely proportional to the temperature, as distinct from the direct proportionality of the linewidth in the H-S-type models, and vanishes as $T\rightarrow\infty$. Third, a numerical estimate of the magnitude of the corrections shows them to be appreciable for singlet excitons in molecular crystals where the exciton bandwidth is greater than the phonon bandwidth (regime (3.26)).[20] We emphasize that the results presented here are restricted to the regime $k_BT>\gamma$.[15]

4. DISCUSSION

In this paper we have extended the usual stochastic formulation of exciton transport[1-6] so as to obtain a description valid at finite temperatures. We have shown that the average exciton-phonon interaction leads to an irreversible dissipative contribution in the equations of motion, a contribution that has been absent from previous stochastic models. The dissipative term is related to the fluctuations via a fluctuation-dissipation relation that ensures thermal equilibration of the excitons with the phonon bath. The dissipative effects, being irreversible, cannot be built into a purely excitonic Hamiltonian.

In order to assess the importance of the dissipative effects, we have considered a specific model valid for many stiuations where the thermal energy k_BT is greater than the phonon bandwidth but possibly smaller than the exciton bandwidth . In the formulation of this model we were guided by the dynamical exciton-phonon system described in the Appendix, but as presented our model is generic and transcends the system discussed in the Appendix. We have shown that the proper inclusion of dissipation brings a new energy parameter into the analysis of exciton transport in additon to the three that are considered in the usual H-S-type infinite temperature models.[1-6] The three parameters usually included are the phonon fluctuation energy D, the phonon bandwidth γ, and the exciton bandwidth B. The parameter that we introduce is the exciton-phonon coupling strength ε, (the dissipation is proportional to ε). The temperature enters the description via the relation $k_BT = D^2/2\varepsilon$ (since in the usual stochastic models ε is in effect set equal to zero, the temperature is implicitly set to infinity).

We have analyzed the effect of the dissipative contribution on exciton lineshapes, leaving the analysis of transport coefficients (e.g. diffusion) to another paper.[15] We find that in the regime $k_BT>\gamma$ (to which our analysis here has been restricted) the dissipative effects are important when the thermal energy is smaller than or of the order of the exciton bandwidth, i.e. when $k_BT \lesssim B$. In this regime we find appreciable corrections to the usual bandwidths of the absorption line, both in the Gaussian and Lorentzian (motionally narrowed) limits. The usual absorption bandwidths are proportional to temperature; our corrections are inversely proportional to temperature. The magnitudes of these corrections are comparable to both the calculated[4] and measured[19] bandwidths for singlet absorption in anthracene. A more careful analysis of our results and their comparison with experiments will be done elsewhere.[15]

APPENDIX

To motivate our equation of motion in Section 3 we illustrate here how it can be obtained explicitly from a model Hamiltonian.

The Hamiltonian for an exciton interacting with a heat bath of phonons that we consider is the "linear-coupling" model[3]

$$H = \sum_k E_k a_k^+ a_k + \sum_{q,\alpha} \omega_{q\alpha} b_{q\alpha}^+ b_{q\alpha} + \sum_k \sum_{k_1} \sum_{q,\alpha} \Gamma_{kq\alpha}^{k_1} (b_{q\alpha}^+ + b_{-q\alpha}) a_k^+ a_{k_1} \quad . \tag{A1}$$

Here a_k^+ and a_k are creation and annihilation operators of a ("dressed") exciton with wave vector k and energy E_k. The operators $b_{q\alpha}^+$ and $b_{q\alpha}$ create annihilate a phonon labeled by (q,α). The last term in the Hamiltonian (A1) expresses the exciton-phonon interaction. For lattice phonons q is a wave vector, α is a branch index, and conservation of lattice momentum requires that $k_1 = k+q+G$ where G is a reciprocal lattice vector. Thus, for lattice phonons the coupling coefficient $\Gamma^{k_1}_{kq\alpha}$ is proportional to $\delta_{k_1,k+q+G}$. For intramolecular modes we need only one labeling index and could therefore dispense with the index q. We retain both indices so that vibrational as well as vibronic modes can be included in (A1). The energy of phonon (q,α) is $\omega_{q\alpha}$. The exciton and phonon operators are assumed to obey Bose-Einstein commutation relations.

The dynamical equations for the exciton operators are obtained from the relations

$$\dot{a}_k = i\,[H,a_k]$$

$$= -\,i\,E_k a_k - \frac{i}{2}\sum_{k_1}\sum_{q,\alpha}\Gamma^{k_1}_{kq\alpha}[(b^+_{q\alpha}+b_{-q\alpha})a_{k_1}+a_{k_1}(b^+_{q\alpha}+b_{-q\alpha})] \quad \text{(A2)}$$

and its hermetian conjugate, for each k. We have symmetrized (A2) in anticipation of the fact that the b's explicitly depend on the a's (cf. below), and therefore the relative order in which the factors appear in the equation is uncertain.

To eliminate the phonon operators from these equations of motion we write the bath dynamical equations

$$\dot{b}_{q\alpha} = -\,i\,\omega_{q\alpha}\,b_{q\alpha} - i\sum_k\sum_{k_1}\Gamma^{k_1}_{kq\alpha}\,a^+_k\,a_{k_1} \quad \text{(A3)}$$

and its hermetian conjugate for each (q,α). The solution of (A3) is

$$b_{q\alpha}(t) = e^{-i\omega_{q\alpha}t}\,b_{q\alpha}(0) - i\sum_k\sum_{k_1}\int_0^t d\tau\,\Gamma^{k_1}_{kq\alpha}e^{-i\omega_{q\alpha}(t-\tau)}\,a^+_k(\tau)a_{k_1}(\tau). \quad \text{(A4)}$$

Substitution of (A4) into (A2) gives the equation of motion

$$\dot{a}_k(t) = -\,i\,E_k a_k(t) - i\sum_{k_1}\sum_{q,\alpha}\Gamma^{k_1}_{kq\alpha}[e^{i\omega_{q\alpha}t}\,b^+_{q\alpha}(0) + e^{-i\omega_{q\alpha}t}\,b_{-q\alpha}(0)]a_{k_1}(t)$$

$$+\tfrac{1}{2}\sum_{k_1}\sum_{k_2}\sum_{k_3}\sum_{q,\alpha}\int_0^t d\tau\,\Gamma^{k_1}_{kq\alpha}\Gamma^{k_3^*}_{k_2q\alpha}e^{i\omega_{q\alpha}(t-\tau)}\left\{[a^+_{k_3}(\tau)a_{k_2}(\tau)][a_{k_1}(t)]\right\}_s$$

$$-\tfrac{1}{2}\sum_{k_1}\sum_{k_2}\sum_{k_3}\sum_{q,\alpha}\int_0^t d\tau\,\Gamma^{k_1}_{kq\alpha}\Gamma^{k_3}_{k_2,-q\alpha}\,e^{-i\omega_{q\alpha}(t-\tau)}\left\{[a^+_{k_2}(\tau)a_{k_3}(\tau)][a_{k_1}(t)]\right\}_s$$

(A5)

where $\{\ \}_s$ represents the symmetrized product or anticommutator, i.e.

$$\{[A][B]\}_s \equiv AB + BA \quad . \tag{A6}$$

Note that (A5) now only contains *initial* values of the phonon operators.

To aid in the interpretation of (A5) it is convenient to reorganize the terms. We first note that a simple change in the dummy summation indices in the integrands allows us to write the last two terms in (A5) as

$$i \sum_{k_1} \sum_{k_2} \sum_{k_3} \sum_{q,\alpha} \int_0^t d\tau \frac{\Gamma^{k_1}_{kq\alpha} \Gamma^{k_3}_{k_2,-q,\alpha}}{\omega_{q\alpha}} \left[\frac{d}{d\tau} \cos \omega_{q\alpha}(t-\tau)\right]$$

$$\cdot \left\{\left[a^+_{k_2}(\tau)\, a_{k_3}(\tau)\right]\left[a_{k_1}(t)\right]\right\}_s \quad . \tag{A7}$$

Integrating (A7) by parts and inserting it into (A5) gives

$$\dot{a}_k(t) = -\, i\, E_k a_k(t) + i \sum_{k_1} \sum_{k_2} \sum_{k_3} \sum_{q,\alpha} \frac{\Gamma^{k_1}_{kq\alpha} \Gamma^{k_3}_{k_2,-q,\alpha}}{\omega_{q\alpha}}$$

$$\cdot \left\{[a^+_{k_2}(t) a_{k_3}(t)][a_{k_1}(t)]\right\}_s - i \sum_{k_1} \hat{F}_{k\,k_1}(t)\; a_{k_1}(t)$$

$$- i \sum_{k_1} \sum_{k_2} \sum_{k_3} \sum_{q,\alpha} \int_0^t d\tau \frac{\Gamma^{k_1}_{kq\alpha} \Gamma^{k_3}_{k_2,-q,\alpha}}{\omega_{q\alpha}} \cos \omega_{q\alpha}(t-\tau)$$

$$\cdot \frac{d}{d\tau} \left\{[a^+_{k_2}(\tau) a_{k_3}(\tau)][a_{k_1}(t)]\right\}_s \tag{A8}$$

where

$$\hat{F}_{kk_1}(t)\; a_{k_1}(t) \equiv F_{kk_1}(t) a_{k_1}(t)$$

$$+ \sum_{k_2} \sum_{k_3} \sum_{q,\alpha} \frac{\Gamma^{k_1}_{kq\alpha} \Gamma^{k_3}_{k_2,-q,\alpha}}{\omega_{q\alpha}} \cos \omega_{q\alpha} t \left\{[a^+_{k_2}(0) a_{k_3}(0)][a_{k_1}(t)]\right\}_s \tag{A9}$$

and

$$F_{kk_1}(t) = \sum_{q,\alpha} \Gamma^{k_1}_{kq\alpha} \left[b^+_{q\alpha}(0)\, e^{i\omega_{q\alpha}t} + b_{-q\alpha}(0)\, e^{-i\omega_{q\alpha}t} \right] \qquad \text{(A10)}$$

Equation (A8) is exact. To proceed further it is necessary to make some approximations. Results to the usual order of approximation (i.e. to second order in the Γ's) are obtained if we make the "weak coupling" assumption in the equations of motion and retain terms to $O(\Gamma)$ in $\hat{F}_{kk_1}(t)$ and to $O(\Gamma^2)$ in the last term in (A8). As we will see shortly, it is necessary to retain terms to different orders in Γ in the equations of motion to ensure a proper fluctuation-dissipation relation.[25,26] To this order of approximation we then have

$$\dot{a}_k(t) = -i E_k a_k(t) + i \sum_{k_1} \sum_{k_2} \sum_{k_3} C^{kk_1}_{k_2k_3} \left\{ \left[a^+_{k_2}(t) a_{k_3}(t) \right] \left[a_{k_1}(t) \right] \right\}_s$$

$$- i \sum_{k_1} F_{kk_1}(t)\, a_{k_1}(t)$$

$$- \sum_{k_1} \sum_{k_2} \sum_{k_3} \left(E_{k_3} - E_{k_2} \right) \int_0^t d\tau\, K^{kk_1}_{k_2k_3}(t-\tau)$$

$$\cdot \left\{ \left[a^+_{k_2}(\tau) a_{k_3}(\tau) \right] \left[a_{k_1}(t) \right] \right\}_s \qquad \text{(A11)}$$

$$C^{kk_1}_{k_2k_3} \equiv \sum_{q,\alpha} \frac{\Gamma^{k_1}_{kq\alpha}\, \Gamma^{k_3}_{k_2,-q,\alpha}}{\omega_{q\alpha}} \qquad \text{(A12)}$$

$$K^{kk_1}_{k_2k_3}(t) = \sum_{q,\alpha} \frac{\Gamma^{k_1}_{kq\alpha}\, \Gamma^{k_3}_{k_2,-q,\alpha}}{\omega_{q\alpha}} \cos \omega_{q\alpha} t \quad . \qquad \text{(A13)}$$

The coupling of the excitons to the phonon bath leads to three contributions in (A11). The first (of $O(\Gamma^2)$) can be viewed as a modification of the exciton portion of the Hamiltonian and will be omitted in the remainder of this discussion (i.e. we neglect $C^{kk_1}_{k_2k_3}$). The second contribution (of $O(\Gamma)$) differs from the others in that it depends on the bath initial conditions. *It is this term that we interpret as the fluctuating portion.*[25-28] The statistical properties of this term are determined by those of the phonon operators in an ensemble of *initially* canonically distributed states. Averages involving $F_{kk_1}(t)$ are thus to be evaluated using the diagonal density matrix with elements

$$P_{eq}(b^+_{q\alpha}, b_{q\alpha}) = Z^{-1}_{ph}(\beta)\,\exp(-\beta\omega_{q\alpha} b^+_{q\alpha}\, b_{q\alpha}) \tag{A14}$$

where $\beta = (k_B T)^{-1}$ and $Z_{ph}(\beta)$ is the phonon partition function. Averages over the phonon bath at t=0 are calculated via the usual prescription

$$\langle f(\{b^+_{q\alpha}, b_{q\alpha}\})\rangle \equiv \mathrm{Tr}(P_{eq}\, f) \quad . \tag{A15}$$

As a result of this definition, $F_{kk_1}(t)$ is zero centered and has Gaussian statistics such that

$$\langle F_{kk_1}(t)\rangle = 0 \tag{A16}$$

and

$$\left\langle F_{kk_1}(t)F_{k_2k_3}(t')\right\rangle = \sum_{q,\alpha} \Gamma^{k_1}_{kq\alpha}\,\Gamma^{k_3}_{k_2,-q,\alpha}\left[(e^{\beta\omega_{q\alpha}}-1)^{-1}\, e^{i\omega_{q\alpha}(t-t')} + (1-e^{-\beta\omega_{q\alpha}})^{-1}\, e^{-i\omega_{q\alpha}(t-t')}\right] \quad . \tag{A17}$$

We note that the fluctuation term can be incorporated into a Hamiltonian of the form (1.1), with

$$V_{nm}(t) = \sum_k \sum_{k_1} e^{ikn}\, e^{-i k_1 m}\, F_{kk_1}(t) \quad . \tag{A18}$$

The last term in (A.11) is fundamentally different from the others in that, it is *irreversible* and therefore cannot be included in a hermetian Hamiltonian involving only exciton operators. It is this term that provides the *dissipation* for the exciton system. The fluctuations and this dissipative term are related by

$$\tfrac{1}{2}\left\langle F_{kk_1}(t)F_{k_2k_3}(t') + F_{kk_1}(t')F_{k_2k_3}(t)\right\rangle$$

$$= \sum_{q,\alpha} \Gamma^{k_1}_{kq\alpha}\,\Gamma^{k_3}_{k_2,-q,\alpha}\;\coth\left(\frac{\beta\,\omega_{q\alpha}}{2}\right)\cos\,\omega_{q\alpha}(t-t'), \tag{A19}$$

where we have used (A17). Equation (A19) is the *fluctuation-dissipation relation* that ensures proper equilibration of the exciton system at temperature T. The dissipative term is absent in the stochastic Hamiltonian formulation. Its importance diminishes relative to the fluctuations as $T\to\infty$, but at ordinary temperatures it cannot be neglected. We note that in order to obtain a thermodynamically consistent fluctuation-dissipation relation it is necessary to retain dissipative terms to $O(\Gamma^2)$ if one retains fluctuating terms to $O(\Gamma)$. We also note that one can extend these results to higher orders in the exciton-phonon coupling coefficients. In doing so one must retain terms up to $O(\Gamma^{2\ell})$ in the dissipation to balance terms up to $O(\Gamma^{\ell})$ in the fluctuations. It may then also be necessary to redefine the phonon distribution (A14) to take into account the equilibration of phonons in the presence of excitons.[27,28]

The fluctuation-dissipation relation (A19) can be rewritten in a form that directly relates the fluctuations and the dissipative kernel $K_{k_2k_3}^{kk_1}(t)$ in (A11). We replace the sum over q,α by an integral over a phonon density of states $g_{ph}(\omega)$ in the usual way and rewrite (A19) as

$$\tfrac{1}{2}\left\langle F_{kk_1}(t)F_{k_2k_3}(t')+F_{kk_1}(t')F_{k_2k_3}(t)\right\rangle \equiv \Phi_{k_2k_3}^{kk_1}(t-t')$$

$$= \int_0^\infty d\omega\; g_{ph}(\omega)\; Q_{k_2k_3}^{kk_1}(\omega)\,\omega \coth\left(\frac{\beta\omega}{2}\right) \cos\omega\,(t-t') \tag{A20}$$

where

$$Q_{k_2k_3}^{kk_1}(\omega) = \frac{\Gamma_k^{k_1}(\omega)\;\Gamma_{k_2}^{k_3}(\omega)}{\omega} \quad . \tag{A21}$$

A Fourier inversion of (A20) with respect to t-t' gives

$$g_{ph}(\omega)\; Q_{k_2k_3}^{kk_1}(\omega) = \frac{\tanh\;(\beta\omega/2)}{\pi\omega} \int_0^\infty d\tau\; \Phi_{k_2k_3}^{kk_1}(\omega)\; \cos\omega\tau \;. \tag{A22}$$

The kernel (A13) can then be written as

$$K_{k_2k_3}^{kk_1}(t) = \int_0^\infty d\omega\; g_{ph}(\omega)\; Q_{k_2k_3}^{kk_1}(\omega)\; \cos\omega t$$

$$= \frac{1}{\pi}\int_0^\infty d\omega\,\frac{\tanh\,(\beta\omega/2)}{\omega}\int_0^\infty d\tau\;\Phi_{k_2k_3}^{kk_1}(\tau)\;\cos\omega\tau\;\cos\omega t$$

$$= \frac{1}{2\pi}\int_0^\infty d\tau\;\ell n\left[\coth\frac{\pi}{2\beta}\,(t+\tau)\;\coth\frac{\pi}{2\beta}\,(t-\tau)\right]\Phi_{k_2k_3}^{kk_1}(\tau)\;. \tag{A23}$$

This is the *fluctuation-dissipation relation* that ensures eventual thermal equilibration of the excitons at temperature T.

If k_BT is greater than the phonon bandwidth, (A23) can be approximated by

$$K_{k_2k_3}^{kk_1}(t) \simeq \frac{\beta}{2}\,\Phi_{k_2k_3}^{kk_1}(t)\qquad . \tag{A24}$$

Since the dissipation represents the effect of the average exciton-phonon interaction, to leading order we take it to be independent of temperature. It therefore follows that the correlation function in this temperature regime is proportional to temperature, i.e. $\Phi \propto k_BT$. If k_BT is smaller than the phonon bandwidth then it is more difficult to determine the β-dependence of Φ.[15] In particular, it is clear that Φ can then not be a simple exponential (as it can at high k_BT). For a particular exciton-phonon interaction such as the one considered here, one can deduce the appropriate form of Φ and hence of K.[15] The form (A24) is valid independently of the detailed model at high temperatures.

ACKNOWLEDGEMENT

One of the authors (B.J. West) would like to thank the Defense Advanced Research Projects Agency for the support of this work.

REFERENCES

1. H. Haken and G. Strobl, Z. Phys. 262, 135 (1973).
2. H. Haken and G. Strobl, in The Triplet State, ed. by A.B. Zahlan (Cambridge University, London, 1967), p. 311.
3. V. M. Kenkre and P. Reineker, Exciton Dynamics in Molecular and Crystal Aggregates, Springer Tracts in Modern Physics Vol. 94 (Springer-Verlag, Berlin, 1982).
4. H. Sumi, J. Chem. Phys. 67, 2943 (1977).
5. A. Blumen and R. Silbey, J. Chem. Phys. 69, 3589 (1978).
6. B. Jackson and R. Silbey, J. Chem. Phys. 75, 3293 (1981).
7. M. Grover and R. Silbey, J. Chem. Phys. 54, 4843 (1971).

8. D. Yarkony and R. Silbey, J. Chem. Phys. 67, 5818 (1977).
9. R. W. Munn and R. Silbey, J. Chem. Phys. 68, 2439 (1978).
10. R. Silbey and R. W. Munn, J. Chem. Phys. 72, 2763 (1980).
11. V. M. Kenkre and R. S.Knox, Phys. Rev. B9, 5279 (1974).
12. V. M. Kenkre and R. S. Knox, Phys. Rev. Lett. 33, 803 (1974).
13. V. M. Kenkre, Phys. Rev. B11, 1741 (1975).
14. V. M. Kenkre, Phys. Rev. B12, 2150 (1975).
15. K. Lindenberg and B. J. West, in preparation.
16. R. Kubo, J. Phys. Soc. Japan 17, 1100 (1962).
17. R. Kubo, J. Math. Phys. 4, 174 (1963).
18. N. G. van Kampen, Stochastic Processes in Physics and Chemistry (North-Holland, Amsterdam, 1981).
19. I.B. Rips and V. Cápek, Phys. Status Solidi B100, 451 (1980).
20. G. C. Morris and M. G. Sceats, Chem. Phys. 3, 342 (1974).
21. A. Einstein, Ann. Physik 33, 1275 (1910).
22. R. Benguria and M. Kac, Phys. Rev. Lett. 46, 1 (1981).
23. R. Silbey, Ann. Rev. Phys. Chem. 27, 203 (1976).
24. See e.g. S. W. Lovesey, Condensed Matter Physics: Dynamic Correlations (Benjamin/Cummings, Reading, 1980).
25. V. Seshadri and K. Lindenberg, Physica 115A, 501 (1982).
26. B. J. West, Phys. Rev. A25, 1683 (1982).
27. R. Zwanzig, J. Stat. Phys. 9, 215 (1973).
28. K. Lindenberg and V. Seshadri, Physica 109A, 483 (1981).

HOPPING CONDUCTION FROM MULTIPLE SCATTERING THEORY AND CONTINUOUS TIME RANDOM WALK TO THE COHERENT MEDIUM APPROXIMATION

M. Lax
Physics Department, City College of the City University of New York, New York, 10031 and Bell Laboratories, Murray Hill, N.J. 07974

T. Odagaki*
Laboratory of the National Foundation for Cancer Research at the Physics Department, City College of the City University of New York, New York, 10031

ABSTRACT

A random-walk in a random environment is studied in connection with hopping conduction in doped semiconductors and with the dynamical percolation problem. A brief review is given of linear response theory, the continuous time random walk method, multiple scattering theory and the coherent medium approximation. The ac hopping conductivity in doped semiconductors is obtained in the coherent medium approximation and compared with experiment. Critical properties of the ac conductivity of the dynamical percolation problem are discussed with the use of some rigorous arguments and the coherent medium approximation.

1. INTRODUCTION

In 1961, a significant experiment on the low-frequency ac-conductivity of doped semiconductors was reported by Pollak and Geballe.[1] A typical dependence of the real part of the ac-conductivity of Si doped with B and P on frequency and temperature is shown in Fig. 1. A noticeable point of the frequency dependence is that the conductivity is an increasing function of frequency at a given temperature, at least in the frequency range shown in Fig. 1. Although impurity conduction had been the accepted mechanism for the conductivity in these doped semiconductors,[2,3] it was not until Scher and Lax[4,5] analyzed the problem using the continuous time random walk (CTRW) that it was possible to explain the dc and ac response of the conductivity with a single theory. In fact, Scher and Lax[4,5] showed that an appropriate scaling of conductivity, frequency and temperature brings data shown in Fig. 1 into a universal curve. Figure 2 is a replot of Fig. 1 by scaling the ac conductivity by $N_A e^2/kT a^2 w_0' \exp(-\Delta/kT)$ and the frequency by $w_0' \exp(-\Delta/kT)$, where N_A is the number density of acceptors, e is the electronic charge, k is Boltzmann's constant, T is the absolute temperature, a is given by $(4\pi/3N_D)^{1/3}$ for a number density N_D of donors, Δ is the activation energy which can be determined from the dc conductivity, and w_0' is a parameter which determines the jump rate of hopping carriers[3,5] and can be estimated from a microscopic theory.[3]

Typical features of the frequency dependence of the scaled ac conductivity are: (i) there is a finite dc conductivity which depends on the donor-doping level, (ii) the ac conductivity increases as the frequency increases, and (iii) there is a frequency range where the real part of the ac conductivity shows a power-law dependence on the frequency and the power is a function of the donor-concentration.

---

*Present address, Physics Dept., Brandeis University, Waltham, Massachusetts, 02254.

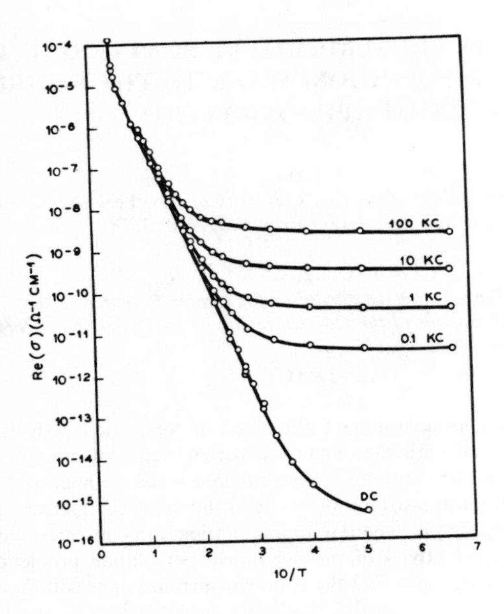

Fig. 1. Temperature and frequency dependence of the real part of the ac conductivity for Si doped with B ($N_A = 0.8 \times 10^{15} cm^3$) and P ($N_D = 2.7 \times 10^{17} cm^3$). [Taken from Fig. 3 of Pollak and Geballe, Ref. 1.]

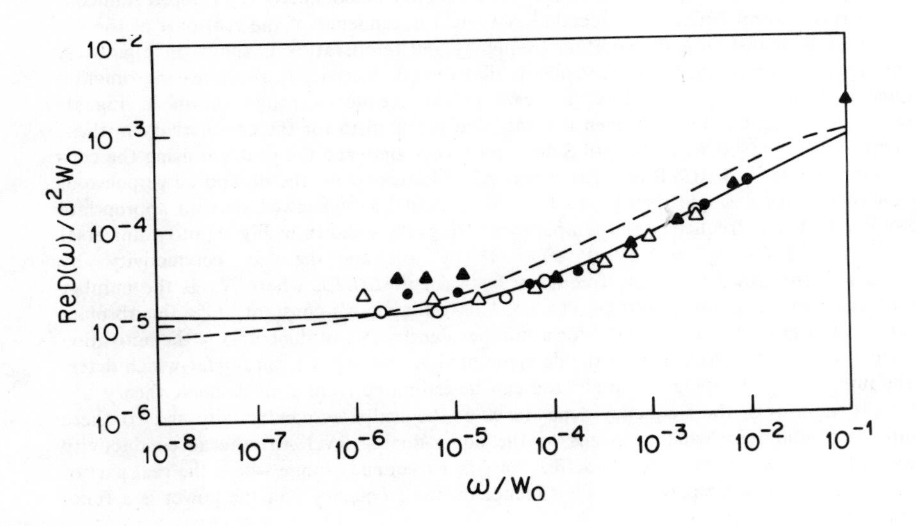

Fig. 2. Replot of data shown in Fig. 1. The conductivity and the frequency are scaled by $N_A e^2/kTa^2 w_0' \exp(-\Delta/kT)$ and $w_0' \exp(-\Delta/kT)$, where $a = (4\pi N_D/3)^{-1/3}$, w_0' is given by Miller and Abrahams[3] and Δ is determined from the dc conductivity.[5] The dashed line is the CTRW result and the solid line is the CMA result. [Taken from Ref. 24.]

The characteristic power-law dependence of the conductivity on the frequency has been observed in various other systems such as doped VO_2,[6] alkaline-doped βAl_2O_3,[7] MoO_3,[8] Te-doped Se,[9] spinel type MnCoNi complex Oxide,[10] $Qn(TCNQ)_2$,[11] and others.

This kind of frequency dependence of the conductivity cannot be explained by a simple Lorentz-Drude type model, because in the Lorentz-Drude model unbound carriers always lead to the conductivity decreasing with increasing frequency, and bound carriers do not give a finite dc conductivity.

For the doped semiconductor discussed earlier, hopping of an electron from a neutral donor to another donor site ionized by compensation has been thought to be the dominant mechanism of electronic conduction.

In fact, Pollak and Geballe[1] analyzed their results by making use of a hopping model between two sites[12] and succeeded in reproducing the power-law behavior of the ac conductivity in a certain frequency range. However, the hopping model between two sites always leads to vanishing dc conductivity, and hence fails to explain the global nature of the ac conductivity observed in the doped semiconductors. Actually, any carrier confined in a finite region in space cannot give rise to a finite dc conductivity. Therefore, we must consider hopping of carriers on an infinitely extended channel.

Suppose a carrier sits on a localized center. The carrier hops to another center with the assistance of a phonon or another dynamical excitation. After staying there for a while, it hops again to another site. The wave function of the carrier will lose phase coherence while the carrier stays on one localized center, and the assistance of other degrees of freedom will make the motion of carriers stochastic. Consequently, we may assume that the carrier performs a *random walk* among the localized centers. The jump rate of a carrier from one center to another depends on the separation between two centers and on the energy difference. Since these centers are usually distributed randomly, the jump rates are random quantities. Thus, we arrive at a random walk with random jump rates to describe hopping conduction in doped semiconductors. Similarly, we can think of a random walk of carriers with random jump rates in each system described earlier.

In connection with percolation theory, there is another motivation for studying a random walk with random jump rates. The percolation theory was originally formulated as a geometrical clustering of percolating sites.[13,14] Although the quantum percolation problem has been studied recently,[15–18] the classical counterpart has not been explicitly formulated except for a descriptive explanation by de Genne[19] and for the bond percolation problem of crystal lattices by the present authors.[20,21] As we will see later, we can formulate the various percolation processes including temperature dependent percolation, directed-bond percolation and continuous percolation by introducing a suitable form for the jump rates and a suitable distribution of localized centers. A random walker performs its random walk among these centers with the specified jump rates, and one can study whether the walker can go infinitely far from the origin on the average. We call this "*the dynamical percolation*" problem.

The random walk in a random environment has been studied by many people. The most successful earlier theory was that due to Scher and Lax[4,5] based on the continuous time random walk (CTRW) technique of Montroll and Weiss.[22] Their theory was quite successful in reproducing experimental results of Pollak and Geballe shown in Figs. 1 and 2. The CTRW method was exploited in various systems.[10,23] When the CTRW method is applied to a system in which any site can be an isolated hopping site the resulting dc conductivity is always zero.[13] But this is not necessarily the case as we will see in the dynamical percolation problem.

Recently, we have developed a new theory, the coherent medium approximation, based on the idea of the coherent potential approximation.[24] This method is found to yield an ac conductivity in a good agreement with experiments for doped semiconductors. It is also shown to work quite well in the dynamical percolation problem where the original CTRW method fails to work.

If one specializes the problem to one-dimension, there is an extensive study by Bernasconi, Alexander, Orbach and their collaborators.[25] They analyzed an integral equation similar to the one introduced by Schmidt[26] for one-dimensional phonon with random mass, by using a scaling hypotheses and a kind of effective medium approximation. Their approach was generalized to the Bethe lattices by Movaghar.[27] He also studied the similarity between the random walk problem and the electronic problem based on the tight binding Hamiltonian.[28]

There has been another type of approach: a Monte Carlo study of the random walk in random environment. Works by Moore,[29] McInnes et al,[30] Richards and Renken,[31] and Vicsek[32] are in this category.

We organize the present paper as follows. In Section 2, we present a short derivation of the generalized Einstein relation from a rather general point of view, which relates the ac conductivity to the random walk. We also introduce a master equation which is assumed to govern the dynamics of the random walker. In Section 3, we derive rather generally some exact results for the ac conductivity, using properties of the master equation. We supply a basic idea of the CTRW method in Section 4 and discuss a difficulty in applying it to a certain system. As we will see in Section 2, we have to evaluate an ensemble average of a kind of propagator in order to obtain the ac conductivity and other quantities. There is a beautiful theory, the multiple scattering theory, developed by one of the present authors (ML)[33] to do this. The multiple scattering theory was originally derived for the wave propagation in random media, but it is applicable to any objects whose description can be couched in a multiple scattering formalism. Actually, the multiple scattering theory was exploited to derive the coherent potential approximation (CPA).[34–36] We give a summary of the multiple scattering theory and its connection to the CPA in Section 5. In section 6 we explain the coherent medium approximation (CMA), which is a generalized application of the CPA to the hopping conduction problem. Section 7 treats applications of the CMA to the impurity conduction problem in one- and three-dimensions. We solve the dynamical percolation problem using the CMA in Section 9, where the bond percolation problem and the percolation in spatially disordered systems are studied.

2. LINEAR RESPONSE THEORY AND THE MASTER EQUATION

Suppose that a sinusoidal external field $E\cos\omega t$ adds a perturbation $-\dot{A}E\cos\omega t$ to an unperturbed system H, and we observe the response by measuring the change $\Delta\dot{A}$ of a physical quantity, $\dot{A} = dA/dt$ as in the case for a conductivity measurement. According to the linear response theory,[24,37,38] the change $\Delta\dot{A}$ is expressed in terms of the susceptibility $\chi_{\dot{A}A}(\omega)$ as

$$\Delta\dot{A} = \mathrm{Re}\,\chi_{\dot{A}A}(\omega)\,E\,e^{-i\omega t}\ , \tag{2.1}$$

and the generalized susceptibility is given by the one-sided Fourier transform of the response function $\phi_{\dot{A}A}(t)$,

$$\chi_{\dot{A}A}(\omega) = \int_0^\infty \phi_{\dot{A}A}(t)\,e^{-i\omega t}\,dt\ \ , \tag{2.2}$$

where the response function $\phi_{\dot{A}A}(t)$ can be written as

$$\phi_{\dot{A}A}(t) = -\frac{1}{2}\frac{\partial^2}{\partial t^2}\int_0^{1/kT} \mathrm{Tr}[\,|A(t)-A(-i\hbar\lambda)|^2\rho]\,d\lambda\ . \tag{2.3}$$

Here, $\rho \equiv e^{-H/kT}$ is the equilibrium density matrix and $A(x) = e^{iHx/\hbar}A\,e^{-iHx/\hbar}$ is the Heisenberg representation of A. When the physical quantity A has a localized nature, or more precisely A has the property

$$\langle \mathbf{s}|A|\mathbf{s}'\rangle = A_{\mathbf{s}}\,\delta(\mathbf{s},\mathbf{s}')\ \ , \tag{2.4}$$

with a local basis function $|\mathbf{s}\rangle$ and a Kronecker δ function $\delta(\mathbf{s},\mathbf{s}')$, the generalized susceptibility is reducible[24] to

$$\chi_{\dot{A}A}(\omega) = -\frac{\omega^2}{2kT}\sum_{\mathbf{s},\mathbf{s}_0}(A_{\mathbf{s}}-A_{\mathbf{s}_0})^2\,\tilde{P}(\mathbf{s},i\omega\,|\,\mathbf{s}_0)\,f(\mathbf{s}_0)\ . \tag{2.5}$$

The quantity $\tilde{P}(\mathbf{s},u\,|\,\mathbf{s}_0)$ is the Laplace transform of the absolute square of the Green's function $P(\mathbf{s},t\,|\,\mathbf{s}_0,0) \equiv |<\mathbf{s}|\exp(-iHt/\hbar)|\mathbf{s}_0>|^2$ defined for $t>0$;

$$\tilde{P}(\mathbf{s},u\,|\,\mathbf{s}_0) = \int_0^\infty e^{-ut}\,P(\mathbf{s},t\,|\,\mathbf{s}_0,0)\,dt\ . \tag{2.6}$$

For the ac conductivity, the physical quantity is the position operator

$$A = \sum_{\mathbf{s}} |\mathbf{s}>e\,s_\alpha<\mathbf{s}|\ , \tag{2.7}$$

where s_α is a Cartesian component of **s**. Therefore, the scalar ac conductivity $\sigma(\omega)$ for an isotropic medium can be reducible to the generalized Einstein relation as follows:

$$\sigma(\omega) = \frac{ne^2}{kT}\,D(\omega)\ , \tag{2.8}$$

$$D(\omega) = -\frac{\omega^2}{2d}\sum_{\mathbf{s},\mathbf{s}_0}<(\mathbf{s}-\mathbf{s}_0)^2\tilde{P}(\mathbf{s},i\omega\,|\,\mathbf{s}_0)f(\mathbf{s}_0)>\ , \tag{2.9}$$

where d is the dimensionality of the system and n is the number density of carriers. The bracket <...> has been used to denote the ensemble average. The average over the initial site $\mathbf{s}_0$ can be omitted in Eq. (2.9) since the ensemble average removes the $\mathbf{s}_0$ dependence of the summand. If the sites $\{\mathbf{s}\}$ reside on a regular lattice, $(\mathbf{s}-\mathbf{s}_0)^2$ can be taken outside of the ensemble average.

Incidentally, we define the dimensionless ac conductivity $\tilde{\sigma}(\omega)$ and dimensionless diffusion constant $\tilde{D}(\omega)$ by

$$\tilde{\sigma}(\omega) = \tilde{D}(\omega) = \sigma(\omega)/(ne^2a^2w_0/kT)\ , \tag{2.10}$$

where a and w_0 are scaling parameters of distance and frequency, respectively, which will show up in each individual problem.

Now, let us remember that the function $P(\mathbf{s},t\,|\,\mathbf{s}_0,0)$ denotes the probability of finding a carrier at site **s** at time t if it left site $\mathbf{s}_0$ at time $t=0$. Note that the definition of $P(\mathbf{s},t\,|\,\mathbf{s}_0,0)$,

$$P(\mathbf{s},t\,|\,\mathbf{s}_0,0) = |<\mathbf{s}|\exp(-iHt/\hbar)|\mathbf{s}_0>|^2\ , \tag{2.11}$$

also includes implicitly a summation over final and average over the initial internal states other than the carrier in consideration. A carrier started at $\mathbf{s}_0$ at time $t = 0$ travels to site **s** in time t and after staying there for a while, it will make a further journey, jumping from **s** to another site **s**′. Actually, Miller and Abraham[3] calculated explicitly the jump rate of an electron between two localized centers in doped semiconductors using the deformation potential. However, it is virtually impossible to study the entire time dependence of $P(\mathbf{s},t\,|\,\mathbf{s}_0,0)$ from first principles. Instead, we introduce a master equation, a random-walk equation, to represent the essential nature of the stochastic motion of carriers. We assume that $P(\mathbf{s},t\,|\,\mathbf{s}_0,0)$ obeys

$$\frac{\partial}{\partial t}P(\mathbf{s},t\,|\,\mathbf{s}_0,0) = -\Gamma_{\mathbf{s}}P(\mathbf{s},t\,|\,\mathbf{s}_0,0) + \sum_{\mathbf{s}'\neq\mathbf{s}} w(\mathbf{s},\mathbf{s}')P(\mathbf{s}',t\,|\,\mathbf{s}_0,0)\ , \tag{2.12}$$

where the decay rate $\Gamma_{\mathbf{s}}$ is given by

$$\Gamma_{\mathbf{s}} = \sum_{\mathbf{s}'\neq\mathbf{s}} w(\mathbf{s}',\mathbf{s})\ , \tag{2.13}$$

to insure the probability conservation $\sum_{\mathbf{s}} P(\mathbf{s},t\,|\,\mathbf{s}_0,0)=1$, and the elementary jump rate $w(\mathbf{s}',\mathbf{s})$ from site **s** to **s**′ is a function of temperature and the distance between two sites **s** and **s**′. The

Laplace transform $\tilde{P}(\mathbf{s},u\,|\,\mathbf{s}_0)$ then obeys

$$(u+\Gamma_\mathbf{s})\,\tilde{P}(\mathbf{s},u\,|\,\mathbf{s}_0)-\sum_{\mathbf{s}'\neq\mathbf{s}} w(\mathbf{s},\mathbf{s}')\,\tilde{P}(\mathbf{s}',u\,|\,\mathbf{s}_0)=\delta(\mathbf{s},\mathbf{s}_0)\ . \tag{2.14}$$

For later convenience, we introduce a matrix $\hat{\mathbf{H}}$ whose elements are

$$\hat{\mathbf{H}}_{\mathbf{s},\mathbf{s}'}=w(\mathbf{s},\mathbf{s}')\ ,\quad \text{for } \mathbf{s}\neq\mathbf{s}'$$

$$\hat{\mathbf{H}}_{\mathbf{s},\mathbf{s}}=-\Gamma_\mathbf{s}\ , \tag{2.15}$$

and a vector

$$\mathbf{P}_\mathbf{s}=\tilde{P}(\mathbf{s},u\,|\,\mathbf{s}_0)\ . \tag{2.16}$$

Then Eq. (2.14) can be written in a compact form

$$(u\hat{\mathbf{E}}-\hat{\mathbf{H}})\,\mathbf{P}=\mathbf{1}_{\mathbf{s}_0}\ , \tag{2.17}$$

with a unit matrix $\hat{\mathbf{E}}$ and a unit vector $\mathbf{1}_{\mathbf{s}_0}$ whose elements are zero except for the $\mathbf{s}_0$ element. Equation (2.17) can be solved formally to yield

$$\tilde{P}(\mathbf{s},u\,|\,\mathbf{s}_0)=\left\{(u\hat{\mathbf{E}}-\hat{\mathbf{H}})^{-1}\right\}_{\mathbf{s},\mathbf{s}_0}\ . \tag{2.18}$$

This restatement of the random-walk equation Eq. (2.14) supports the analogy between a classical random-walk and quantum mechanical propagation. We call $\hat{\mathbf{H}}$ a random-walk matrix and $\hat{\mathbf{P}}\equiv(u\hat{\mathbf{E}}-\hat{\mathbf{H}})^{-1}$ a random-walk propagator.

3. SOME EXACT PROPERTIES

Because of the characteristics of the random-walk matrix $\hat{\mathbf{H}}$ given in Eq. (2.15), some properties of the ac hopping conductivity can be derived rigorously.

First, we list three properties of the random-walk matrix;

$$\text{(1)}\quad \det\hat{\mathbf{H}}=0\ . \tag{3.1}$$

(2) $\hat{\mathbf{H}}$ is negative semidefinite .

If the number of localized centers $\{\mathbf{s}\}$ is finite, i.e. $\hat{\mathbf{H}}$ is of finite dimensions

(3) zero is a nondegenerate eigenvalue of $\hat{\mathbf{H}}$.

The proof of these properties can be found elsewhere.[39]

Suppose that the system can be partitioned into a set of clusters containing a finite number of hopping sites and each site in such a cluster is connected by a non-zero jump rate to at least one of other sites in the same cluster. We also assume that no pairs of sites belonging to different clusters are connected by non-zero jump rate. Then the random-walk matrix can be reducible to a set of matrices of finite dimensions. A random-walk matrix of finite dimensions has always a nondegenerate eigenvalue zero separated from other eigenvalues by a finite gap.[39] Using these properties, we can prove without difficulty that the real and imaginary parts of the ac conductivity for this case vanishes quadratically and linearly, respectively, as the frequency goes to zero. In particular, the diffusion constant can be written as

$$D(\omega)\approx i\xi^2\omega+O(\omega^2)\ , \tag{3.2}$$

and the correlation length ξ is given by

$$\xi^2 = \frac{1}{2d} < \sum_{\mathbf{s}_0} \sum_{\mathbf{s}} \frac{1}{Z} (\mathbf{s}-\mathbf{s}_0)^2 f(\mathbf{s}_0) f(\mathbf{s}) > , \tag{3.3}$$

where $Z = \sum_{\mathbf{s}}' f(\mathbf{s})$ and the summation $\sum_{\mathbf{s}}'$ is taken all sites in the cluster designated by $\mathbf{s}_0$.

The expression for the high frequency limit of the ac conductivity can be derived more easily. First we expand the random-walk propagator in a power series of $1/u$:

$$(u\hat{\mathbf{E}}-\hat{\mathbf{H}})^{-1} = \frac{1}{u}\hat{\mathbf{E}} + \frac{1}{u^2}\hat{\mathbf{H}} + \frac{1}{u^3}\hat{\mathbf{H}}^2 + \cdots . \tag{3.4}$$

Noting that the terms satisfying $\mathbf{s} = \mathbf{s}_0$ in Eq. (2.9) are automatically excluded, and thus the first term on the right hand side of Eq. (3.4) does not contribute to the conductivity, one obtains the generalized diffusion constant at the high frequency limit as

$$D(\infty) = \frac{1}{2d} < \sum_{\mathbf{s},\mathbf{s}_0} (\mathbf{s}-\mathbf{s}_0)^2 w(\mathbf{s},\mathbf{s}_0) f(\mathbf{s}_0) > . \tag{3.5}$$

4. THE CONTINUOUS TIME RANDOM WALK

We can write down a formal solution of Eq. (2.12) or Eq. (2.14) in the form of the Continuous Time Random Walk (CTRW). First we define the probability density $R(\mathbf{s},t\,|\,\mathbf{s}_0,0)$ that a carrier arrives at $\mathbf{s}$ at time t if it started $\mathbf{s}_0$ at time $t=0$, and the probability $\Phi(\mathbf{s},t)$ that a carrier stays on the site $\mathbf{s}$ in the time interval $[0,t]$. Then, the probability $P(\mathbf{s},t\,|\,\mathbf{s}_0,0)$ can be written in terms of the convolution of $R(\mathbf{s},t\,|\,\mathbf{s}_0,0)$ and $\Phi(\mathbf{s},t)$;

$$P(\mathbf{s},t\,|\,\mathbf{s}_0,0) = \int_0^t \Phi(\mathbf{s},t-\tau)\, R(\mathbf{s},\tau\,|\,\mathbf{s}_0,0)\, d\tau . \tag{4.1}$$

The Laplace transform $\tilde{P}(\mathbf{s},u\,|\,\mathbf{s}_0)$ is then related to the Laplace transforms of $\Phi(\mathbf{s},t-\tau)$ and $R(\mathbf{s},t\,|\,\mathbf{s}_0,0)$;

$$\tilde{P}(\mathbf{s},u\,|\,\mathbf{s}_0) = \tilde{\Phi}(\mathbf{s},u)\, \tilde{R}(\mathbf{s},u\,|\,\mathbf{s}_0) , \tag{4.2}$$

where

$$\tilde{\Phi}(\mathbf{s},u) = \int_0^\infty \Phi(\mathbf{s},t) \exp(-ut)\, dt , \tag{4.3a}$$

$$\tilde{R}(\mathbf{s},u\,|\,\mathbf{s}_0) = \int_0^\infty R(\mathbf{s},t\,|\,\mathbf{s}_0,0) \exp(-ut)\, dt . \tag{4.3b}$$

Furthermore, we introduce a probability $\psi(\mathbf{s},\mathbf{s}',t)dt$ that a single jump from site $\mathbf{s}'$ to $\mathbf{s}$ occurs in $[t,t+dt]$. Using the elementary jump probability density $\psi(\mathbf{s},\mathbf{s}',t)$, the quantities $R(\mathbf{s},t\,|\,\mathbf{s}_0,0)$ and $\Phi(\mathbf{s},t)$ are expressed as

$$R(\mathbf{s},t\,|\,\mathbf{s}_0,0) = \sum_{\mathbf{s}'} \int_0^t d\tau \psi(\mathbf{s},\mathbf{s}',t-\tau) R(\mathbf{s}',\tau\,|\,\mathbf{s}_0,0) d\tau$$

$$+ \delta(\mathbf{s},\mathbf{s}_0)\, \delta(t-0^+) , \tag{4.4}$$

and

$$\Phi(\mathbf{s},t) = 1 - \sum_{\mathbf{s}'\neq\mathbf{s}} \int_0^t \psi(\mathbf{s}',\mathbf{s},\tau)\, d\tau . \tag{4.5}$$

We now take Laplace transform of Eqs. (4.4) and (4.5) to obtain

$$\tilde{R}(\mathbf{s},u\,|\,\mathbf{s}_0) - \sum_{\mathbf{s}'}\tilde{\psi}(\mathbf{s},\mathbf{s}',u)\tilde{R}(\mathbf{s}',u\,|\,\mathbf{s}_0) = \delta(\mathbf{s},\mathbf{s}_0) \quad , \tag{4.6}$$

and

$$\tilde{\Phi}(\mathbf{s},u) = \frac{1}{u}\left\{1 - \sum_{\mathbf{s}'\neq\mathbf{s}}\tilde{\psi}(\mathbf{s}',\mathbf{s},u)\right\} , \tag{4.7}$$

where

$$\tilde{\psi}(\mathbf{s},\mathbf{s}',u) = \int_0^\infty \psi(\mathbf{s},\mathbf{s}',t)e^{-ut}dt \quad . \tag{4.8}$$

Equation (4.6) can be solved formally by using a matrix notation; let us define matrices $\hat{\mathbf{R}}$, $\hat{\psi}$ and $\hat{\Phi}$ by

$$\hat{\mathbf{R}}_{\mathbf{s},\mathbf{s}'} = \tilde{\mathbf{R}}(\mathbf{s},u\,|\,\mathbf{s}') \quad , \tag{4.9}$$

$$\hat{\psi}_{\mathbf{s},\mathbf{s}'} = \tilde{\psi}(\mathbf{s},\mathbf{s}',u) \quad , \tag{4.10}$$

and

$$\hat{\Phi}_{\mathbf{s},\mathbf{s}'} = \delta(\mathbf{s},\mathbf{s}')\,\tilde{\Phi}(\mathbf{s},u) \quad . \tag{4.11}$$

Then, from Eq. (4.6) $\hat{\mathbf{R}}$ is given by

$$\hat{\mathbf{R}} = (\hat{\mathbf{E}} - \hat{\psi})^{-1} \quad , \tag{4.12}$$

and hence, the probability $\tilde{P}(\mathbf{s},u\,|\,\mathbf{s}_0)$ in Eq. (4.2) is the $\mathbf{s},\mathbf{s}_0$ matrix element of

$$\hat{\mathbf{P}} = \hat{\Phi}\cdot\hat{\mathbf{R}} = \hat{\Phi}(\hat{\mathbf{E}} - \hat{\psi})^{-1} \quad . \tag{4.13}$$

Comparing Eq. (4.13) with Eq. (2.18) and noting that $\hat{\Phi}$ is diagonal, one obtains

$$\tilde{\Phi}(\mathbf{s},u) = \frac{1}{u+\Gamma_{\mathbf{s}}} \quad , \tag{4.14}$$

and

$$\tilde{\psi}(\mathbf{s},\mathbf{s}',u) = \frac{w(\mathbf{s},\mathbf{s}')}{u+\Gamma_{\mathbf{s}'}} \quad , \tag{4.15}$$

in agreement with the expression given by Scher and lax.[5,40]

Up to this point, all the manipulations are exact. To obtain an ensemble average of the probability $\tilde{P}(\mathbf{s},u\,|\,\mathbf{s}_0)$ appeared in the expression for the ac conductivity Eq. (2.9), Scher and Lax[4,5,40] introduced a Hartree approximation

$$\tilde{\Phi}(\mathbf{s},u) \approx \tilde{\Phi}(u) \equiv < \frac{1}{u+\Gamma_{\mathbf{s}}} > \quad , \tag{4.16}$$

and

$$\tilde{\psi}(\mathbf{s},\mathbf{s}',u) \approx \tilde{\psi}(\mathbf{s}-\mathbf{s}',u) \equiv < \frac{w(\mathbf{s},\mathbf{s}')}{u+\Gamma_{\mathbf{s}'}} > \quad . \tag{4.17}$$

Then, the generalized diffusion constant $D(\omega)$ is reducible to

$$D(\omega) = \frac{1}{2d}\,\sigma^2_{\mathrm{rms}}(\omega) < \Gamma_{\mathbf{s}}/(i\omega+\Gamma_{\mathbf{s}}) > / < 1/(i\omega+\Gamma_{\mathbf{s}}) > \quad , \tag{4.18}$$

where

$$\sigma_{\rm rms}^2(\omega) = \sum_{\mathbf{s}} (\mathbf{s}-\mathbf{s}')^2 \tilde{\psi}(\mathbf{s}-\mathbf{s}', i\omega) / \sum_{\mathbf{s}} \tilde{\psi}(\mathbf{s}-\mathbf{s}', i\omega) \ . \tag{4.19}$$

In particular, when the sites $\{\mathbf{s}\}$ reside on a hyper-cubic regular lattice with lattice constant a and the jump rate $w(\mathbf{s},\mathbf{s}')$ is zero unless $\mathbf{s}$ and $\mathbf{s}'$ are the nearest neighbors, then the diffusion constant is given by

$$D(\omega) = \frac{a^2}{2d} < \Gamma_{\mathbf{s}}/(i\omega+\Gamma_{\mathbf{s}}) > / < 1/(i\omega+\Gamma_{\mathbf{s}}) > \ . \tag{4.20}$$

The Scher-Lax procedure is quite successful in reproducing the global characteristics of the ac conductivity observed in doped semiconductors as shown in Fig. 2.

In this procedure, it is easy to see that $D(0) = 0$ if we have a finite probability of having $\Gamma_{\mathbf{s}}=0$, since the denominator always carries, then, the term $1/i\omega$ which diverges, and hence $D(\omega)$ vanishes as ω approaches zero. However, $D(0)$ is not always equal to 0, even if $\Gamma_{\mathbf{s}}$ vanishes with a finite probability, as we will see in Section 8.

5. MULTIPLE SCATTERING THEORY AND THE CPA

As we have seen in Section 2, the calculation of the ac hopping conductivity can be expressed as an average of the random-walk propagator. Similar ensemble averages of propagators have been investigated by many authors.[33,36,41] We present here the essential feature of the multiple scattering theory and the derivation of an exact coherent potential from which the coherent potential approximation (CPA) can be deduced.

We are concerned with the solution of a problem of the form

$$(E - H_0 - \sum_j v_j)\ \Psi = 0 \ . \tag{5.1}$$

In multiple scattering language, it is appropriate to regard

$$H_0 = \frac{\mathbf{p}^2}{2m} = -\frac{\hbar^2}{2m} \nabla^2 \ , \tag{5.2}$$

as the free wave Hamiltonian,

$$v_j = v(\mathbf{r}-\mathbf{r}_j) \ , \tag{5.3}$$

as the potential associated with a scatterer at $\mathbf{r}_j$, and Ψ as the total wave in the presence of all the scatterers $j=1,2,\ldots$. Although our derivation of a coherent potential was motivated by the multiple scattering picture, it is important to note that our proof[33,36] depends only on the algebraic structure of Eq. (5.1) and not on the physical interpretation just given. Indeed, we have used this coherent potential procedure when Ψ is not a wave but a probability density and v_j is not a potential but the jump rate associated with a bond.[24] Each v_j can even represent a cluster of objects, provided only that $\sum v_j$ represents the total.

To describe the motion in a coherent or effective medium, V_c, we introduce a modified "free wave" operator

$$H = H_0 + V_c \ , \tag{5.4}$$

and correct each of the N scatterers to

$$V_j = v_j - v_j^c \ , \tag{5.5}$$

where

$$V_c = \sum_j v_j^c \ ,$$

so that, with no approximation

$$H_0 + \sum_j v_j \equiv H + \sum_j V_j \quad , \tag{5.6}$$

even with an arbitrary choice of the "coherent potential" V_c.

Often the problem is to calculate the density of states:

$$n(E) = \mathrm{tr} < \delta(E - H - \sum_j V_j) >$$

$$= - \frac{1}{\pi} \mathrm{Im}\, \mathrm{tr} < G > \quad , \tag{5.7}$$

when $< >$ denotes an average over the ensemble of scatterer configuration and G, the Green's function, obeys

$$(E - H - \sum_j V_j)\, G = 1 \quad , \tag{5.8}$$

or in integral form

$$G = G_c + \sum_j \frac{1}{E-H} V_j\, G \equiv G_c + \sum_j L_j \; , \tag{5.9}$$

where

$$G_c \equiv (E-H)^{-1}, \tag{5.10}$$

is the coherent propagator and $L_j \equiv (E-H)^{-1} V_j G$ can be interpreted as the wave that leaves scatterer j. We may then define

$$G^i = G_c + \sum_{j \neq i} L_j \; , \tag{5.11}$$

as the wave that excites scatterer i. The relation between the exciting and total fields

$$G - L_i = G^i \quad , \tag{5.12}$$

provides us with the integral equation

$$G - (E-H)^{-1} V_i\, G = G^i \quad . \tag{5.13}$$

Let us seek a solution to this equation in the form

$$G = (1 + R_i)\, G^i = G^i + L_i \quad , \tag{5.14}$$

which expresses the leaving wave L_i as $R_i G^i$, and the total as the striking wave G^i plus the scattered wave $R_i G^i$. Thus R_i is found to obey the equation

$$R_i - (E-H)^{-1} V_i R_i = (E-H)^{-1} G^i \quad . \tag{5.15}$$

Since the righthand side contains $(E-H)^{-1}$ as a factor, let

$$R_i = (E-H)^{-1}\, T_i \quad . \tag{5.16}$$

The transition operator T_i then obeys the integral equation

$$T_i = V_i + V_i (E-H)^{-1}\, T_i \quad . \tag{5.17}$$

With the help of $L_i = R_i\, G^i$ and Eq. (5.16), the total field (5.9) and the exciting field (5.11) can be written in the form

$$G = G_c + \sum_i (E-H)^{-1}\, T_i\, G^i \; , \tag{5.18}$$

$$G^i = G_c + \sum_{j \neq i} (E-H)^{-1} \, T_j \, G^j \ . \tag{5.19}$$

Equations (5.18) and (5.19) constitute a set of multiple scattering equations. For the case of a bare propagator ($V_c = 0$), these equations were first proposed as intuitively obvious.[33] Note that the above derivation is valid even for scatterers with overlapping potentials!

Not only is the above derivation exact, it is valid for any choice of coherent potential V_c. The *exact* coherent potential can be defined by the requirement that the average propagator

$$< G > = G_c = (E-H_0-V_c)^{-1} \ , \tag{5.20}$$

be given exactly by G_c. The ensemble average of Eq. (5.18) then yields an *exact* condition

$$\sum_i < T_i \, G^i > = 0 \ , \tag{5.21}$$

for the coherent potential. In order to implement this condition, however, one needs a solution to the multiple scattering equation (5.19).

The widely used coherent potential approximation can be obtained in this formalism by neglecting the difference between the exciting field and the average field:

$$< G^i >_i \approx G_c \ , \tag{5.22}$$

where the subscript i implies that an average over the random variables (e.g. positions) of all scatterers but the ith have been averaged over. Equation (5.21) then leads to the usual statement

$$\sum_i < T_i > \approx 0 \ , \tag{5.23}$$

of the coherent potential approximation; namely that the coherent potential V_c is so chosen that the additional scattering due to the deviation V_i of each V_i from V_c produces a vanishing transition operator T_i on the average.

The CPA is sometimes formulated as the following single site approximation. Consider a single site i, with a scattering potential V_i embedded in the average medium H. Let

$$K^i \equiv (E-H-V_i)^{-1} \ , \tag{5.24}$$

be the Green's function of this single site problem. Require that the average single site Green's function be identical to the coherent Green's function of the embedding medium G_c:

$$< K^i > \equiv < (E-H-V_i)^{-1} > = (E-H)^{-1} \, . \tag{5.25}$$

Equation (5.24) can be algebraically reexpressed as

$$K^i - (E-H)^{-1} \, V_i \, K^i = (E-H)^{-1} \equiv G_c \ . \tag{5.26}$$

But this is precisely Eq. (5.13) with G^i replaced by G_c. Its solution via Eq. (5.14) is simply

$$K^i = (1 + R_i) \, G_c = G_c + (E-H)^{-1} \, T_i \, G_c \ . \tag{5.27}$$

With all sites understood to be equivalent,

$$< \sum_i T_i > = N < T_i > = 0 \ , \tag{5.28}$$

in the CPA so that

$$< K^i > = G_c \ , \tag{5.29}$$

just the single site criterion, Eq. (5.25).

To implement the exact condition, (5.21), for the coherent potential, or the approximate condition, (5.23), we need to evaluate T_j. For the case of an A - B type alloy, $v_j = v(\mathbf{r}-\mathbf{j})$ takes one of two values $v_A(\mathbf{r}-\mathbf{j})$ or $v_B(\mathbf{r}-\mathbf{j})$. Each kind of atom has its own T matrix, e.g., from Eq. (5.17) (with the site **j** shifted to the origin)

$$T_A = v_A(\mathbf{r}) - v_c(\mathbf{r}) + [v_A(\mathbf{r}) - v_c(\mathbf{r})]\, e^{-1}\, T_A \ , \tag{5.30}$$

and

$$T_B = v_B(\mathbf{r}) - v_c(\mathbf{r}) + [v_B(\mathbf{r}) - v_c(\mathbf{r})]\, e^{-1}\, T_B \ , \tag{5.31}$$

where

$$e \equiv E - H_0 - V_c \ .$$

In an alloy with concentrations C_A and C_B, the coherent potential approximation Eq. (5.23) reduces to

$$C_A\, T_A + C_B\, T_B = 0 \ , \tag{5.32}$$

as a condition for determining the site coherent potential v_c with the result

$$v_c = C_A\, v_A + C_B\, v_B - (v_A - v_c)\, e^{-1}\, (v_B - v_c) \ . \tag{5.33}$$

For the case of a liquid-like disorder, the operator V_c will be diagonal in **k** space because of translational invariance. Thus we can write

$$v_j^c = V_c/N \ , \tag{5.34}$$

independent of j. Equation (5.17) can be solved formally, with the result

$$T_j = v_j - (V_c/N) + \left(v_j - \frac{V_c}{N} \right) \left(E - H_0 - \frac{N-1}{N}\, V_c - v_j \right)^{-1} \left(v_j - \frac{V_c}{N} \right) \ . \tag{5.35}$$

In the "thermodynamic" limit in which N approaches infinity

$$T_j = \hat{t}_j - V_c/N \tag{5.36}$$

where

$$\hat{t}_j = v_j + v_j\, (E - H_0 - V_c - v_j)^{-1}\, v_j \ , \tag{5.37}$$

and terms of order $1/N$ have been omitted in (5.37). Equation (5.37), however, is the solution of the one-body scattering problem

$$\hat{t}_j = v_j + v_j\, (E - H_0 - V_c)^{-1}\, \hat{t}_j \ , \tag{5.38}$$

using the modified propagator.

The coherent potential approximation, (5.23), then leads to the "simple" expression

$$V_c = V_{\mathrm{CPA}} = \sum_j < \hat{t}_j > \tag{5.39}$$

whose matrix elements in **k** space take the diagonal form:

$$(V_c)_{ba} = \int n(\mathbf{j})\, d\mathbf{j} \exp[i\,(\mathbf{a}-\mathbf{b})\cdot\mathbf{j}]\, \hat{t}_{ba} = n\, \delta_{ba}\, \hat{t}_{ba} \ . \tag{5.40}$$

Thus the coherent potential in a spatially disordered system is given by the forward scattered amplitude, for one-body scattering, calculated, self-consistently, in the modified medium.

6. THE COHERENT MEDIUM APPROXIMATION

First, we note that the random-walk matrix defined by Eq. (2.15) can be written as a sum of matrices $\hat{\mathbf{W}}^i$ of infinite dimensions in which only the four elements are nonzero

$$\hat{\mathbf{W}}^i_{\mathbf{s},\mathbf{s}'} = w(\mathbf{s},\mathbf{s}') \; ; \; \hat{\mathbf{W}}^i_{\mathbf{s}',\mathbf{s}'} = -w(\mathbf{s},\mathbf{s}') \; ; \hat{W}^i_{\mathbf{s}',\mathbf{s}} = w(\mathbf{s}',\mathbf{s}) \; ; \; \hat{W}^i_{\mathbf{s},\mathbf{s}} = -w(\mathbf{s}',\mathbf{s}) \; . \tag{6.1}$$

and all other matrix elements are zero: namely,

$$\hat{\mathbf{H}} = \sum_i \hat{\mathbf{W}}^i \; , \tag{6.2}$$

where the index i runs over all pairs of sites which are connected by nonzero jump rate. Therefore, the multiple scattering method described in Section 5 can be applied to obtain the ensemble average of the random walk propagator $\hat{\mathbf{P}} = (u\hat{\mathbf{E}}-\hat{\mathbf{H}})^{-1}$.

First we consider the case where the sites {s} form a regular cubic lattice and the elementary jump rates $w(\mathbf{s},\mathbf{s}')$ and $w(\mathbf{s}',\mathbf{s})$ are zero unless sites $\mathbf{s}$ and $\mathbf{s}'$ are nearest neighbors. Further, we assume $w(\mathbf{s},\mathbf{s}')=w(\mathbf{s}',\mathbf{s})$ for simplicity. Then, each pair of nearest neighbor sites is associated with a nonzero symmetric matrix $\hat{\mathbf{W}}^i$ localized on the bond connecting these two sites and vice versa. Taking the jump rate between a *pair* of nearest neighbor sites, say 1 and 2, as our elementary perturbation V_j we apply the single "site" formulation discussed in Section 5. After a simple algebraic manipulation, we obtain a self-consistency equation for the coherent jump rate w_c

$$< \frac{w_c - w_{12}}{1 - 2(\bar{P}_{11}-\bar{P}_{12})(w_c-w_{12})} > = 0 \; . \tag{6.3}$$

Here, $< \cdots >$ denotes an average over all possible w_{12}, and $\bar{P}_{11}$ and $\bar{P}_{12}$ are, respectively, the (1,1) and (1,2) matrix of the coherent random walk propagator $(u\hat{\mathbf{E}}-\hat{\mathbf{H}}_c)^{-1}$. The coherent random walk matrix $\hat{\mathbf{H}}_c$ is given by

$$\hat{\mathbf{H}}_c = \sum_i \hat{\mathbf{W}}^i_c \; , \tag{6.4}$$

with the coherent jump rate matrix $\hat{\mathbf{W}}^i_c$. For the bond of two site 1 and and 2, for example, $\hat{\mathbf{W}}^i_c$ has nonzero matrix elements w_c at (1,2) and (2,1) and $-w_c$ at (1,1) and (2,2), and the remaining elements of $\hat{\mathbf{W}}^i_c$ are zero. The coherent jump rate to be determined self-consistently in Eq. (6.3) is a function of the Laplace parameter u. The dimensionless ac-conductivity in this approximation is simply given by

$$\tilde{\sigma}(\omega) = w_c(i\omega)/w_0 \; . \tag{6.5}$$

In order to apply the present procedure to spatially disordered systems, we must set up a tractable model. We first pick up a site, say $\mathbf{s}$, and then divide the space around $\mathbf{s}$ into z-equivalent parts with an arbitrary constant z. We assume that the nearest site to $\mathbf{s}$ in each of z-parts contributes dominantly to the jump of the random-walking carrier from site $\mathbf{s}$, and jumps to other sites are negligible. We take one of these neighboring pair of sites as a random unit embedded in a coherent medium. The coherent medium is assumed to be an appropriate lattice of coordination number z with a nonzero constant nearest neighbor jump rate w_c. Jump rates between further neighbors are assumed to be zero. Then it is not difficult to show that the condition to determine the coherent jump rate reduces to Eq. (6.3) with a suitable distribution for the jump rate w_{12}.[24]

In passing, we make a few remarks. The self-consistency equation (6.3) was derived here with emphasis on its relation to the multiple scattering theory. It was originally presented by the present authors in a slightly different manner.[24] A similar self-consistency condition has been used in a random-resistor network problem[42] and electronic problem.[43] Equation (6.3) was also rederived by other authors.[44–46]

Finally, we use the terminology "the coherent medium" instead of "the coherent potential" since a more general quantity such as a jump rate can be treated on the same footing as a potential.

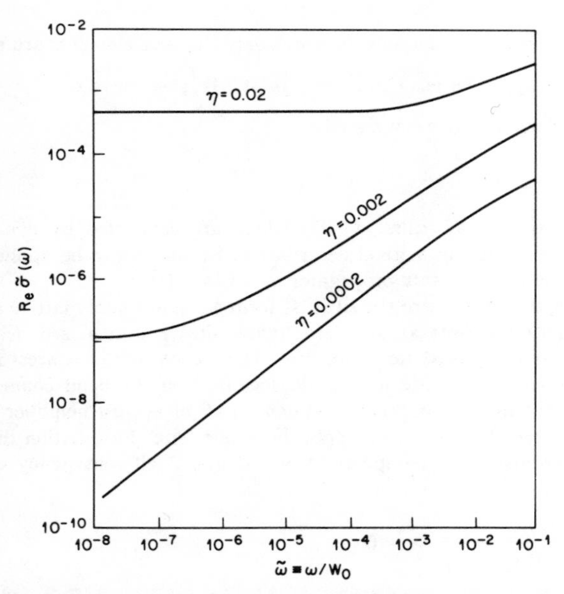

Fig. 3. Typical behavior of the real part of the ac conductivity for various donor levels obtained by the CMA method. $\eta = 4\pi N_D R_d^3$ is the reduced number density of donor.

7. IMPURITY CONDUCTION

First we apply the procedure given in Section 6 to the impurity conduction discussed in the introduction. We replace $|\mathbf{s}-\mathbf{s}_0|^2$ in the expression of the ac conductivity Eq. (2.9) by a^2, a being given in Section 1, and employ the method discussed in the previous section to determine the average of the random walk propagator. The average in Eq. (6.3) is taken over the distribution of the nearest neighbor distance r in one of the z-parts in the three dimensional space,

$$N(r) = \frac{4\pi N_D}{z} r^2 \exp\{-\frac{4\pi}{3z} N_D r^3\} . \tag{7.1}$$

We assume a simple exponential dependence of jump rate $w_{\mathbf{s},\mathbf{s}'}$ on the distance $|\mathbf{s}-\mathbf{s}'|$

$$w_{\mathbf{s},\mathbf{s}'} = w_0 \exp\{-|\mathbf{s}-\mathbf{s}'|/R_d\} , \tag{7.2}$$

where w_0 is a constant dependent on the temperature and R_d is half the effective Bohr radius. As the structure of the coherent medium, we use a pseudo-lattice with the coordination number z=6 and the semi-elliptic density of states.[24] From the numerical solution of Eq. (6.3) with Eqs. (7.1) and (7.2) we obtained the frequency dependence of the ac conductivity which is shown as the solid line in Fig. 2. The agreement of the present theory with experiment is excellent as is the CTRW method. Typical frequency dependence of the ac conductivity for various value of n_D is shown in Fig. 3, where $\eta = 4\pi N_D R_d^3$.

An apparent difference between the CTRW method and the CMA method appears at the dependence of the dc part of the conductivity on the donor concentration. Indeed, the CMA method[24] predicts

$$\tilde{\sigma}(0) \sim \frac{1}{2} \exp\{-(18\ln\frac{3}{2})^{1/3}\eta^{-1/3}\} , \tag{7.3}$$

for small η, while the CTRW method[5] yields

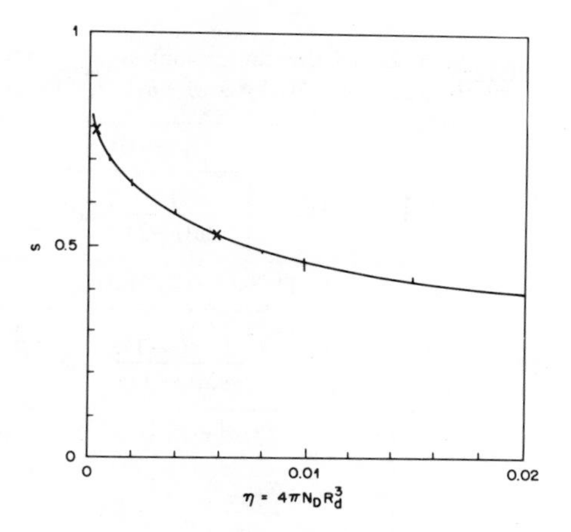

Fig. 4. The donor density dependence of the exponent s predicted by the CMA method. Vertical bars denote fluctuations due to the numerical determination of the exponent. The curve is well fitted by $\exp[-3.5\eta^{1/3}]$. Crosses are experiments for doped silicon.[1,5]

$$\tilde{\sigma}(0) \sim e^{\gamma} \eta^{1/4} \exp\{-2\eta^{-1/2}/3\}/6\sqrt{\pi} \ , \tag{7.4}$$

where γ=0.5772.

As one can see in Fig. 3, the real part of the ac conductivity appears to have a power-law dependence, ω^s, on the frequency in a certain frequency range. The exponent s is a function of the donor concentration N_D. Our numerical data yield Fig. 4 for the N_D dependence of the exponent s. The curve in Fig. 4 is fitted quite well by

$$s \sim \exp\{-3.5\eta^{1/3}\} \ . \tag{7.5}$$

Two experimental points for doped semiconductors are also shown in Fig. 4, which are in a good agreement with the theoretical value of s.

The one-dimensional analogy to the doped semiconductor is of interest in connection with quasi one-dimensional conductors.[11,25] In completely random one dimensional chains, the distribution of the nearest neighbor distance is simply given by the Poisson distribution

$$N(r) = N_D \exp(-N_D r) \ . \tag{7.6}$$

Using the same method as above in one-dimension, we obtained five regimes for the low-frequency behavior of the conductivity shown in Table I.[47–49] The real part of the ac conductivity near the static limit shows again a power-law behavior and the exponent depends on the donor concentration. Figure 5 shows the explicit dependence of the exponent on the reduced density $\rho = N_D R_d$.

In passing, the five regimes in Table I are the consequence of the different properties of the first and second inverse moments of random jump rate.[47] The expression on the first row in Table I, where both the first and second inverse moments of jump rate exist, has been shown to be exact.[50] The system with nondivergent first and second inverse moments of jump rate is the simplest case in the problem[25,47,48] and has been discussed by others[46,51,52] in a more general form.

TABLE I. The low frequency behavior of the dimensionless ac conductivity for the one-dimensional analogy to doped semiconductors. $\tilde{\sigma}(\omega) \equiv \tilde{\sigma}(0) + \tilde{\sigma}_1(\omega)$, $\tilde{\omega} \equiv \omega/w_0$ and $\rho \equiv N_D R_d$.

ρ	$\tilde{\sigma}(0)$	$\tilde{\sigma}_1(\omega) \quad (\omega \sim 0)$
$2 < \rho$	$\frac{\rho-1}{\rho}$	$\left(\frac{\rho-1}{\rho}\right)^{1/2} \frac{1}{2\rho(\rho-2)} (i\tilde{\omega})^{1/2}$
$\rho = 2$	$1/2$	$-(1/8\sqrt{2})(i\tilde{\omega})^{1/2} \ln(i\tilde{\omega})$
$1<\rho<2$	$\frac{\rho-1}{\rho}$	$\left(\frac{\rho-1}{\rho}\right)^{\frac{\rho+1}{2}} \frac{2^{1-\rho}(\rho-1)\pi}{\sin(\rho-1)\pi} (i\tilde{\omega})^{\frac{\rho-1}{2}}$
$\rho = 1$	0	$-2/\ln\{-i\tilde{\omega}/\ln(i\tilde{\omega})\}$
$0<\rho<1$	0	$\left(\frac{2^{1-\rho}\rho\pi}{\sin\rho\pi}\right)^{-\frac{2}{1+\rho}} (i\tilde{\omega})^{\frac{1-\rho}{1+\rho}}$

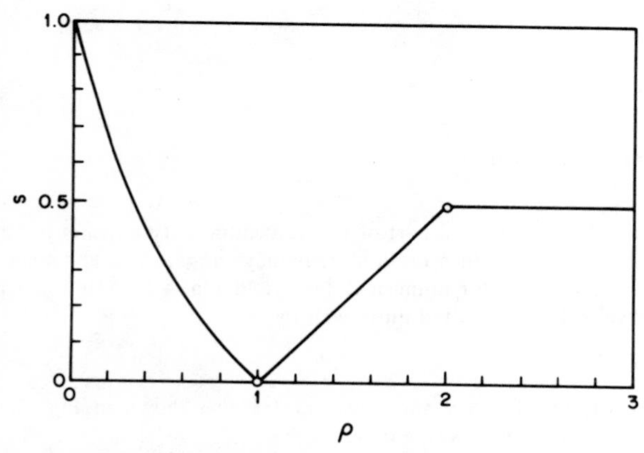

Fig. 5. The dependence of exponent s on the reduced density $\rho = N_D R_d$ for one-dimensional model of doped semiconductors. Open circles denote points where a logarithmic term appears in the ac conductivity. The CMA method is used to determine the conductivity. [Taken from Ref. 47.]

8. AC CONDUCTIVITY FOR A PERCOLATION MODEL

Let us first consider a bond percolation model in a d-dimensional hyper-simple-cubic lattice. [The sites {s} form a regular cubic lattice in d-dimensions.] The nearest neighbor jump rate $w(\mathbf{s},\mathbf{s}') = w(\mathbf{s}',\mathbf{s})$ takes as value either a nonzero constant w_0 or zero with probability p and $1-p$ respectively. The self-consistency equation (6.3) with this distribution is reducible to

$$\{z - 2 + 2u\bar{P}_{11}\} w_c(u) = \{pz - 2 + 2u\bar{P}_{11}\} w_0 \, . \tag{8.1}$$

We are primarily interested in the low-frequency behavior of the ac conductivity. Analyzing the properties of the diagonal elements $\bar{P}_{11}$ of the coherent random walk propagator near $u = 0$, we obtained the expressions for the low frequency behavior of the ac part of the conductivity $\tilde{\sigma}(\omega) - \tilde{\sigma}(0)$ shown in Tables II-IV.[21]

TABLE II. The ac part of the conductivity $\tilde{\sigma}(\omega)-\tilde{\sigma}(0)$ for small ω is tabulated versus dimension d, above the percolation threshold.

d	$\tilde{\sigma}(\omega) - \tilde{\sigma}(0)$
$1<d<2$	$\dfrac{2h(1)\Gamma(\frac{d}{2})\Gamma(1-\frac{d}{2})(z-2)^{2-d/2}(1-p)}{z^{d-1}(p-p_c)^{d/2}}[\cos\frac{d\pi}{4}+\sin\frac{d\pi}{4}i]\omega^{d/2}$
$d=2$	$\dfrac{2h(1)(1-p)}{z(z-2)(p-p_c)}(-\omega\ln\omega)i + \dfrac{\pi h(1)(1-p)}{(z-2)(p-p_c)}\omega$
$2<d<4$	$\dfrac{2m_1(1-p)}{z(z-2)(p-p_c)}\omega i + \dfrac{2h(1)\Gamma(\frac{d}{2}-1)\Gamma(2-\frac{d}{2})(1-p)}{z^{d-1}(z-2)^{2-d/2}(p-p_c)^{d/2}}\cos\frac{d\pi}{4}\omega^{d/2}$
$d=4$	$\dfrac{2m_1(1-p)}{z(z-2)(p-p_c)}\omega i - \dfrac{2h(1)(1-p)}{z^3(p-p_c)^2}\omega^2\ln\omega$
$4<d$	$\dfrac{2m_1(1-p)}{z(z-2)(p-p_c)}\omega i + \dfrac{2(1-p)}{z^3(p-p_c)^2}[\dfrac{2m_1^2}{z(p-p_c)} + m_2]\omega^2$

TABLE III. The ac part of the conductivity is tabulated versus dimension d, at the percolation threshold.

d	$\tilde{\sigma}(\omega) - \tilde{\sigma}(0)$
$1<d<2$	$\frac{1}{z}\{\frac{2z}{z-2}h(1)\Gamma(\frac{d}{2})\Gamma(1-\frac{d}{2})\}^{\frac{2}{d+2}}\{\cos\frac{d\pi}{d+2}+\sin\frac{d\pi}{d+2}i\}\omega^{\frac{d}{d+2}}$
$d=2$	$\left\{\dfrac{h(1)}{2z(z-2)}\right\}^{\frac{1}{2}}(1+i)(-\omega\ln\omega)^{\frac{1}{2}}$
$2<d$	$\left\{\dfrac{m_1}{z(z-2)}\right\}^{\frac{1}{2}}(1+i)\omega^{\frac{1}{2}}$

TABLE IV. Asymptotic form of $A(z,d,p)$ and $B(z,d,p)$ for $p \leqslant p_c$. The ac conductivity $\tilde{\sigma}(\omega)$ is written as $A(z,d,p)\,\tilde{\omega}i + B(z,d,p)\tilde{\omega}^2$ near the static limit $\omega=0$.

d	$A(z,d,p)$	$B(z,d,p)$
$1<d<2$	$\dfrac{1}{z}\left[\dfrac{2h(1)\Gamma(\frac{d}{2})\Gamma(1-\frac{d}{2})}{z(p_c-p)}\right]^{2/d}$	$\dfrac{2}{dz^2}\left\{\dfrac{h(1)\Gamma(d/2)\Gamma(1-d/2)}{z}\right\}^{4/d}\dfrac{1-p}{(p_c-p)^{1+4/d}}$
$d=2$	$\dfrac{-2h(1)\ln(p_c-p)}{z^2(p_c-p)}$	$\dfrac{4h(1)^2(1-p)[\ln(p_c-p)]^2}{4\,(p_c-p)^3}$
$2<d$	$\dfrac{2m_1}{z^2(p_c-p)}$	$\dfrac{4m_1^2(1-p)}{z^4(p_c-p)^3}$

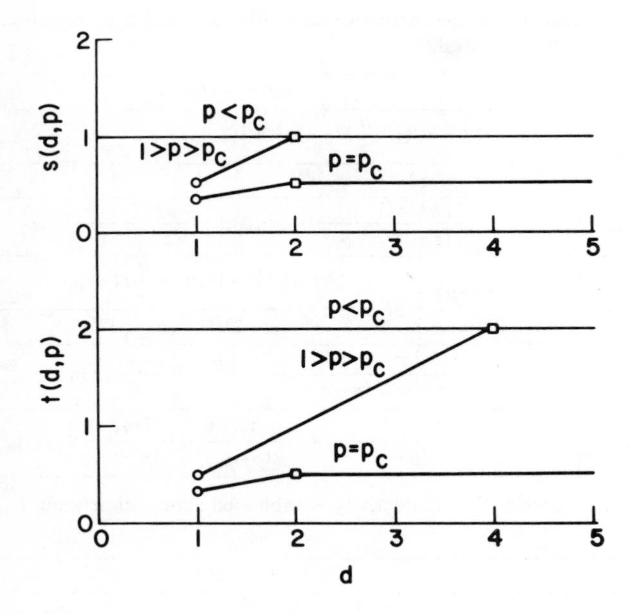

Fig. 6. Probability and dimensionality dependence of the exponents $s(d,p)$ and $t(d,p)$, where the ac part $\hat{\sigma}(\omega)-\hat{\sigma}(0)$ of the conductivity for the d-dimensional bond percolation model is written as $A(z,d,p)\omega^{s(d,p)}\,i+B(z,d,p)\omega^{t(d,p)}$ near the static limit. The open squares denote points where the conductivity has a logarithmic term. The open circles show nonexisting point. The conductivity is determined by the CMA method. [Taken from Ref. 21.]

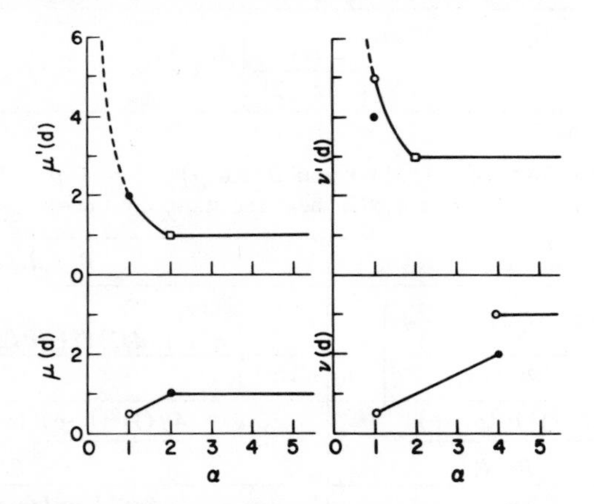

Fig. 7. Dimensionality dependence of critical exponents $\mu(d)$, $\mu'(d)$, $\nu(d)$ and $\nu'(d)$ defined in Eqs. (8.10) and (8.11) in the text. The open squares denote that a logarithmic term appears and the open circles show nonexistent points. The closed circles emphasize the existence of points. The exponents are determined by the CMA method. [Taken from Ref. 21.]

In these Tables, $h(1)$ and m_K are defined in terms of the density of states $n(x)$ as

$$h(1) = \lim_{x \to 1} n(x)/(1-x)^{d/2-1} , \tag{8.2}$$

and

$$m_K = \int_{-\infty}^{1} \frac{n(x)}{(1-x)^K} dx . \tag{8.3}$$

The density of states $n(x)$ is defined by

$$n(x) = \lim_{N \to \infty} \frac{1}{N} \sum_{\mathbf{k}} \delta[x - f(\mathbf{k})] , \tag{8.4}$$

where

$$f(\mathbf{k}) = \frac{1}{z} \sum_{\substack{\mathbf{s}' \\ (\text{n.n. of } \mathbf{s})}} \exp[i\mathbf{k}\cdot(\mathbf{s}'-\mathbf{s})] , \tag{8.5}$$

N is the total number of sites in the system and z is the number of nearest neighbor sites. In the coherent medium approximation used here, the dc conductivity $\tilde{\sigma}(0)$ is always given by

$$\tilde{\sigma}(0) = \begin{cases} \dfrac{p-p_c}{1-p_c} & p \geqslant p_c \\ 0 & p < p_c \end{cases} , \tag{8.6}$$

with $p_c=2/z$ regardless of the dimensionality. We identify p_c as the critical percolation probability. Thus, the critical percolation probability p_c agrees with exact result for a chain and with the empirical relation $zp_c=2$ for two-dimensional lattices,[53] but it is higher than the values estimated by other methods in three-dimensions.[54]

As one can see in Tables II-IV, the ac part of the conductivity shows various critical behaviors at the percolation threshold. Let us write

$$\tilde{\sigma}(\omega)-\tilde{\sigma}(0) = A(z,d,p) f(\omega) i + B(z,d,p) g(\omega) \tag{8.7}$$

near the static limit, and define indices $s(d,p)$ and $t(d,p)$ by

$$f(\omega) = \omega^{s(d,p)} , \tag{8.8}$$

$$g(\omega) = \omega^{t(d,p)} , \tag{8.9}$$

when $f(\omega)$ and $g(\omega)$ obey a power law. We also define the critical indices for the coefficients $A(z,d,p)$ and $B(z,d,p)$ by

$$A(z,d,p) \propto \begin{cases} (p-p_c)^{-\mu(d)} & p>p_c \\ (p_c-p)^{-\mu'(d)} & p<p_c \end{cases} , \tag{8.10}$$

$$B(z,d,p) \propto \begin{cases} (p-p_c)^{-\nu(d)} & p>p_c \\ (p_c-p)^{-\nu'(d)} & p<p_c \end{cases} . \tag{8.11}$$

Figure 6 shows the dimensionality dependence of $s(d,p)$ and $t(d,p)$. The critical exponents $\mu(d)$, $\mu'(d)$, $\nu(d)$, and $\nu'(d)$ are plotted against the dimensionality in Fig. 7. In these figures, the open squares denote that a logarithmic term appears in the quantity discussed, the open circles show nonexistent points and the closed circles emphasis existing points.

In a one-dimensional chain, the bond percolation model has been solved exactly by the present authors.[20] According to this exact solution, $A(2,1,p)$ and $B(2,1,p)$ are given by

$$A(2,1,p) = \frac{p}{2(1-p)^2} \ , \tag{8.12}$$

$$B(2,1,p) = \frac{p(1+p)^2}{4(1-p)^4} \ . \tag{8.13}$$

The coherent medium approximation yields

$$A(2,1,p) = \frac{p(2-p)}{4(1-p)^2} \ , \tag{8.14}$$

$$B(2,1,p) = \frac{p(2-p)}{8(1-P)^4} \ . \tag{8.15}$$

Therefore, for one-dimension, the exponents $\mu'(1)$ and $\nu'(1)$ given by the CMA agree with the exact values, though the coefficients in $A(2,1,P)$ and $B(2,1,p)$ are different.

As we have discussed in Section 3, the imaginary and real parts of the ac conductivity below the percolation threshold vanish linearly and quadratically, respectively, with the frequency, because all the clusters are of finite size when the probability p is less than the percolation threshold. Consequently, the exponents $s(d,p)$ and $t(d,p)$ defined in Eqs. (8.8) and (8.9) are exact when $p < p_c$.

The coefficient $A(z,d,p)$ for $p < p_c$ is directly related to the correlation length. The critical index $\mu'(d)$ defined in Eq. (8.10) coincides with the critical exponent of the correlation length estimated in the conventional percolation theory[55] for the dimensionalities $d=1$ and $d \geqslant 6$.

Finally, we show an application to the percolation problem in spatially disordered systems. In the master equation (2.12) we set

$$w(\mathbf{s}',\mathbf{s}) = w(\mathbf{s},\mathbf{s}') = \begin{cases} w_0 & \text{when } |\mathbf{s}-\mathbf{s}'| \leqslant r_0 \\ 0 & \text{otherwise} \end{cases} \ . \tag{8.16}$$

This choice of the jump rate defines completely the dynamical percolation process in spatially disordered systems. The sites can be distributed in any way. We assume here the sites $\{\mathbf{s}'\}$ are distributed uniformly. We applied the procedure presented in Section 6 to obtain the critical percolation density in three-dimensions

$$\rho_c^{1/3}\, r_0 = 0.834 \ , \tag{8.17}$$

in a good agreement with the result 0.874 given by Pike and Seager[56] by a Monte Carlo simulation. Here, ρ denotes the number density of the percolating sites. More detailed results will be published elsewhere.[57]

ACKNOWLEDGMENTS

This work at City College of The City University of New York was supported in part by grants from Department of Energy, Army Research Office, National Science Foundation and Professional Staff Congress - City University of New York Research Award Program. One of the authors (TO) benefited from support from the National Foundation for Cancer Research. This work has also been done in conjunction with the binational agreement between the National Science Foundation and the Japanese Society for the Promotion of Science under contract number INT- 7918591.

REFERENCES

[1]M. Pollak and T. H. Geballe, Phys. Rev. **122**, 1742 (1961).
[2]N. F. Mott, Can. J. Phys. **34**, 1356 (1956).
[3]A. Miller and E. Abrahams, Phys. Rev. **120**, 745 (1960).
[4]H. Scher and M. Lax, Phys. Rev. **B7**, 4491 (1973).
[5]H. Scher and M. Lax, Phys. Rev. **B7**, 4502 (1973).
[6]T. M. Reyes, M. Sayer, A. Mansingh and R. Chen, Can. J. Phys. **54**, 413 (1976).
[7]A. S. Barker Jr., J. A. Ditzenberger and J. P. Remeika, Phys. Rev. **B14**, 4254 (1976).
[8]M. Sayer, A. Mansingh, J. B. Webb and J. Noad, J. Phys. **C11**, 315 (1978).
[9]R. M. Mehra, P. C. Mathur, A. K. Kathuria and R. Shyam, Phys. Rev. **B18**, 5620 (1978).
[10]M. Suzuki, J. Phys. Chem. Solid, **41**, 1253 (1980).
[11]S. Alexander, J. Bernasconi, W. R. Schneider, R. Biller, W. G. Clark, G. Grüner, R. Orbach and A. Zettl, Phys. Rev. **B24**, 7474 (1981).
[12]M. Lax, unpublished.
[13]S. R. Broadbent and J. M. Hammersley, Proc. Camb. Phil. Soc., **53**, 629 (1957).
[14]V. K. S. Shante and S. Kirkpatrick, Adv. Phys. **20**, 325 (1971).
[15]T. Odagaki, N. Ogita and H. Matsuda, J. Phys. **C13**, 189 (1980).
[16]M. M. Pant and B. Y. Tong, J. Phys. **C13**, 1237 (1980).
[17]R. Raghavan and D. C. Mattis, Phys Rev. **B23**, 4791 (1981).
[18]Y. Shapir, A. Aharony and A. B. Harris, Phys. Rev. Lett. **49**, 486 (1982).
[19]P. G. de Gennes, La Recherche, **7**, 919 (1976).
[20]T. Odagaki and M. Lax, Phys. Rev. Lett. **45**, 847 (1980).
[21]T. Odagaki, M. Lax and A. Puri, Phys. Rev. **B** (to be published).
[22]E. W. Montroll and G. H. Weiss, J. Math. Phys. **6**, 167 (1965).
[23]For example, H. Scher and E. W. Montroll, Phys. Rev. **B12**, 2455 (1975); G. Pfister and H. Scher, Phys. Rev. **B15**, 2062 (1977); A. Blumen, J. Klafter and R. Silbey, J. Chem. Phys. **72**, 843 (1980).
[24]T. Odagaki and M. Lax, Phys. Rev. **24**, 5284 (1981); M. Lax and T. Odagaki, in Lecture Notes in Physics, Vol. 154, *Proceedings of the Conference on the Macroscopic Properties of Disorder Media,* edited by R. Burridge, S. Childress and G. Papanicoloau, (Springer, N.Y. 1982), p. 148.
[25]S. Alexander, J. Bernasconi, W. R. Schneider and R. Orbach, Rev. Mod. Phys. **53**, 175 (1981).
[26]H. Schmidt, Phys. Rev. **105**, 425 (1957).
[27]B. Movaghar, J. Phys. **C13**, 4915 (1980).
[28]B. Movaghar, B. Pohlmann and W. Schirmacher, Phil. Mag. **B41**, 49 (1980).
[29]E. J. Moore, J. Phys. **C7**, 1840 (1974).
[30]J. A. McInnes, P. N. Butcher and J. D. Clark, Phil Mag. 1 (1980).
[31]P. M. Richards and R. L. Renken, Phys. Rev. **B21**, 3740 (1980).
[32]T. Vicsek, Z. Phys. **B45**, 153 (1981).
[33]M. Lax, Rev. Mod. Phys. **23**, 287 (1951); Phys. Rev. **85**, 621 (1952).
[34]P. Soven, Phys. Rev. **156**, 809 (1967).
[35]D. W. Taylor, Phys. Rev. **156**, 1017 (1967).
[36]M. Lax, in *Stochastic Differential Equations,* SIAM-AMS Proceedings Vol. 6 (American Mathematical Society, Providence, R.I. 1973), p. 35.
[37]R. Kubo, J. Phys. Soc. Japan, **12**, 570 (1957).
[38]M. Lax, Phys. Rev. **109**, 1921 (1958).
[39]T. Odagaki and M. Lax, Phys. Rev. **B26**, 6480 (1982).
[40]M. Lax and H. Scher, Phys. Rev. Lett. **39**, 781 (1977).
[41]K. M. Watson, Phys. Rev. **105**, 1388 (11957).
[42]S. Kirkpatrick, Rev. Mod. Phys. **45**, 574 (1973).

[43]T. Odagaki and F. Yonezawa, J. Phys. Soc. Japan, **47**, 379 (1979).
[44]I. Webmann, Phys. Rev. Lett. **47**, 1496 (1981).
[45]V. Halpern, J. de Phys. (Paris), **42**, C4-119 (1981).
[46]J. W. Haus, K. W. Kehr and K. Kitahara, Phys. Rev. **B25**, 4918 (1982).
[47]T. Odagaki and M. Lax, Phys. Rev. **B25**, 2301 (1982).
[48]T. Odagaki and M. Lax, Phys. Rev. **B25**, 2307 (1982).
[49]T. Odagaki and M. Lax, Mol. Cryst. Liq. Cryst. **85**, 129 (1982).
[50]R. Zwanzig, J. Stats. Phys. **28**, 127 (1982).
[51]J. Machta, Phys. Rev. **B24**, 5260 (1981).
[52]I. Webmann and J. Klafter, Phys. Rev. **B26**, 5950 (1982).
[53]W. A. Vyssotsky, S. B. Gordon, H. L. Frisch and J. M. Hammersley, Phys. Rev. **123**, 1566 (1961).
[54]J. M. Ziman, J. Phys. **C1**, 532 (1968).
[55]H. Nakanishi and H. E. Stanley, Phys. Rev. **B22**, 2466 (1980).
[56]G. E. Pike and C. H. Seager, Phys. Rev. **B10**, 1421 (1974).
[57]T. Odagaki and M. Lax, to be published.
[58]After this manuscript was completed, an elaborate review paper on random walks appeared: G. H. Weiss and R. J. Rubin, Adv. Chem. Phys. **52**, 363 (1983).

RANDOM WALK THEORY OF GEMINATE RECOMBINATION

H. Scher
Xerox Webster Research Center, Webster, NY 14580*

S. Rackovsky
Dept. of Physics and Astronomy, University of Rochester, Rochester, NY 14627

ABSTRACT

We have developed the first comprehensive model of geminate recombination which depends on molecular parameters and focuses on the competition between the inter- and intramolecular rates. The model is random walk on a lattice in a combined Coulomb and external field. The model is solved exactly and is computationally straightforward. The analytic method we have used is quite general and can easily be extended to include a broad class of problems involving large numbers of (correlated) "special sites". The main feature of the computation needed to solve these problems, is the evaluation of the lattice Green's functions (G) in the presence of E, the electric field. We will show that the computation of the E-dependent G-functions are greatly simplified with the use of a newly derived symmetry relation. We will elaborate this approach and discuss our results for the quantum efficiency as a function of external field, temperature and molecular concentration, η (E,T, c). In general, within this framework, one can study the influence on η of such factors as dimensionality, lattice structure, disorder, tunneling transition rates, intramolecular rates and intrinsic energy level differences.

*Now at: Sohio Research Center, Cleveland, Ohio 44128

0094-243X/84/1090155-18 $3.00

INTRODUCTION

The theory of geminate recombination has been dominated by a phenomenological theory due to Onsager.[1] It has been successful in fitting photogeneration data in a number of different types of condensed phases [2-5] in terms of two parameters: φ_0 , a scaling factor representing the high field saturation of the quantum efficiency η and r_0 the initial separation of thermalized carrier pairs. There, however, is no clear relation of φ_0 , r_0 to the molecular properties of the system. We have developed a model that incorporates these properties and the dependence of η on them as well as the electric field (E), temperature (T) and molecular concentration (c). The model is a random walk on a lattice in a combined Coulomb and external field and is solved exactly. The solution we have obtained is in qualitative agreement with the Onsager results but exhibits an expanded range of behavior. For a fixed initial separation r_0 , one can vary the span of the field dependence of η by more than one decade and considerably change the temperature dependence by different choices of important molecular variables and by changes in the form of the transition rates. These results will be discussed below and is a considerable condensation of a complete work[6] which will appear elsewhere. The main emphasis, however, in this paper will be on the method of solution which follows the "defect" method.[7]

The analytical method we have used is quite general and can easily be extended to include a broad class of problems involving large numbers of (correlated) "special sites". In the context of the present problem, one can also include dimensionality effects and site disorder. The solution involves the numerical evaluation of large determinants whose elements contain lattice Green's functions[8], which must be evaluated in the presence of the applied electric field, E. There is a large and fascinating literature on some types of these Green's functions[9,10]. However, no attention has been paid to the calculation of Green's function in the presence of E. In addition, there is little work on these functions for the lattice we have chosen, the fcc. The computation involved a good deal of effort. However, once having evaluated these functions for the perfect lattice in the presence of E, one is enabled to consider the large group of problems mentioned above. In other words, the "entry fee" for solving diffusion problems on discrete lattices, with site specified properties, is the analytic or numerical evaluation of the lattice Green's functions. From a general computational point of view, working with discrete lattices avoids the inherent difficulties involved in the simultaneous limits of long wavelength (continuum limit w.l. $\to \infty$) and $t \to \infty$ that complicate transport problems.[8] It is certainly preferable to numerical solution of the resulting partial differential equations with singular potentials.

MODEL

The dynamics or diffusion of a charge carrier (cc) on a (fcc) lattice, under the influence of a force field can be described by a generalization of the continuous-time random walk (CTRW).[7,12] A site position can be expressed as

$$\mathbf{r} = \ell_1 \hat{\mathbf{a}}_1 + \ell_2 \hat{\mathbf{a}}_2 + \ell_3 \hat{\mathbf{a}}_3 \tag{1}$$

with ℓ_i equal to an integer and a_i, the primitive translation vectors[$\hat{a}_1 = a(j + k)/2$, $\hat{a}_2 = a(i + k)/2$, $\hat{a}_3 = a(i + j)/2$]. The walks are restricted to finite lattices (N^3 unit cells) with periodic boundary conditions (we shall eventually let $N \to \infty$) and for brevity, $\ell \equiv \{\ell_1, \ell_2, \ell_3,\}$.

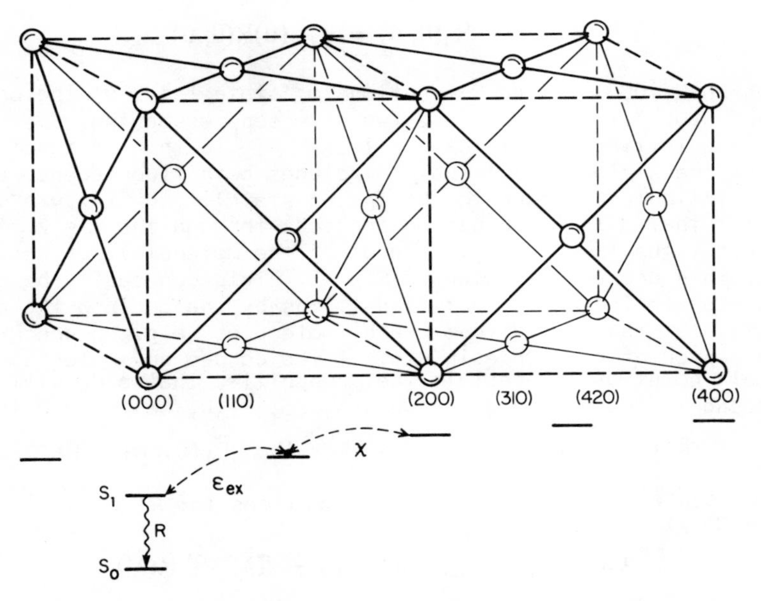

Fig. 1 The face-centered cubic lattice and the energy levels associated with the recombination site (000) and some of its neighbors. The kinetic scheme used in this work is illustrated by arrows, and the parameters associated with the different types of transition are noted at appropriate arrows.

Let $\tilde{R}(\ell, t)$ be the probability per unit time for the cc to just arrive at ℓ at time t, if it started at ℓ_0 at t = 0+ . The sites of the lattice are of two kinds: those in a set L (designated λ) where the transition rates to leave the site are affected by both a potential V(r) and an external dc electric field E and the rest of the sites where the transition rates to leave the site are affected only by E. The difference in the sites are reflected in two types of distribution functions $\tilde{\psi}(\ell, t)$, which denote the probability/unit time that the time interval between successive arrivals is t and the displacement is ℓ. The ψ-function depends only on relative displacement $\ell - \ell'$ if $\ell' \notin$ L, and on both $\ell - \ell'$ and ℓ' if $\ell' \varepsilon$ L.

The equation for $\tilde{R}(\ell, t)$ is set up in the same way as in Refs. (7,13) and the method of solution is the same. The special features of this problem are that we now consider a countercharge at the origin as the source of a Coulomb potential and a number of states, i = 1, 2, . . ., on this site. The state i = 1 will designate the ground state in which recombination occurs when occupied by the cc, and i = 2, 3 ..., higher lying excited states of the molecule. An additional(LT) equation is included to specify the occupation of the different states at the origin,

$$R_1(0, u) = \psi_{12}(u)R(0, u) \tag{2}$$

where $R_1(0,u)$ is the R-function for state i = 1 at the origin and R(0,u) for state i = 2 (we have suppressed the subscript 2 and initially restrict i = 1,2), ψ_{12} is the ψ-function for the transition 2 $\rightarrow$ 1 (which includes both fluorescence and radiationless return to the ground state). Eq. (2) express the fact that all recombination proceeds through the i = 2 state (e.g., the lowest singlet level of the molecule). A schematic of this process is shown in Fig. 1. This can easily be generalized to include direct transitions by the cc into the ground state at the origin from nearby molecules (e.g., a donor-acceptor pair). Once in state 1 the cc can never leave. A main point of interest is the probability that a cc will be found in the ground state when $t \rightarrow \infty$, $\tilde{P}_1(0, \infty)$, and using the fact that $\tilde{\psi}(0_1, t) = 0$, one can show that $\tilde{P}_1(0, \infty) = R_1(0,0)$.

One can generate [7,13] a set of equations for $R(\lambda, u)$ $(\equiv R(\lambda))$

$$G(\lambda - \ell_0) = \sum_{\lambda'} [G(\lambda - \lambda') - f(\lambda; \lambda')] R(\lambda') \tag{3}$$

where $G(\ell)$ is the Green's function for the perfect lattice including the E-field and

$$f(\ell;\lambda) \equiv \sum_{\ell'} G(\ell-\ell')\,\psi(\ell'-\lambda;\lambda). \tag{4}$$

In general, to solve for $R(\lambda)$ one must invert a matrix M with elements

$$M_{\lambda,\lambda'} \equiv G(\lambda-\lambda') - f(\lambda;\lambda'). \tag{5}$$

For a reasonable range of V(r) (e.g., up to and including 4th nearest neighbors of the fcc lattice)the dimension of the M matrix can be large (55 x 55). Obtaining the solution for $R(\lambda, u)$ for all λ, one can determine the time dependence of site occupancy, i.e., $\tilde{R}(\lambda,t)$. This last step would involve numerical inversion of the LT. Both the matrix inversion of M and the LT numerical inversion represent a modest amount of computation by present standards. Fortunately one can simplify the computation even further if one is primarily interested in the probability of recombination measured in a typical collection experiment.[3]

Using the relation between $\tilde{P}(\ell,t)$, the probability of finding a cc on ℓ at time t, and the R-function,[13] and $\psi(0_1,t)=0$ we can obtain

$$\tilde{P}_1(0,\infty) = \int_0^\infty \tilde{R}_1(0,t)\,dt = R_1(0,0). \tag{6}$$

from which it follows using Eq.(2)

$$\tilde{P}_1(0,\infty) = \psi_{12}(0)R(0,0), \tag{7}$$

thus the determination of one value of the LT of $\tilde{R}(\lambda,t)$, for $\lambda=0$ and for $u=0$, in Eq. (3) determines, directly (i.e., without a LT inversion), the probability of recombination.

We can use Kramer's rule and solve Eq. (3) for R(0,u),

$$R(0,u) = D_0(M)/D(M) \tag{8}$$

where D(M) is the determinant of M and D_0 (M) is D(M) with the $\lambda = \{0,0,0\}$ column replaced by $G(-\ell_0, u)$. We have thus reduced the problem to calculating the ratio of two determinants.[8] Further by taking the limit $u \to 0$ in Eq. (8) one can obtain directly $\tilde{P}_1(0,\infty)$ using the relation in Eq.(7)

The Green's function in the $N^3 \to \infty$ limit can be replaced by an integral,

$$G(\ell, u) = (2\pi)^{-3} \int\int\int_0^{2\pi} \frac{d^3\theta e^{i\theta.\ell}}{1-\Lambda(\theta, u)} \quad . \tag{9}$$

In order to specify the structure factor Λ we must define the form of the transition rate.

We assume nearest neighbor transitions only, and for the perfect lattice in the presence of a dc field,

$$W(\ell) = W \exp(-\beta(\Delta - eE.\ell)) = W_o\, e^{-\beta eE\cdot\ell} \tag{10}$$

where $\beta = (k_BT)^{-1}$, Δ is a barrier energy (e.g., small polaron activation energy) and E is the electric field (Eq.22) is a simplified version of the transition rate in Ref. (17). If we assume an applied field along one of the cube edges, i.e., $\underline{E} = E\hat{z}$, the structure factor has the form

$$\Lambda(\theta, u) = [1+2\kappa]^{-1}[(e^{-i\theta_3}e^{\gamma} + e^{i\theta_3}e^{-\gamma})(c_1+c_2)/2 + c_1c_2] \tag{11}$$

where $c_i \equiv \cos\theta_i$, $\kappa \equiv \cosh\gamma + u/8W_o$ and $\gamma \equiv e\beta Ea/2$.

It is expedient to define

$$\bar{G}(\ell, u) \equiv G(\ell, u)/(1+2\kappa) = 4w_oP(\ell, u). \tag{12}$$

a dimensionless LT of $\tilde{P}(\ell,t)$.

We have been able to derive an important symmetry relation for the Greens function [6]

$$\bar{G}(m,n,\ell) = e^{\ell\gamma} F(mn\ell) \tag{13}$$

where we have suppressed the dependence on u, and

$$F(mn\ell) \equiv \pi^{-3} \int\int\int_0^{\pi} d^3\theta \cos m\theta_1 \cos n\theta_2 \cos \ell\theta_3 [1+2\kappa - c_1c_2 - c_2c_3 - c_1c_3]^{-1}. \tag{14}$$

The F functions have the <u>full</u> symmetry of the fcc lattice and depend on the E-field through κ. The symmetry relation in Eq.(13) has resulted in a considerable simplification of the evaluation of the dc field-dependent Greens functions.

The first two integrations can be performed analytically, [6]

$$F(000) \equiv F_o = \pi^{-2} \int_o^{\pi} d\theta \bar{K}(k) \tag{15}$$

with $\bar{K}(k) \equiv K(k)/(1+\kappa)$, where K(k) is the complete elliptic integral of the first kind, with modulus

$$k=[1+2\kappa+\cos^2\theta]^{1/2}/(1+\kappa). \qquad (16)$$

In general all the Green's functions can be expressed as an integral of an algebraic combination of $\bar{K}(k)$, E(k) (the complete elliptic integral of the second kind) and $Z(\psi,k)$, the Jacobian zeta function[15], which is defined as

$$Z(\psi,k) \equiv E(\psi,k) - (E/K)\ F(\psi,k), \qquad (17)$$

where $F(\psi,k)$, $E(\psi,k)$ are incomplete elliptic integrals of the first and second kind, respectively and $\psi \equiv \sin^{-1}[(1-\cos\theta)^{1/2}/(1+\kappa)^{1/2}k]$. The k-dependence of K(k) and E(k) is assumed understood. The important properties of $Z(\psi,k)$ are listed in Ref.(15), p.33. A list of initially useful Green's functions is:

$$F(m00) = \pi^{-2}\int_0^{\pi} d\theta \cos(m\theta)\,\bar{K} \qquad (18)$$

$$F(m10) = \pi^{-2}\int_0^{\pi} d\theta \cos(m\theta)\,(2KZ(\psi,k)/\sin\theta - \bar{K}) \qquad (19)$$

$$F(m20) = \pi^{-2}\int_0^{\pi} d\theta \cos(m\theta) I(\theta), \qquad (20)$$

with

$$I(\theta) \equiv \bar{K} - 4(1+\kappa)(2\,\mathrm{ctn}\theta\, KZ(\psi,k) + E - (\kappa+\cos\theta)\bar{K}). \qquad (21)$$

For values of m not compatible with fcc lattice points the F functions vanish (by symmetry). The rest of the initial set of F's can all be determined in terms of Eqs.(18-20) by using the recursion relations for $G(\ell,u)$[12] and other relations enumerated in Ref.(6).

An important aid and check on our numerical evaluation of Eqs.(18-20) is the availability of an analytic expression for F_o ($\equiv$ F(000)) developed by Joyce[10] in another context,

$$F_o = 2\pi^{-2}K(k_+)K(k_-)/(\kappa+1) \tag{22}$$

with

$$(\kappa+1)^{3/2}k_\pm^{\,2} = ½((\kappa+1)^{3/2} - \kappa(\kappa-1)^{½}) \pm (\kappa+½)^{½} \tag{23}$$

In the limit $\kappa \sim 1$

$$F_o = \sqrt{3}\,(K_1/\pi)^2 - \sqrt{3}\,(\kappa-1)^{½}/(2\pi(\kappa+½)^{½}) + \ldots \tag{24}$$

where $K_1 \equiv K(\sin(\pi/12))$. In general it is possible to develop an analytic form for F(mnl) similiar to the one for the bcc lattice (Ref.(16) in Ref.(9)),and we hope to report on this in a later work. The use of F_o allows for the elimination of all end point limit problems in the numerical integration of the integrals in Eqs.(18-20). We note that the branch point singularity of F_o at $\kappa = 1$ results in a linear dependence on E for small fields, i.e., $(\kappa-1)^{1/2} \propto \gamma/\sqrt{2}$, as $\gamma \to 0$. This leasds to the same type of low field dependence as in the Onsager solution. The low field behavior of all the F functions and $\tilde{P}_1$ (0, oo) will be pursued in a future study, in which we will analytically develop the "slope-to-intercept" ratio (cf. discussion of this ratio in Ref.(5)).

For large γ (e.g., $\gamma \geq 2$) a more reliable and convenient method to calculate F(mnl), especially for m (or n, l) ≥ 3, involves the use of a series representation. In ref.(6) an expansion of F(mnl) in powers of $(1+2\kappa)^{-1}$ is derived. The use of the series expansion also allows us to develop an analytic expression for $\tilde{P}(m,n,l;\,t)$.[6]

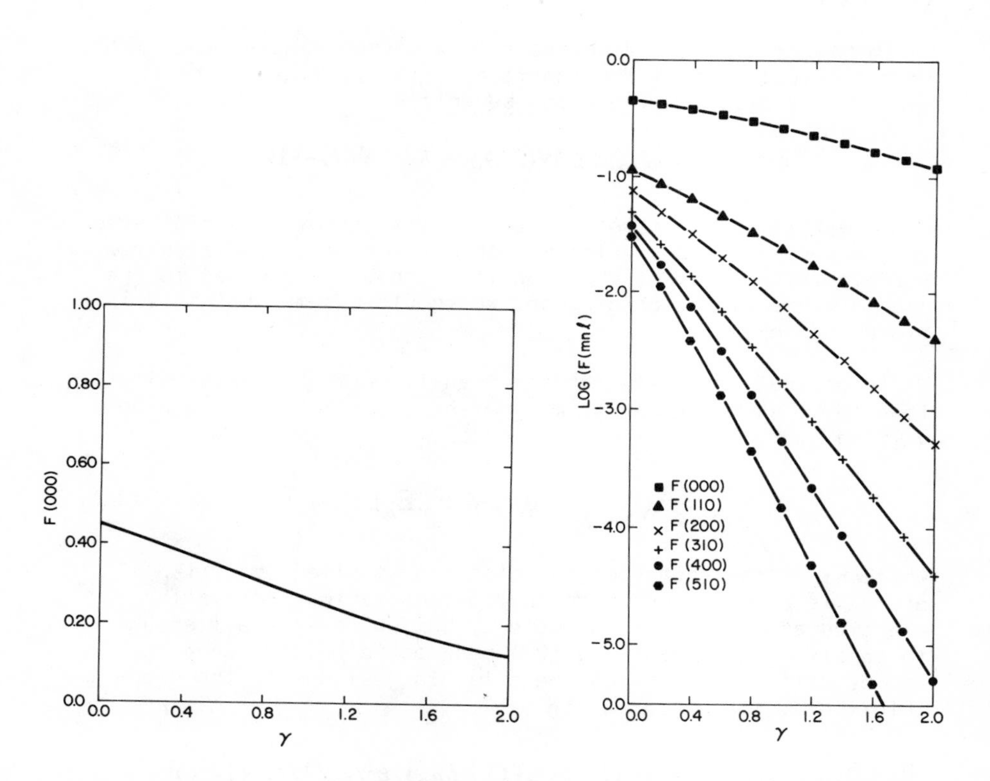

Fig.2 a) Linear plot of F(000) vs. γ (see text for definitions).

b) Semilog plot of F(mnl) vs. γ for various (mnl)values.

In Fig.2 we show the results of our numerical integration evaluation of F(mnl). In Fig.2a we have a linear plot of F(000) vs. γ , and in Fig. 2b we show a semi log plot of F(mnl) as a function of γ for representative points (m,n,l). For l=0, the G and F functions are equal and the field enters the F function only in the combination $2\cosh\gamma + u/4w_o$. Thus, with an appropriate scale change, Fig.2b can also indicate a plot of F(mnl) vs. u. For $\gamma > 0.8$ the log (F(mnl)) is linear in γ with a slope $\sim -(1 + m)$. The value for the slope is in close agreement with that obtained from the leading term in the series representation of F(mnl) in Ref.(6) with the approximation $(1 + 2\kappa)^{-1} \sim e^{-\gamma}$.

The final step yielding the matrix elements $M_{\lambda,\lambda'}$ (Eq.(5)) is the calculation of the function $f(\lambda;\lambda')$, i.e., we must specify the $\psi(\ell;\lambda)$. These can be written

$$\psi(\ell;\lambda) = W(\ell;\lambda)/[\Sigma\, W(\ell\,';\lambda) + \delta_{\lambda,o}\, W\,(2\rightarrow 1)] \qquad (25)$$

where $W(\ell;\lambda)$ is the rate of the transition from a special site λ to a site at $\lambda+\ell$, which may or may not be a special site. $W(2\rightarrow 1)$ is the rate of the transition from the excited to the ground state at the origin, and we should assume that this is field independent.

In gereral, the rate for the transition from special site to site λ, where ℓ may or may not be a special site, can be written

$$W_{\lambda\rightarrow\ell} = W_o e^{-\beta(E_\ell - E_\lambda)} \qquad (26)$$

where E_i is the energy of the electron at site i, $\beta = (k_B T)^{-1}$ and W_o is a prefactor which may contain various activation energies and molecular parameters, but which we shall assume to be site-independent. Eq.(26) reduces to Eq.(10) when the energy difference between the sites is just due to the electric field. Now we can write

$$E_\ell - E_\lambda = -e(r_\ell - r_\lambda)\,E - (1 - \delta_{\lambda,o})\,(1 - \delta_{\ell,o})\;e^2/\varepsilon\,[(1/r_\ell)-(1/r_\lambda)] \qquad (27)$$

$$+ (\delta_{\lambda,o} - \delta_{\ell,o})\,E_{ex}$$

where ϵ is a dielectric constant and E_{ex} represents the difference in energy between the first excited state at the origin and a state with a hole at the origin and an excess electron at a nearest neighbor. This energy can be written as

$$E_{ex} = E_I^{(1)} - E_I^{(2-)} - \sqrt{2}e^2/a\epsilon \tag{28}$$

where $E_I^{(1)}$ is the ionization potential of the first excited state at the origin, $E_I^{(2-)}$ is the ionization potential of the anion at a nearest neighbor site, and the third term is the Coulomb contribution (remembering that the nearest-neighbor distance in a f.c.c. lattice is $a/\sqrt{2}$).

Multiplying Eq.(27) by β clearly leads to three dimensionless parameters:

$$\gamma = \beta eEa/2, \tag{29}$$

$$\chi = 2\beta e^2/\epsilon a, \tag{30}$$

and

$$\epsilon_{ex} = \beta E_{ex} \tag{31}$$

Note that from Eq.(28),

$$\epsilon_{ex} = \beta[E_I^{(1)} - E_I^{(2-)}] - \chi/\sqrt{2} \equiv \epsilon_I - \chi/\sqrt{2} \tag{32}$$

Each of these parameters scales an independent aspect of the problem. The external field effects are governed by γ , transitions from special sites not at the origin under the influence of the Coulomb field are goverened by χ , and transitions to and from the excited state at the origin are governed by ϵ_{ex}.

The assumption that transitions from second neighbors to more distant sites are not affected by the Coulomb potential is equivalent to the assumption that the difference between the Coulomb energy at the second neighbor sites nearest the origin and that at their furthest neighbors (fifth nearest neighbors from the origin) is less than kT. If we denote the distance between the Coulomb energy of the second neighbors nearest to the origin by $E_c^{(2)}$ and that of their nearest neighbors most distant from the origin by $E_c^{(3)}$,

$$E_c^{(3)} - E_c^{(2)} \leq k_B T. \tag{33}$$

Since the distance of the nearest second neighbor sites from the origin in units of a/2 is 2, we find the following condition to hold:

$$2e^2/\varepsilon a[1/2 - 1/\sqrt{10}] \leq k_B T \tag{34}$$

But this implies

$$x \leq 5.44 \tag{35}$$

We have therefore defined the parameter region in which the approximation is reasonable.

If we define the exponential in Eq.(25) by

$$\Omega(\lambda,\lambda') = e^{-\beta(E_{\lambda'} - E_\lambda)} \tag{36}$$

then we can write Eq.(25) as

$$\psi(\ell;\lambda) = \Omega(\lambda+\ell,\lambda)/[\Sigma_{\ell'}\Omega(\lambda+\ell',\lambda) + \delta_{\lambda,o}R] \tag{37}$$

The parameter

$$R = W(2\rightarrow 1)/W_o \tag{38}$$

is a dimensionless measure of the rate of deexcitation of the 1st excited state at the origin.

Note that the $\psi(\ell,\lambda)$ are normalized; i.e.,

$$\Sigma_\ell \, \psi(\ell;\lambda) = 1 \tag{39}$$

This implies that

$$\psi_{21} = \psi(0_1 \, ; 0_2) = R/[\Sigma_\ell \, \Omega(\ell,0) + R] \tag{40}$$

It should be emphasized that the specification of the problem as geminate recombination occurs only through the inputs in $\psi(\ell;\lambda)$. To insert diagonal disorder into the site properties would simply mean using a different $\psi(\ell;\lambda)$. The Green's functions, once determined, can be used for a wide variety of problems.

RESULTS

We now present a selection of our results for $\eta(\Upsilon)$ as a function of ϵ_{ex}, χ, R (Eqs.(32,30,38)). We group the η-curves into small R and large R regimes. First we review what precisely is the physical meaning of varying each of the molecular parameters. Variation of ϵ_{ex} with fixed (χ,R) corresponds to changing the ϵ_l term in Eq. (32) - i.e., to changes in the energy level structure of the molecule at the origin and/or its nearest neighbors. Variation of χ with fixed (ϵ_{ex}, R) can be seen, from Eqs. (29) and (31), to correspond to variation of both ϵa and ϵ_l. ϵa can be varied in two ways. The dielectric constant can be varied, for example by changing materials, or by local fluctuation in non-homogeneous materials. The lattice constant (a) can be varied by changing materials, by local fluctuations in amorphous or doped systems or by concentration changes in doped systems. Variation of R with fixed (ϵ_{ex}, χ) corresponds to a change of the rate $W(2\rightarrow1)$ for deexcitation to S_o at the origin. Thus, for example, an experiment in which doping concentration is varied in a molecularly doped polymer[4] would correspond to a path in a parameter space along which χ and ϵ_{ex} vary in opposite directions. Study of a series of donors as dopants at a given concentration in a given matrix would correspond to a path in parameter space in which ϵ_{ex} and R vary simultaneously, with fixed χ.

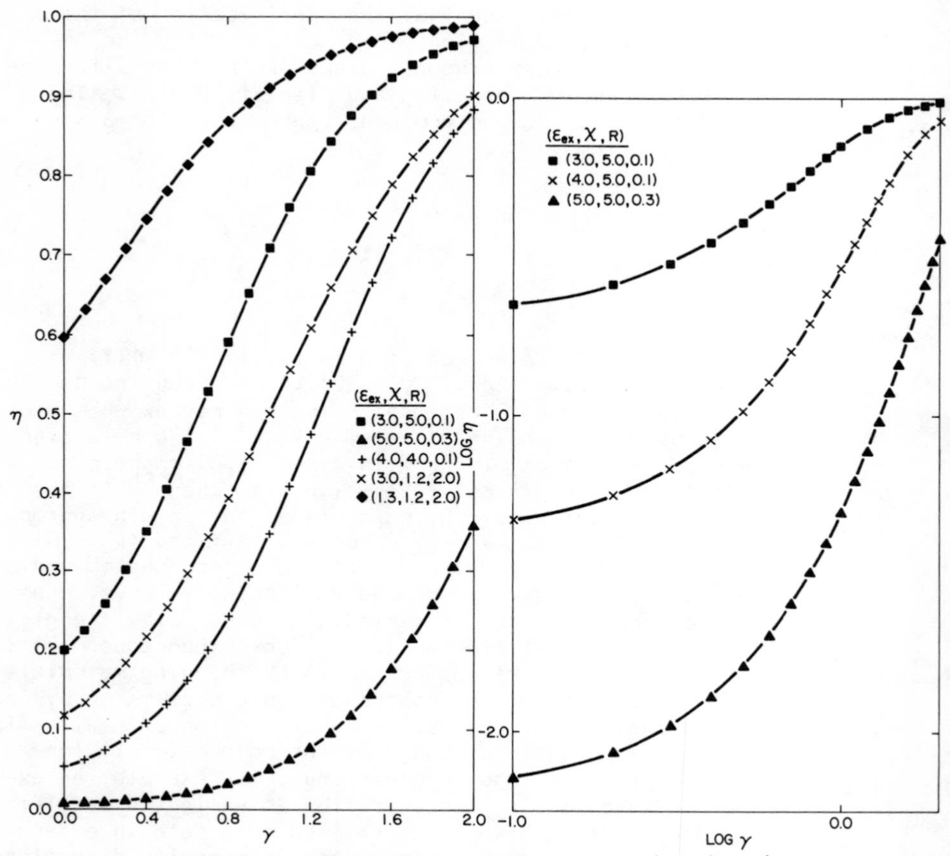

Fig.3 η(Boltzmann-weighted) vs. γ for several sets of (ϵ_{ex}, χ, R).

Fig.4 Log-log plot of η illustrating the effect of varying ϵ_{ex} with (χ, R) essentially fixed.

Increase of any of (ϵ_{ex}, χ, R) leads to a decrease in η at low fields, which is to be expected, since such an increase corresponds to an increase in the effective barrier the electron faces in escaping from the hole. This can be seen in Fig.3.

A striking feature of these curves in Fig.3 is the wide range of behavior, both qualitative and quantitative, which can be obtained by varying the molecular parameters. Both in terms of low field properties and rate of saturation, η shows substantial variation.

In Fig.4 we reproduce on a log plot the top and bottom η curves of Fig.3 with the addition of a third, intermediate curve. These curves show the dramatic effect on the yield of changing just one parameter ϵ_{ex} (in steps of k_BT) or more precisely, with fixed χ, of changing ϵ_I (Eq.(32)). ϵ_I can change because of a change in $E_I^{(1)}$, the ionization potential of the first excited state at the origin, and/or a change in $E_I^{(2)}$, the ionization potential of the anion at a nearest neighbor site. More generally, a change in ϵ_I is due to a change in the intrinsic energy levels of the molecule at the origin or the anion of the molecule at a nearest neighbor site, without the Coulomb effect of the hole at the origin molecule. The low field value of η changes by nearly two decades for a <u>fixed</u> initial distribution. In the Onsager curves this shape change can only occur by changing the initial separation r_o. The variation in ϵ_I can be caused by disorder, e.g., local energy level fluctuations in amorphous semiconductors or molecularly doped polymers. Fig.4 shows the strong effect these rather mild fluctuations can have on the low field η values. Thus, if fit by the Onsager theory, in this case r_o would be parametizing a variation in molecular energy levels!

The results of Fig.4 illustrate another strength of the lattice theory: the ability to treat an arbitrary initial distribution of electrons about the hole. The results of Fig.4 are for an initial distribution in which the electron is on a nearest-neighbor site of the hole, with angular distribution weighted by the applied field (Boltzmann average). This type of distribution has not previously been considered. The effect of varying the initial distribution (including a Frenkel exciton at the origin) is considered in Ref.(6).

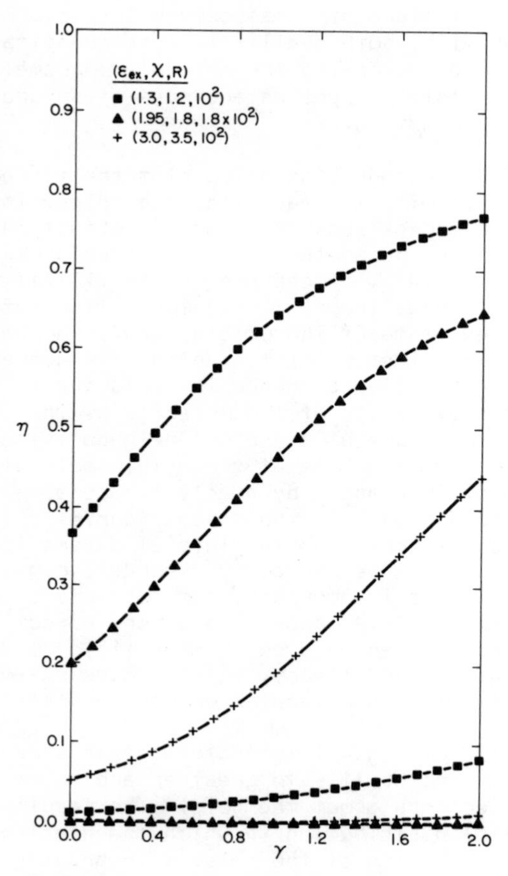

Fig.5 η vs. γ for three parameter sets in the large-R regime. Top three curves are for nearest neighbor initial position; bottom three curves are for Frenkel exciton (S_1) initial position.

We now turn to some results in the large-R regime. It should be noted that the value of R specifies the type of material we are considering. Large values correspond to materials with small transfer integrals; e.g., molecularly doped polymers. Moderate R values define materials with larger overlap integrals, such as amorphous semiconductors. In Fig.5 we show η (γ) for several parameter sets, with two types of initial distribution. The upper three are for nearest neighbor initial position, the lower three are for Frenkel exciton (S_1) initial position.

In contrast to the η curves in Figs.3-4, the high field saturation values are not nearly unity. The large R results emphasize the intrinsic loss in the recombination.

In the Onsager theory (continuum approximation) the high field saturation of the yield is controlled by φ_0. Here we see that the molecular parameters (notably R) control the saturation, and not a scaling factor inserted to represent the "primary quantum yield". The success of the continuum approach in explaining experimental data can be understood by considering the complimentary roles played by φ_0 and the "initial displacement" r_0. The high field saturation is controlled by events which occur in the neighborhood of the recombination site (the origin). These events are parametrized, in the continuum approximation, by φ_0. The principal contribution to the field dependence of η is from electrons which have succeeded in reaching a separation large enough that they are much-less influenced by the recombination center. r_0 represents some average value of this separation. It is easy to understand, from this viewpoint, why typical experimental r_0 values are several to many times the characteristic intermolecular separation. It is also clear that φ_0 contains all the information on the early dynamics of charge separation in the continuum approach. [In Ref.(6) other results for large R clearly show that changes in R also change the field dependence. Thus, $\eta(\infty)$ can be controlled by the same molecular parameters and rate processes that determine the field dependence of η.]

Details of the model-temperature dependence, low field behavior, dependence on molecular parameters, etc., as well as a comparison with the continuum approximation, are explored in Ref.(6) and in forthcoming publications. It is expected that this approach will provide a useful laboratory in which these and other factors such as site energy disorder, dimensionality and different forms for the transition rates, can be studied. In addition a large emphasis will be placed on the transient solution to model recent picosecond regime experiments.[16]

REFERENCES

1. L. Onsager, J. Chem. Phys. 2, 599 (1934); Phys. Rev. 54, 554 (1938).

2. D. Mauzerall and S.G. Ballard, Ann. Rev. Phys. Chem. 33, 377 (1982) and references therein.

3. D.M. Pai and R.C. Enck, Phys. Rev. B 11, 5163 (1975).

4. P.M. Borsenberger, L.E. Contois, and D.C. Hoesterey, J. Chem. Phys. 68, 637 (1978).

5. R.R. Chance and C.L. Braun, J. Chem. Phys. 64, 3573 (1976) and references therein.

6. H. Scher and S. Rackovsky, submitted to J. Chem. Phys.

7. E.W. Montroll, J. Math. Phys. 10, 753 (1969); E.W. Montroll and H. Scher, J. Stat. Phys. 9, 101 (1973).

8. F.T. Hioe, Statistical Mechanics and Statistical Methods in Theory and Application, ed. by U. Landman (Plenum Press, New York, 1977), p.165.

9. F.T. Hioe, J. Math. Phys. 19, 1064 (1978) and references therein.

10. G.S. Joyce, J. Phys. C: Solid State Phys. 4, L53 (1971).

11. H. Scher, in Photoconductivity and Related Phenomena, ed. by J. Mort and D.M. Pai, (Elsevier, Amsterdam, 1976), p.107.

12. E.W. Montroll and G.H. Weiss, J. Math. Phys. 6, 167 (1966); H. Scher and M. Lax. Phys. Rev. B7, 4491 (1973).

13. H. Scher and C.H. Wu, Proc. Natl. Acad. Sci. 78, 22 (1981); H. Scher and C.H. Wu, J. Chem. Phys. 74, 5366 (1981)

14. H. Scher and T. Holstein, Phil. Mag. B44, 343 (1981).

15. P.F. Byrd and M.D. Friedman, Handbook of Elliptic Integrals for Engineers and Scientists (Springer-Verlag, 2nd ed., NY, 1971).

16. D.E. Ackley, J. Tauc, and W. Paul, Phys. Rev. Lett. 43, 715 (1979).

CONTINUOUS TIME RANDOM WALK ASPECTS IN REACTION AND TRANSPORT

J. Klafter and A. Blumen*
Corporate Research Science Laboratories
Exxon Research and Engineering Company
P. O. Box 45, Linden, N.J. 07036

ABSTRACT

We present a study of transport and reaction in disordered materials ased on the continuous time random walk (CTRW) formalism. We focus on he ensemble averaged probability that no transfer occurs during a given mount of time, and derive from this quantity the distribution of tepping times. The distribution of stepping times is then used to imic the random walk on a disordered system. We derive first in the TRW framework the diffusion coefficient D(t) for a random system devoid f traps, and make connection to other approximate approaches used to alculate D(t). Then we introduce trapping centers and establish the ecay laws due to irreversible trapping showing that different decay ehaviors obtain for different classes of stepping time distributions. s an example for applications we use our results to analyze electron cavenging in glassy materials.

I. INTRODUCTION

Transport in disordered materials has been an important subject of tudy in recent years.[1-7] The problem arises in the fields of energy ransfer,[2,3,5-10] of spin dynamics,[11] of conduction in amorphous aterials[1,4,12-16] and of electron scavenging.[17-20] The disordered ature of the medium over which the transport takes place strongly nfluences experimentally accessible basic quantities like exciton life imes, the electron-scavenger recombination rates and the a.c. and d.c. onductivities. The fundamental theoretical question in this respect s in how far may models based on ordered lattices be extended so as to e able to qualitatively describe effects due to disorder, and when such xtensions break down.[21]

Applications of random-walk models have encountered considerable uccess in many fields, see Refs. (22) and (23) for extensive reviews. his is due to the fact that the theory of random walks on *regular attices* has obtained a high degree of maturity, which greatly facilitates he calculation of exact solutions for particular models.

In this paper we will consider the random environment which arises rom the *substitutional* occupancy by active centers of sites on an *rdered* lattice. The random walk will then take place over the active enters only. When treating trapping by impurities, these will be taken o replace active centers. The trapping problem will thus include *two ypes of disorder*: the randomness of the carrier medium (active centers) nd the random distribution of traps. In the diffusion problem one, owever, faces only the first type of disorder, namely that due to the andom distribution of the active centers. This can be viewed as a roblem of a random walk on *random* lattices. In this case the situation s considerably more complex than for regular lattices and one is forced o make use of approximating schemes.

An approach which has been shown to be powerful for the general problems of transport in disordered systems is the continuous time random walk (CTRW). The CTRW ideas were suggested by Montroll and Weiss[24] who introduced the distribution function $\psi(t)$ into the usual random walk formalism to discuss a random walk with random stepping times. This approach was elaborated later by Scher and Lax[12,13] who applied the CTRW to transport in disordered systems by generalizing the functions $\psi(t)$ to forms $\psi(\underset{\sim}{r},t)$ which couple the spatial and temporal disorder aspects. Applications of a decoupled scheme (i.e. $\psi(\underset{\sim}{r},t) = f(\underset{\sim}{r})\psi(t)$)was also proposed by Scher and Montroll[14]. The connection of the CTRW to the master equation, which is the basic equation to describe the transport of a particle in a random medium was discussed by Bedeaux, Lakatos-Lindenberg and Shuler[25] and the connection to the generalized master equation was pointed out by Kenkre, Montroll and Shlesinger[26] and by Kenkre and Knox.[27] This later connection was studied only in a decoupled scheme, which corresponds to the above mentioned Scher-Montroll approximation.[14]

Recently Klafter and Silbey[28,29] proved that the original random walk problem on a random lattice can be rigorously reduced to a CTRW formulation of the Scher-Lax type. The derivation makes use of projection operator techniques, and $\psi(\underset{\sim}{r},t)$ is related to a self-energy (or a memory) function. However, the calculation of the self-energy is difficult and one has to look for the best possible approximations.

The purpose of this paper is to provide concise framework for the problems of diffusion and trapping in random systems in terms of the CTRW by starting from a single site type of approximation. We focus on the ensemble averaged probability $\Psi(t)$ that no transfer from an initially occupied active center occurs during time t[30]. As we will show in Sec. II this quantity can be used to obtain forms for the stepping time distributions $\psi(t)$ and $\psi(\underset{\sim}{r},t)$. These approximate forms were used quite successfully to mimic the random walk on a random system[12,13] for which the exact $\psi(\underset{\sim}{r},t)$ is not known. In Sec. III we will treat the transport in the absence of traps and will establish the diffusion coefficient $D(t)$ which follows from $\psi(\underset{\sim}{r},t)$ in the CTRW framework. There we also make connection to other approaches[31,32,33] which use $\Psi(t)$ to determine $D(t)$. In Sec. IV we introduce the traps and calculate the CTRW decay law due to trapping showing that different behaviors for the decay obtain for different classes of the stepping time distribution $\psi(t)$. As an example for applications we make use, in Sec. V, of our results to analyze experimental findings in the field of electron scavenging[17-20].

II. THE FUNCTION $\Psi(t)$

In this section we present the ensemble averaged transfer from a donor molecule to randomly distributed acceptor molecules. We start our considerations by assuming that the positions of the donor and of the acceptors are fixed, so that the acceptors form a certain configuration $\underset{\sim}{K}$ around the donor. The probability $\Psi(\underset{\sim}{K};t)$ that the donor did not transfer its energy during time t to any of the acceptors is exponential.

$$\Psi(\underset{\sim}{K};t) = \exp[-t\sum_{\underset{\sim}{r}\varepsilon\underset{\sim}{K}} w(\underset{\sim}{r})] = \prod_{\underset{\sim}{r}\varepsilon\underset{\sim}{K}} \exp[-tw(\underset{\sim}{r})] \tag{2.1}$$

In Eq. (2.1) $w(\mathbf{r})$ denotes the transfer rate to the acceptor at position $\mathbf{r}$ and the sum and product extend over all acceptors. Often encountered forms for $w(\mathbf{r})$ are

$$w(\mathbf{r}) = \frac{1}{\tau}\left(\frac{d}{r}\right)^{s} \tag{2.2}$$

for multipolar interactions and

$$w(\mathbf{r}) = \frac{1}{\tau} \exp\left[\gamma(d - r)\right] \tag{2.3}$$

for exchange interactions. The later form is often used for electron scavenging models (Sec. V). The parameter s in Eq. (2.2) equals 6 for dipole-dipole, 8 for dipole-quadrupole and 10 for quadrupole-quadrupole interactions. The quantity γd in Eq. (2.3) is a measure of the range of the exchange interaction and is larger for shorter range interactions. We have chosen to scale (2.2) and (2.3) in terms of the transfer rate τ^{-1} of the donor-acceptor pair at nearest distance d. Forms (2.2) and (2.3) are only approximations to realistic transfer rates which may have a considerably more complex structure.

The quantity of interest experimentally is not $\Psi(\mathbf{K};t)$, which is due to a particular acceptor configuration, but the ensemble average of Ψ over all acceptor configurations

$$\Psi(t) = \sum_{\mathbf{K}} p(\mathbf{K})\Psi(\mathbf{K},t) \tag{2.4}$$

where $p(\mathbf{K})$ denotes the probability with which configuration $\mathbf{K}$ occurs.

We now consider a system in which active molecules occupy the lattice sites in a random, non-correlated way with probability p. Eq. (2.1) thus reads with the donor located at the origin:

$$\Psi(\mathbf{K};t) = \exp[-t\Sigma\zeta(\mathbf{r})w(\mathbf{r})] \tag{2.5}$$

where $\zeta(\mathbf{r})$ are random variables which take only the values 1 and 0 with probability p and 1-p respectively, depending on whether the site at distance $\mathbf{r}$ is occupied or not. The configurational average of Eq. (2.5) is (See Ref. (34) for other derivations):

$$\begin{aligned}\Psi(t) \equiv \langle \Psi(\mathbf{K};t\rangle_{\{\zeta(\mathbf{r})\}} &= \langle \Pi'\exp[-t\zeta(\mathbf{r})w(\mathbf{r})]\rangle_{\{\zeta(\mathbf{r})\}} \\ &= \prod_{\mathbf{r}}{}' \langle\exp[-t\zeta(\mathbf{r})w(\mathbf{r})]\rangle_{\zeta(\mathbf{r})} \\ &= \prod_{\mathbf{r}}{}' \left[1-p + pe^{-tw(\mathbf{r})}\right]\end{aligned} \tag{2.6}$$

Here use was made of the fact that the $\zeta(\mathbf{r})$ are uncorrelated. In Eq. (2.6) the product extends over all lattice sites and the final form depends only on p and not on the configurations $\mathbf{K}$. Eq. (2.6) is exact. It is valid for all times and all concentrations of active molecules and it explicitly includes the structure of the underlying lattice. It also holds regardless of the particular form of $w(\underset{\sim}{r})$ and of the detailed structure of the transfer law for each donor-acceptor pair, $\exp[-tw(\underset{\sim}{r})]$. This fact leads to useful generalizations.[30] For an arbitrary number A of <u>different</u> acceptor species, which interact with the donor via $w_\alpha(\underset{\sim}{r})$ $\alpha = 1 \ldots A$, and occupy the lattice sites with probability p_α, the transfer law is

$$\Psi(t) = \prod_{\underset{\sim}{r}}{}' \sum_{\alpha=0}^{A} p_\alpha \exp[-tw_\alpha(\underset{\sim}{r})] \tag{2.7}$$

where α=o denotes the nonaccepting molecules. Thus $p_o = 1 - \sum_{\alpha=1} p_\alpha$ and $w_{\alpha=0}(\underset{\sim}{r})=0$. The proof of relation (2.7) follows similarly to Eq. (2.6), see also Ref. (34). In the case that the index varies continuously Eq. (2.7) takes the form[35]

$$\Psi(t) = \prod_{\underset{\sim}{r}}{}' \int d\alpha\, p(\alpha) \exp[-tw_\alpha(\underset{\sim}{r})] \tag{2.8}$$

where now $p(\alpha)$ is the corresponding probability density.

Under the assumption of a low acceptor concentration a number of approximate forms for the donor transfer law have been derived using different procedures[36-41]. We retrieve now these forms from the exact equations by taking the logarithm of both sides of Eq. (2.6)-(2.8) and by expanding with respect to p. From eq. (2.6)[34,42]:

$$\begin{aligned} \ln \Psi(t) &= \sum_{\underset{\sim}{r}} \ln(1-p\{1-\exp[-tw(\underset{\sim}{r})]\}) = \\ &= -\sum_{\underset{\sim}{r}} \sum_{k=1}^{\infty} \frac{p^k}{k} \{1-\exp[-tw(\underset{\sim}{r})]\}^k \end{aligned} \tag{2.9}$$

where the last line is valid for small p ($p \ll 1$). For a densely packed medium the sum in Eq. (2.9) may be replaced by an integral (continuum approximation), thus obtaining

$$\Psi(t) = \exp(-p\rho \int \{1-\exp[-tw(\underset{\sim}{r})]\} d\underset{\sim}{r}) \tag{2.10}$$

where ρ is the density of molecules in the medium. Thus $p\rho$ is the concentration of active molecules. Interestingly, the approximate expression (2.10) occurs frequently in many fields, see Eq. (8.16) of Ref. (1). According to Lax this result was known to Markoff and may have been known to Laplace.

For short times the exclusion of the origin from the sums in Eq. (2.9) is important and may be taken into account in Eq. (2.10) through a cut-off at small $\underset{\sim}{r}$[34] here we will not consider this point and will start integration at the origin. For isotropic interactions $w(\underset{\sim}{r})$ we have therefore,[30]

$$\Psi(t) = \exp \{-\Delta V_\Delta \rho p \int [1-\exp(-tw(r))] r^{\Delta-1} dr\} \tag{2.11}$$

where Δ is the dimensionality of the lattice and V_Δ the volume of a unit sphere in Δ-dimensions. We insert now the forms(2.2) and (2.3) into Eq. (2.11) and obtain for isotropic multipolar interactions[34]

$$\Psi(t) = \exp [-V_\Delta p \rho d^\Delta \Gamma(1-{}^\Delta/s)\ (t/\tau)^{\Delta/s}] \tag{2.12}$$

where Γ (x) is Euler's gamma function, and for isotropic exchange interactions[35]

$$\Psi(t) = \exp[-V_\Delta p \rho \gamma^{-\Delta} g_\Delta(\frac{t}{\tau} \exp(\gamma d))] \tag{2.13}$$

where the function $g_\Delta(z)$ is defined through

$$g_\Delta\ (z) = \Delta \int \{1-\exp[-z\exp(-y)]\} y^{\Delta-1} dy \tag{2.14}$$

A list of the properties of the function $g_\Delta(z)$ is given in the appendix of Ref. (35). Formula (2.12) for Δ=3 and s=6 was derived in the energy transfer field by Förster,[36] for Δ=3 and s arbitrary the result was found by Inokuti and Hirayama; [39] the case of s=6 and Δ=1,2 is given in Ref. (40). Eq. (2.13) for Δ=3 was obtained by Inokuti and Hirayama[39] and the function $g_3(z)$ is their function g(z); for arbitrary Δ the result is given by Blumen and Silbey.[41]

We note that identical expressions for $\Psi(t)$, calculated by inserting Eq. (2.2) or Eq. (2.3) into Eq. (2.11) were also obtained in Ref. (12) in the field of conductivity in disordered systems and for the electron scavenging problem.[43]

We now use the above expression calculated for $\Psi(t)$ in order to derive the distribution of stepping times $\psi(t)$. We will follow the approximation suggested by Scher and Lax[12,13] to relate $\psi(t)$ to Ψ:

$$\psi(t) = -\frac{d}{dt}\Psi(t) \tag{2.15}$$

From Eq. (2.6) follows,

$$\psi(t) = \sum_{\underset{\sim}{r}} pw(\underset{\sim}{r}) e^{-tw(\underset{\sim}{r})} \{\prod_{\substack{\underset{\sim}{r}'\neq \underset{\sim}{r} \\ \underset{\sim}{r}'\neq \underset{\sim}{0}}} [1-p+pe^{-tw(\underset{\sim}{r}')}]\}$$

$$\equiv \sum_{\underset{\sim}{r}} \psi(\underset{\sim}{r},t) \tag{2.16}$$

In Eq. (2.16) $\psi(\underset{\sim}{r},t)$ is the probability of an excitation jumping a distance $\underset{\sim}{r}$ with stepping time t. In the continuum approximation and for low p, p << 1,

$$\psi(\underset{\sim}{r},t) \simeq p \exp[-tw(\underset{\sim}{r})] \Psi(t) \tag{2.17}$$

where we have replaced the product in Eq. (2.16) by its form (2.10). Interestingly, Eq. (2.17) represents an example for a coupled stepping time distribution.(12,13,31,32,44)

III. DIFFUSION IN THE FRAMEWORK OF THE CTRW

We now consider diffusion over a disordered system in the CTRW framework. As in Sec. II we follow the Scher-Lax model,(12,13) namely the transport occurs on an ordered lattice, while the disorder is contained in the distribution of stepping times $\psi(t)$, Eq. (2.16).

Within this model the probability $P(\underset{\sim}{r},t)$ of finding the walker at site $\underset{\sim}{r}$ at time t is given in terms of its Fourier and Laplace transforms:(12,28)

$$P(\underset{\sim}{k},u) = \frac{1-\psi(u)}{u} \cdot \frac{1}{1-\psi(\underset{\sim}{k},u)} \qquad (3.1)$$

where $\psi(k,u)$ is the Fourier-Laplace transform of $\psi(\underset{\sim}{r},t)$ defined in Eq. (2.17). The Laplace transform of the mean squared displacement $\langle r^2(t)\rangle$ of the walker is given by (L represents the Laplace transform)

$$\langle r^2(u)\rangle = L\,\langle r^2(t)\rangle = \sum_{\underset{\sim}{r}} r^2 P(\underset{\sim}{r},u) =$$

$$= -\nabla^2_{\underset{\sim}{k}}\, P(\underset{\sim}{k},u)\big|_{\underset{\sim}{k}=o} \qquad (3.2)$$

from which with (3.1) it follows

$$\langle r^2(u)\rangle = \frac{1}{u[1-\psi(u)]} \sum_{\underset{\sim}{r}} r^2\psi(\underset{\sim}{r},u) \qquad (3.3)$$

The time dependent diffusion coefficient D(t) is defined as

$$D(t) \equiv \frac{1}{2\Delta}\,\frac{d}{dt}\,\langle r^2(t)\rangle \qquad (3.4)$$

So that

$$D(u) = \frac{1}{2\Delta[1-\psi(u)]} \sum_{\underset{\sim}{r}} r^2\psi(\underset{\sim}{r},u) \qquad (3.5)$$

Thus $\psi(\underset{\sim}{r},t)$ is sufficient to determine in the CTRW approach the diffusion coefficient D(t) uniquely. In the continuum approximation we obtain by using (2.17)

$$\sum r^2\psi(\underset{\sim}{r},t) = L\,[\Psi(t)\Omega(t)] \qquad (3.6)$$

with

$$\Omega(t) = p\rho \int d\underset{\sim}{r}\; r^2 w(\underset{\sim}{r}) e^{-tw(\underset{\sim}{r})} \qquad (3.7)$$

from which Eq. (3.5) can be evaluated as:

$$D(u) = L\,[\Psi(t)\Omega(t)]/_{[2\Delta u\Psi(u)]} \qquad (3.8)$$

$\Psi(t)$ was extensively studied in Sec. II. The form (3.8) for D(u) is the main result of this section and we now consider for example multipolar interactions in 3 dimensions. With Eq. (2.2) one has:

$$\Omega(t) = C_1 t^{5/s-1} \tag{3.9}$$

with

$$C_1 = \Gamma(1-5/s)\frac{4\pi p\rho}{s} \left(\frac{d^s}{\tau}\right)^{5/s} \tag{3.10}$$

From Eq. (2.12) one has:

$$\Psi(t) = \exp[-C_2 t^{3/s}] \tag{3.11}$$

with

$$C_2 = \Gamma(1-3/s)\,\frac{4\pi}{3}\,p\rho \left(\frac{d^s}{\tau}\right)^{3/s} \tag{3.12}$$

In order to obtain the diffusion coefficient one generally has to evaluate the inverse Laplace transform of Eq. (3.8). In our case for short and long times the behavior of D(t) can be found directly from $\Omega(t)$ and $\Psi(t)$ using Tauberian theorems.[45]

For short times $\Psi(t)$ is a slowly varying function of t, so that

$$\Psi(u) \sim \frac{1}{u} \tag{3.13}$$

and, using Eq. (3.9)

$$L\,[\Psi(t)\Omega(t)] \sim \frac{C_1\Gamma(5/s)}{u^{5/s}} \tag{3.14}$$

From Eq. (3.8) it now follows:

$$\begin{aligned} D(t) &\sim L^{-1}\left[\frac{C_1\Gamma(5/s)}{6u^{5/s}}\right] \sim \frac{C_1}{6}\, t^{5/s-1} \\ &= \Gamma(1-5/s)\,\frac{2\pi p\rho}{3s}\left(\frac{d^s}{\tau}\right)^{5/s} t^{5/s-1} \end{aligned} \tag{3.15}$$

This result concurs with the expression obtained by expanding the functions in powers of $\frac{1}{u}$ by Godzik and Jortner.[32]

For long times (small u) one has

$$L\,[\Psi(t)] = \sum_{j=o}^{\infty} \frac{(-u)^j}{j!}\,\tau_{j+1} \tag{3.16}$$

and

$$L\ [\Omega(t)\ \Psi(t)] = C_1 \sum_{j=o}^{\infty} \frac{(-u)^j}{j!} \tau_{j+5/s} \tag{3.17}$$

with

$$\tau_k \equiv \int dt t^{k-1}\ \Psi(t) = \frac{s}{3}\,\Gamma(ks/3)\ C_2^{-ks/3} \tag{3.18}$$

Since the moments τ_j of Eqs. (3.16) and (3.17) exist, one has from the Tauberian theorems for longer times:(45)

$$D(t) \sim \frac{C_1 \tau_{5/s}}{6\tau_1} + \Theta(1/t) \tag{3.19}$$

typical for time independent diffusion.

Another approximation for calculating diffusion coefficients was suggested by Gochanour Andersen and Fayer(46) and studied further by Blumen, Klafter and Silbey.(33) The basic equation in this approach is:(46)

$$D(u) = \frac{p\rho}{2\Delta u} \int d\underset{\sim}{r}\ \frac{r^2 w(\underset{\sim}{r})}{1+2\Psi(u)w(\underset{\sim}{r})} \tag{3.20}$$

Here again $\Psi(u)$ (The Laplace transform of $\Psi(t)$ discussed in Sec. II) play a dominant role. One should note the difference by factor u between Eq. (3.20) and Eq. (2.15) of Ref. (33). Eq. (3.20) obtains in the pair approximation and is similar to the expression derived by Haan and Zwanzig(47) except that in their case $1/u$ replaces the function $\Psi(u)$. In the derivation of Eq. (3.20) $\Psi(u)$ implicitly allows for multiple step transfers and is generally taken to include back transfer, see Ref. (46). We note that Eq. (3.20) is also similar to an effective medium approximation by Movaghar.(48) As in Eq. (3.13), for short times, large u, $\Psi(u) \sim 1/u$. Replacing $\Psi(u)$ by $1/u$ in Eq. (3.20) results in having $D(u)$ equal the Laplace transform of $\Omega(2t)/\Delta$ as may be readily verified from Eq. (3.7); this is nothing else but the Haan-Zwanzig expression.(47) Hence the same relation between $D(t)$ and $\Omega(t)$ obtains in the short time limit as in the CTRW case, Eq. (3.15), apart from a factor 2 in $\Omega(2t)$, which stems from the backtransfer assumption.(33) For multipolar interactions both τ_1 and the integral (3.20) with $\Psi(u)$ replaced by τ_1 exist. In the small u case the integral in Eq. (3.20) is then a slowly varying function of u. Using again the Tauberian theorem one has

$$D(t) = \frac{p\rho}{2\Delta} \int d\underset{\sim}{r}\ \frac{r^2 w(\underset{\sim}{r})}{1+2\tau_1 w(\underset{\sim}{r})} + \Theta(\frac{1}{\tau}) \tag{3.21}$$

i.e. a non-vanishing time independent diffusion coefficient for longer times.

IV. TRAPPING(RECOMBINATION) IN THE CTRW MODEL

In this section we consider the problem of trapping (recombination) of a walker by traps (or recombination centers) randomly distributed on a lattice. We will use trapping as a generalized concept which may apply in the case of the capture of an optical excitation by low lying traps,[6-10] in the case of carrier recombination in disordered materials[49-52] and for analyzing the process of electron scavenging in organic glasses.[17-20] As an example we will consider charge carriers in a disordered system. The recombination of two oppositely charged carriers A and B (say electrons and holes) is supposed to occur predominantly at particular recombination centers (rc). A general theoretical treatment of the process has to include both the motion of the carriers and also their encounter at the rc. For simplicity one assumes that one sort of carriers (say B) reaches the rc very quickly, so that the recombination is determined by the motion of the A type carriers to the already B-occupied rc. Furthermore one assumes that the A carriers recombine at the first encounter of a rc and describes their motion as a random walk. In the amorphous material the distribution of sites involved in the carrier transport is random, and we face the problem of a random walk on a random lattice. As for the experimental behavior in such cases, for a wide class of materials the decay law due to trapping (recombination) is nonexponential in time. An algebraic decay law provides a good description of the behavior at longer times.[50-52]

$$N(t) \sim ct^{-\beta} \tag{4.1}$$

where $N(t)$ is the survival probability of a carrier at time t, and β is a parameter of the process, $0 < \beta < 1$.

Since no exact approach to the general trapping problem in disordered systems is available, we have to use approximate schemes, such as the <u>decoupled</u> CTRW. As mentioned in the previous sections we consider nearest neighbor random walks on a regular lattice and let the individual steps occur at random times. The time between two consecutive steps is taken to be distributed according to $\psi(t)$ of Sec. II and III. This version decouples in our trapping case the temporal process from the spatial one. Furthermore, we assume that traps are placed at random on the regular lattice, occupying the lattice sites with probability q. The transfer to a certain site should not depend on whether the site is a trap or not. We also take the walker to be quenched instantaneously at the first encounter of a trap.

For a particular realization of the random walk on the perfect (trap-free) lattice, let R_n denote the number of distinct sites visited in n steps. For the same realization of the walk let F_n denote the probability (over the ensemble of lattices doped with traps) that trapping has not occurred up to the n-th step. The quantities R_n and F_n are stochastic variables, related through[10,54]

$$F_n = (1-q)^{R_n - 1} \tag{4.2}$$

where we assume the origin of the walk not to be a trap. The measurable survival probability at time t is then the average of Eq. (4.2) with respect to all realizations of the random walk in <u>space</u> and in <u>time</u>:[10]

$$\Phi(t) = << F_n >> = \sum_{n=o}^{\infty} <(1-q)^{R_n -1}> \phi_n(t) \qquad (4.3)$$

In Eq. (4.3) the symbol < > denotes average with respect to the realizations of the random walk on the lattice, whereas << >> includes also the temporal behavior. The quantity $\phi_n(t)$ in Eq. (4.3) is the probability density for having performed exactly n steps in time t, which in the CTRW is related to the distribution function $\psi(t)$ through[53,55]

$$\phi_n(u) = [\psi(u)]^n[1-\psi(u)]/u \qquad (4.4)$$

Equations (4.2) and (4.3) are exact. Analytically, however, the distribution of R_n values is not generally known. One has therefore to resort to approximations. A canonical way to proceed is to use a cumulant expansion of Eqs. (4.2) and (4.3) in terms of $\lambda \equiv -\ln(1-q)$. This approach was discussed in Ref. (53) where it was pointed out that the expansion works best for higher dimensional lattices. Here we restrict ourselves to a three dimensional simple cubic lattice; in this case for random walks with fixed stepping time frequency and for low trap concentration ($q \ll 1$) already the first cumulant, i.e. the mean number of distinct sites visited, $S_n \equiv \langle R_n \rangle$ allows a satisfactory description of the decay:

$$\Phi_n \simeq (1-q)^{S_n -1} \approx e^{-qS_n} \quad (q \ll 1) \qquad (4.5)$$

A major advantage is using Eq. (4.5) rests in the fact that much information on S_n is known from the generating function formalism;[24] higher cumulants are much more difficult to evaluate.[10,54,56] For three dimensional random walks, S_n behaves asymptotically as $S_n = a.n + O(\sqrt{n})$, where for nearest neighbor random walks on cubic lattices the constant a is given by the inverse of the corresponding Watson integral.[9,24]

In the CTRW representation we obtain therefore for low trap concentrations and three dimensional lattices

$$\Phi(t) \approx \sum_{n=o}^{\infty} e^{-qS_n}\phi_n(t) \quad (q \ll 1) \qquad (4.6)$$

where we used Eqs. (4.3) and (4.5)

Two approximations to Eq. (4.6) are now possible. One is to view the sum over n as a time average and to define a cumulant expansion in time.[53] Here the first cumulant is $S(t) = \sum_{n=o}^{\infty} S_n\phi_n(t)$ and in this case one has:

$$\Phi(t) \simeq e^{-qS(t)} \qquad (4.7)$$

a form which has been used extensively.[53,55,57-61] This is best seen by defining the decay rate k(t) of Φ(through $k(t) \equiv -\dot{\Phi}(t)/\Phi(t)$) and noticing that, from Eq. (4.7):

$$k(t) \approx q\frac{dS}{dt} \qquad (4.8)$$

the first passage time of Ref. (61).

The second approximation to Eq. (4.6) obtains by making use of $S_n \propto a \cdot n$ which together with Eq. (4.4) allows the direct summation of Eq. (4.6):

$$L\,[\Phi(t)] \approx \frac{1-\psi(u)}{u} \sum_{n=o}^{\infty} [e^{-qa}\psi(u)]^n$$

$$= [1-\psi(u)]/\{u[1-e^{-qa}\psi(u)]\} \qquad (4.9)$$

Both forms, Eq. (4.7) and Eq. (4.9) have their respective applications. Eq. (4.7) is certainly superior in describing the short time behavior of $\Phi(t)$. As pointed out in Ref. (53) the approximation (4.7) is convenient for an exponential stepping time distribution, where it approximates well the time decay form, Eq. (4.3). The approximation (4.7) gets poorer, however, if the stepping time distribution gets broader.[53] The disadvantage of Eq. (4.9) is that it is not readily extendable to lower dimensionalities. Its main advantage rests in the long time tail description of the decay law, as we now show.

It is evident from Eq. (4.9) that the decay law depends on $\psi(u)$. We study the decay in Eq. (4.9) due to two different stepping time densities:

$$\psi(t) = c\alpha t^{\alpha-1} \exp(-ct^{\alpha}) \quad o < \alpha < 1 \qquad (4.10)$$

and

$$\psi(t) \propto t^{-1-\gamma} \quad o < \gamma < 1 \qquad (4.11)$$

One should note that $\psi(t)$ in Eq. (4.10) is obtained by using Eq. (2.15) for a generalization of the multipolar $\Psi(t)$ in Eq. (2.12) where Δ/s is replaced by α. Eq. (4.10) has been obtained also by Ngai[58] assuming low frequency fluctuations. Eq. (4.11) which displays a long time tail was suggested by Scher and Montroll[14] to model transport in amorphous media.

In Eq. (4.9) the long time behavior is determined by the first moment τ_1 of the $\psi(t)$ distribution, $\tau_1 = \int^{\infty} dt t\psi(t)$. If τ_1 is finite then $\psi(t) \approx 1-\tau_1 u$ for small u. In this case one has from Eq. (4.9) for longer times $L[\phi(t)] \approx [u+qa/\tau_1]^{-1}$. We remark, furthermore that all moments τ_i of the distribution function $\psi(t)$ exist. Hence, a time domain $\Phi(t)$ compatible with this behavior is:

$$\Phi(t) \approx e^{-qat/\tau_1} \qquad (4.12)$$

i.e. an exponential. For very long times, deviations from Eq. (4.12) may occur due to long time tails in the form of $\psi(t)$. For the stepping time distribution of Eq. (4.11), τ_1 is infinite. One obtains then $\psi(u) \approx 1-\Gamma(1-\gamma)u^{\gamma}/\gamma$ and $L[\Phi(t)] \approx [\Gamma(1-\gamma)/\gamma]u^{\gamma-1}L(u)$, with $L(u) \sim \{1-e^{-qa}[1-\Gamma(1-\gamma)u^{\gamma}/\gamma]\}^{-1}$ L(u) being a slowly varying function[45] of u, for $u \to o$. Using the Tauberian theorem[45] one obtains:

$$\Phi(t) \approx t^{-\gamma}/(qa\gamma) \tag{4.13}$$

for long times. For $\gamma = 1/2$, Eq. (4.13) was already established through an exact inverse Laplace transform.[55,62]

Consider now the approximate form, Eq. (4.7). We remark that S(u), the Laplace transform of S(t) is given by:[24,53]

$$S(u) = \{u(1-\psi(u))P(0;\psi(u))\}^{-1} \tag{4.14}$$

where P(0,z) is the generating function of the walk and $\lim_{z\to 1} P(0,z) = a^{-1}$ for three dimensional lattices. For $\psi(u) \approx 1-\tau_1 u$ one obtains $S(t) \sim at/\tau_1$. Thus Eq. (4.7) leads in this case also to Eq. (4.12). For $\psi(u) \approx 1-\Gamma(1-\gamma)u^{\gamma}/\gamma$ the situation is, however, very different. Using Eq. (4.14) and the same Tauberian theorem leads to $S(t) \sim At^{\gamma}/[\Gamma(\gamma)\Gamma(1-\gamma)]$; therefore, Eq. (4.7) implies

$$\Phi(t) \sim \exp\{-qat^{\gamma}/[\Gamma(\gamma)\Gamma(1-\gamma)]\} \tag{4.15}$$

Note the difference between Eq. (4.15) and Eq. (4.13). From our numerical experience we view Eq. (4.13) to be the better approximation at longer times. This is due to the fact that the assumption S_n=an works very well in three dimensions for moderate and for large n[9]. Eq. (4.15) might still be valid for shorter times. The summation in Eq. (4.9) then takes care of all temporal cumulants of $\phi_n(t)$, whose influence is significant for broad distributions, and which are neglected in (4.7).

We should emphasize that in the derivation of Eq. (4.13) no use has been made of the particular form of S(t), the determining factor being $[1-\psi(u)]/u$ in Eq. (4.9). Nevertheless, one may express Eq. (4.13) in the form

$$\Phi(t) \sim [q\Gamma(1+\gamma)\Gamma(1-\gamma)S(t)]^{-1} \tag{4.16}$$

a fact noted already by Scher.[62] The extension of this result to other dimensionalities calls for further attention.

V. AN EXAMPLE: REACTIONS OF TRAPPED ELECTRONS IN ORGANIC GLASSES

Reactions of trapped electrons (e^-_{tr}) with electron acceptors (scavengers, S) in glasses have been a continuous object of study in recent years.[17-20, 43, 55, 57, 63] The theoretical treatments of the e^-_{tr}+S reaction have evolved around two distinct, limiting assumptions. The first is that e^-_{tr} reacts directly with a scavenger.[19, 20, 43, 63] The second is that the electron hops from one trap to another, until encountering a scavenger.[18,55,57] In fact, much effort has been devoted to the controversial issue of finding out the correct model.[19]

In this section we analyze this problem of electron scavenging and make a comparative study of the direct reaction of electrons with scavengers vs. a multiple-step electron migration over the traps prior to capture. For the direct reaction we obtain expressions for the survival probability based on Sec. II. The electron migration is modelled in the CTRW formalism as discussed in Sections II and IV, by obtaining stepping time distributions from direct transfer expressions, namely the Scher-Lax suggestions of Section II.

In the direct reaction model we assume e^-_{tr} and S to be on a lattice. The survival probability of e^-_{tr} averaged over all S configurations is given, for low S concentrations by Eq. (2.10) where a continuum description has been used. The parameter p in Eq. (2.10) corresponds here to the site occupation probability by the scavengers. In the electron scavenging problem one usually assumes for $w(\underset{\sim}{r})$ of (2.10) the form (2.3) which may be envisaged to arise from the overlap of the wave functions of e^-_{tr} and S, or from tunneling through a potential barrier. Inserting Eq. (2.3) into Eq. (2.10) we obtain for the three dimensional case, $\Delta=3$:

$$\Psi(t) = \exp[-pAg_3(\frac{t}{\tau} e^{\gamma d})] \tag{5.1}$$

where $A={}^{4\pi\rho}/(3\gamma^3)$. The more general forms for $\Psi(t)$ and the function $g_3(z)$ are given by Eqs. (2.13) and (2.14). The approximate expression of $\Psi(t)$, Eq. (5.1), was derived for the case of electron scavenging also in Refs. (43) and (63). For large values of the variable z, $g_3(z)$ admits an asymptotic expansion in lnz:

$$g_3(z) = \ln^3 z + 1.73\ln^2 z + 5.93\ln z + 5.44 \tag{5.2}$$

As noted in Refs. (64) and (65) a good approximation to Eq. (5.2) is $g_3(z)=\ln^3(Cz)$ with $C\simeq1.9$. This leads to a convenient form for Eq. (5.1):

$$\Psi(t) = \exp[-pA\ln^3(C\frac{t}{\tau} e^{\gamma d})] \tag{5.3}$$

As shown in Refs. (66) and (67), the time behavior of electron scavenging is well reproduced in a wide variety of materials through a survival law of the form of Eq. (5.3); this seems to apply that the dynamics of e^-_{tr} + S is determined by the direct mechanism.

Recently claims have been made that a decay form similar to Eq. (5.3) might also be obtained from multiple step processes, by which e^-_{tr} migrates from trap to trap.[18,55,58] The formalism used to describe the migration was the CTRW as described in the previous sections. The $\psi(t)$ distribution functions, which mimic the disorder were derived through Eq. (2.15) using Eq. (5.3),

$$\psi(t) = 3pA[\ln^2(C\frac{t}{\tau} e^{\gamma d})/t] \exp[-pA\ln^3(C\frac{t}{\tau} e^{\gamma d})] \tag{5.4}$$

This form was discussed by Tachiya.[20] An approximate form to Eq. (5.4) valid for a large time domain was introduced by Scher and Montroll[14] and is given by Eq. (4.11). This form was applied to the present problem by Hamill and Funabashi.[18] We can now transfer our results of Sec. IV for trapping to the present problem of electron scavenging with $\psi(t)$ given by Eq. (5.4) or by Eq. (4.11); here q introduced in Eq. (4.2) is now $q = p/(p+p_{tr})$; where p_{tr} is the probability that sites of the original substance are traps (here: sites over which the migration takes place!). As before we consider the case of small q, $q \ll 1$, i.e. $p \ll p_{tr}$.

As is evident from our conclusions in Sec. IV, the stepping time distribution of Eq. (5.4) has finite moments and so the decay law due to the multistep process should approach (4.12) for larger times (i.e., $\Phi(t) \sim e^{-qat/\tau_1}$). For $\psi(t)$ in Eq. (4.11), where the first moment is infinite, one obtains the decay law of (4.13) (i.e., $\Phi(t) \sim t^{-\gamma}/(qa\gamma)$). Both forms obtained in the CTRW framework differ from the decay law due to direct reaction, Eq. (5.3), which has been found to describe the experimental results accurately. Thus, the asymptotically valid forms for the multiple-step CTRW electron scavenging are less apt to reproduce the experimental findings. We thus favor a single step (or, eventually, a few steps) mechanism over the multiple step mechanism, in order to describe electron scavenging in glasses.

ACKNOWLEDGEMENTS

The collaboration with Dr. G. Zumofen in establishing many of the results presented is gladly acknowledged. A. B. is grateful to Exxon Corporation for the hospitality extended to him and acknowledges the support of the Deutsche Forschungesgeneiuschaft and of the Fonds der Chemischen Industrie.

*On leave from Lehrstuhl für Theoretische Chemie Technische Universität München, West Germany.

REFERENCES

1. M. Lax, in Stochastic Differential Equations, edited by J. B. Keller and H. P. McKean (Amer. Math. Soc., Providence, R.I., 1973), p. 35.
2. R. Kopelman, In Radiationless Processes in Molecules and Condensed Phases, edited by F. K. Fong (Springer, Berlin, 1976), p. 297.
3. J. C. Wright, in Radiationless Processes in Molecules and Condensed Phases, edited by F. K. Fong (Springer, Berlin, 1976), p. 239.
4. G. Pfister and H. Scher, Adv. Phys. 27, 747 (1978).
5. V. M. Agranovich and M. D. Galanin, Electronic Excitation Energy Transfer in Condensed Matter (North-Holland, Amsterdam (1982).
6. D. L. Huber, in Laser Spectroscopy of Solids, edited by W. M. Yen and P. M. Selzer (Springer, New York, 1981), p. 83.
7. S. Alexander, J. Bernasconi, W. R. Schneider and R. Orbach, Rev. Modern Phys. 53, 175 (1981).
8. R. C. Powell and Z. G. Soos, J. Lumin. 11, 1 (1975).
9. A. Blumen and G. Zumofen, J. Chem. Phys. 75, 892 (1981).
10. G. Zumofen and A. Blumen, J. Chem. Phys. 76, 3713 (1982).
11. B. E. Vugmeister, Fiz. Tverd. Tela 18, 819 (1976) [English translation: Sov. Phys. Solid State 18, 469 (1976)].
12. H. Scher and M. Lax, Phys. Rev. B7, 4491 (1973).
13. H. Scher and M. Lax, Phys. Rev. B7, 4502 (1973).
14. H. Scher and E. W. Montroll, Phys. Rev. B12, 2455 (1975).
15. K. W. Kehr and J. W. Haus, Physica 93A, 412 (1978).
16. S. Kivelson, Phys. Rev. B21, 5755 (1980).
17. W. P. Helman and K. Funabashi, J. Chem. Phys. 66, 5790 (1977).
18. W. H. Hamill and K. Funabashi, Phys. Rev. B16, 5523 (1977).
19. S. A. Rice and G. A. Kenney-Wallace, Phys. Rev. B21, 3748, (1980).
20. M. Tachiya, Radiat. Phys. Chem. 17, 447 (1981).
21. P. W. Anderson, in Ill-Condensed Matter, edited by R. Balian, R. Maynard and G. Toulouse (North Holland, Amsterdam, 1979) p. 162.
22. M. N. Barber and B. W. Ninham, Random and Restricted Walks; Theory and Applications (Gordon and Breach, New York, 1970).
23. G. H. Weiss and R. J. Rubin, Adv. Chem. Phys. 52, 363 (1983).
24. E. W. Montroll and G. H. Weiss, J. Math. Phys. 6, 167 (1965).
25. D. Bedeaux, K. Lakatos-Lindenberg and K. E. Shuler, J. Math. Phys. 12, 2116 (1971).
26. V. Kenkre, E. Montroll and M. Shlesinger, J. Stat. Phys. 9, 45 (1973).
27. V. Kenkre and R. Knox, Phys. Rev. B9, 5279 (1974).
28. J. Klafter and R. Silbey, J. Chem. Phys. 72, 843 (1980).
29. J. Klafter and R. Silbey, Phys. Rev. Lett. 44, 55 (1980).
30. A. Blumen, Nuovo Cimento 63B, 50 (1981).
31. K. Godzik and J. Jortner, Chem. Phys. Lett. 63, 428 (1979).
32. K. Godzik and J. Jortner, J. Chem. Phys. 72, 4471 (1980).

33. A. Blumen, J. Klafter and R. Silbey, J. Chem. Phys. 72, 5320 (1980).
34. A. Blumen and J. Manz, J. Chem. Phys. 71, 4694 (1979).
35. A. Blumen, J. Chem. Phys. 72, 2632 (1980).
36. T. Förster, Z. Naturforsch. Teil A4, 321 (1949).
37. D. L. Dexter, J. Chem. Phys. 21, 836 (1953).
38. M. D. Galanin, Zh. Eksp. Teor. Fiz. 28, 485 (1955) [English translation: Sov. Phys. JETP 1, 317 (1955)].
39. M. Inokuti and F. Hirayama, J. Chem. Phys. 43, 1978 (1965).
40. M. Hauser, U.K.A. Klein and U. Gosele, Z. Phys. Chem. (Wiesbaden) 101, 255 (1976).
41. A. Blumen and R. Silbey, J. Chem. Phys. 70, 3707 (1979).
42. V. P. Sakun, Fiz. Tverd. Tela 14, 2199 (1972) [Sov. Phys. Solid State 14, 1906 (1973)].
43. M. Tachiya and A. Mozumder, Chem. Phys. Lett. 28, 87 (1974).
44. D. G. Thomas, J. J. Hopfield and W. M. Augustyniak, Phys. Rev. 140, A202 (1965).
45. W. Feller, An Introduction to Probability Theory and Its Applications (wiley, New York, 1968) Vol. II, p. 445.
46. C. R. Gochanour, H. C. Andersen, and M. D. Fayer, J. Chem. Phys. 70, 4254 (1979).
47. S. W. Haan and R. Zwanzig, J. Chem. Phys. 68, 1879 (1978).
48. B. Movaghar, J. Phys. C. Solid State Phys. 13 4915 (1980).
49. J. M. Hvam and M. H. Brodsky, Phys. Rev. Lett. 46, 371 (1981).
50. Z. Vardeny, P. O'Connor, S. Ray and J. Tauc, Phys. Rev. Lett. 44, 1267 (1980).
51. J. Mort, I. Chen, A. Troup, M. Morgan, J. Knights and R. Lujan, Phys. Rev. Lett. 45, 1348 (1980).
52. P. B. Kirby, W. Paul, S. Ray and J. Tauc, Solid State Commun. 42, 533 (1982).
53. A. Blumen and G. Zumofen, J. Chem. Phys. 77, 5127 (1982).
54. G. W. Weiss, Proc. Nat. Acad. Sci. U.S.A. 77, 4391 (1980).
55. W. P. Helman and K. Funabashi, J. Chem. Phys. 71 2458 (1979).
56. G. Zumofen and A. Blumen, Chem. Phys. Lett. 88, 63 (1982).
57. M. F. Shlesinger, J. Chem. Phys. 70, 4813 (1979).
58. K. L. Ngai and F. S. Liu, Phys. Rev. B24, 1049 (1981).
59. R. D. Wieting, M. D. Fayer and D. D. Dlott, J. Chem. Phys. 69, 1996 (1978).
60. J. Klafter and R. Silbey, J. Chem. Phys. 72, 849 (1980).
61. J. Klafter and R. Silbey, J. Chem. Phys. 74, 3510 (1980).
62. H. Scher, J. Phys. (Paris) 42, C4-547 (1981).
63. J. R. Miller, J. Chem. Phys. 56, 5173 (1972), Chem. Phys. Lett. 22, 180 (1973); J. Phys. Chem. 79, 1070 (1975).
64. R. K. Huddleston and J. R. Miller, J. Phys. Chem. 85, 2292 (1981); 86, 200 (1982); 86, 1347 (1982).
65. A. B. Doktorov, R. F. Khairutdinov and K. I. Zamaraev, Chem. Phys. 61, 351 (1981).
66. K. I. Zamaraev, R. F. Khairutdinov and J. R. Miller, Chem. Phys. Lett. 57, 311 (1978).
67. J. V. Beitz and J. R. Miller, J. Chem. Phys. 71, 4579 (1979).

RANDOM WALKS, TRANSPORT, AND DISPERSION IN POROUS MEDIA

Muhammad Sahimi
Department of Chemical Engineering and Materials Science
University of Minnesota, Minneapolis, Minnesota 55455

ABSTRACT

Dispersion results from different flow paths and consequent different first passage times available to tracer particles crossing from one plane to another in a porous medium. A continuous-time random walk model of dispersion is proposed and is investigated by means of Monte Carlo simulations in square and cubic networks of random conductances. Disperson is found to be diffusive and longitudinal dispersion coefficient (i.e. the effective diffusion coefficient in the direction of mean flow) is found to be one order of magnitude larger than transverse dispersion coefficient. As the percolation threshold is approached, dispersivities, i.e., the ratio of dispersion coefficients and the mean flow velocity, increase dramatically. This is attributed to the increase in tortuosity of the backbone of the percolation cluster as the percolation threshold is approached, with the appearance of numerous loops which provide alternate particle paths that are highly effective in dispersing a concentration front of tracer particles. This results in a broad distribution of the first passage times and thus large dispersivities. A possible connection between dispersion process near a percolation threshold and directed percolation is investigated and it is shown how Monte Carlo simulations of dispersion near the percolation threshold yield a great deal of information about the structure of the backbone.

INTRODUCTION

The phenomenon of dispersion refers to the irreversible mixing of two miscible fluids displacing one another. If two miscible fluids are in contact, with an initially sharp interface, they will slowly diffuse into one another. As time passes, the sharp interface between the two fluids will become a diffuse mixed zone grading from one pure fluid to the other. This diffusion arises because of random motion of the molecules. During this process there can be no equilibrium except that of one phase uniformly distributed throughout the second. If there is no change in volume upon mixing the two fluids, then the net transport of one of the constituents across any arbitrary plane can be represented by the Fick diffusion equation. This mixing takes place whether or not the two fluids are moving through the medium. However, if the fluids are flowing through the medium, then there may be some additional mixing taking place. This increased mixing caused by the structure of the medium and the fluid flow condition will be designated dispersion. This phenomenon is important to a wide variety of problems and has witnessed considerable research activities during the past decades. Some examples where dispersion is important are: the intrusion of salt water into the freshwater lens beneath oceanic islands, the displacement of oil by solvents in petroleum reservoir, and the contamination of ground water resources by the disposal of wastes such as atomic wastes. In such a case dispersion is an important physical phenomenon that must be taken into account when evaluating the environmental consequences of transport of a hazardous subtance by a ground water system. A helpful aspect of dispersion is its reduction of possible peak contaminant concentrations arriving at the biosphere. However, detrimental aspects are that a contaminant will be dispersed over larger regions and will have earlier arrival into the environment than predicted by mean ground-water velocities alone. Dispersion phenomenon also occurs during the flow in packed towers; these have been studied intensively by chemical engineers for a long time.

Disperson can take place in microstructures as simple as a capillary tube and in media as complex as a porous medium. In order to understand dispersion phenomenon, we will first consider dispersion in a single capillary tube. We then turn to dispersion processes in a porous medium which is not only a much more scientifically interesting problem but also more important technologically.

DISPERSION IN CAPILLARY TUBES

Let us consider dispersion in a capillary tube. Suppose the tube is filled with one fluid and a second fluid is injected at one end of the tube. If the two fluids are of the same constant physical properties (viscosity, density, etc.), if diffusion effects are not important, and if flow is laminar, then the concentration of displacing fluid in the effluent stream is easily determined by integrating the flow equation for laminar conditions. There will be no injected fluid appearing in the effluent stream until one-half of the tube volume has been injected. For continued injection, the effluent concentration is given by

$$C_i = 1 - [V_p/2V]^2 \quad V \geqslant V_p/2 \tag{1}$$

where C_i is the volume fraction of injected fluid in the effluent, V_p the total volume of the tube, and V the volume injected. This equation which gives the instantaneous average concentration of the *effluent* was first derived by Fowler and Brown.[1]

In reality, molecular diffusion will cause mixing along the interface, so that the net result will be a mixed zone growing at a more rapid rate than would obtain from diffusion alone, but less than the rate as given by equation (1). Griffiths[2] in 1911 was the first one to report some experimental work which demonstrated the essence of the dispersion process in a capillary tube with diffusional effects present, without mathematical treatment. He observed that a tracer injected into a system of water spreads out in a symmetrical manner about a plane in the cross section which moves with the mean speed of flow. He commented that: "It is obvious that the movement of the center of the column of the tracers must measure the mean speed of flow." It turned out that this was not as obvious as Griffiths had thought! Forty two years later, G.I. Taylor[3] pointed out that this is a rather startling result for two reasons. First, because the water at the center of the capillary tube moves with twice the mean speed of the flow (the Hagen-Poiseuille flow), the water at (or near) the center must approach the column of tracer, absorb the tracer as it passes through the column, and then reject the tracer as it leaves on the other side of the column. Second, although the velocity is unsymmetrical about the plane moving at the mean speed, the column of tracer spreads out symmetrically. The concentration of the tracer material is described by the two-dimensional unsteady convection-diffusion equation, a special case of the Fokker-Planck equation:

$$\frac{\partial C}{\partial t} + u(r)\frac{\partial C}{\partial x} = D_r\left(\frac{\partial^2 C}{\partial r^2} + \frac{1}{r}\frac{\partial C}{\partial r}\right) + D_a\frac{\partial^2 C}{\partial x^2}\,. \tag{2}$$

Here C is the point concentration, u the parabolic laminar velocity profile: $u(r) = 2u_m\left(1 - \frac{r^2}{R^2}\right)$, where u_m is the mean speed of flow, R the radius of the capillary, x and r the axial and radial coordinator, and D_r and D_a the radial and axial molecular diffusion coefficients respectively. Taylor[3] ignored longitudinal diffusion and showed that for a large enough time the process could be described by a one-dimensional model,

$$\frac{\partial C_m}{\partial t} + u_m\frac{\partial C_m}{\partial z} = D_L\frac{\partial^2 C_m}{\partial z^2} \quad , \; z = r/R \tag{3}$$

Here C_m is the mean value of the concentration over a cross-section of the tube; i.e. C_m is given by

$$C_m = \frac{2}{R^2} \int_0^R Crdr \; . \tag{4}$$

If we define a coordinate which moves with the mean speed of flow as

$$x_1 = z - u_m t \; , \tag{5}$$

then equation (3) becomes

$$\frac{\partial C_m}{\partial t} = D_L \frac{\partial^2 C_m}{\partial x_1^2} \; . \tag{6}$$

Thus the mixed zone would travel with the mean speed of the injected fluid and would be dispersed as if there were a constant diffusion coefficient and the process were a pure molecular diffusion. Equation (6) is simply the one-dimensional diffusion equation to which solutions are readily available under a variety of conditions. The radial molecular diffusion coefficient D has been replaced by a dispersion or an effective diffusion coefficient which, in the absence of axial molecular diffusion, was shown by Taylor to be

$$D_L = \frac{R^2 u_m^2}{48 D_r} \; . \tag{7}$$

This simple equation provides a great deal of physical insight into the nature of the dispersion process if we interpret the numerator of equation (7) to be a measure of axial convection and the denominator to reflect the intensity of transverse mixing rather than just transverse molecular diffusion. Then it is easily seen that the dispersion coefficient, which is a measure of the rate at which material will spread out axially in the tube, is enhanced by having large differences in velocity exist across the flow. In contrast, any mechanism which increases transverse mixing reduces the dispersion coefficient. This is qualitatively true even if the medium is as complex as a porous medium.

Aris[4] later extended the analysis to include axial molecular diffusion and demonstrated that the dispersion coefficient for this case contains Taylor's result with an additive term due to axial diffusion

$$D_L = D_a + \frac{R^2 u_m}{48 D_r} C \; . \tag{8}$$

This result is now known as the Taylor-Aris theory. We emphasize the importance of the restriction placed on the Taylor-Aris theory with respect to time. Unless sufficient time has elapsed, the mean concentration C_m is not described by equation (3). There is an intermediate region where neither equation (1) or equation (7) is valid. This region was studied by Van Deemter *et al.*[5] but their study did not result in a quantitative representation of behavior over the full intermediate range. Very recently Van den Broeck[6] exploited the equivalence of Langevin and Fokker-Planck formalisms and derived the the Taylor-Aris results.

Taylor-Aris theory describes a very extreme and rare mechanism of dispersion that takes place almost solely in single capillary tubes. However, the phenomenon of dispersion occurs in many other systems, of which porous media are perhaps the most important. In this article, we examine some aspects of the problem of dispersion in a porous medium. The inapplicability of Taylor-Aris type models to dispersion in porous media can be easily seen by observing that dispersion occurs both parallel to and perpendicular to the mean flow: a porous medium exhibits a longitudinal dispersion coefficient D_L, which in general is different from the one given by equation (7) and a transverse dispersion coefficient D_T. In Taylor-Aris theory there is no room for transverse dispersion phenomenon and $D_T = 0$. Now that a procedure has been developed[7,8] for systematically reducing a porous medium to an equivalent random network, we can turn our attention to mechanisms of dispersion in flow through a porous medium and contrast these mechanisms with that of Taylor-Aris dispersion process.

DISPERSION IN FLOW THROUGH POROUS MEDIA

Dispersion in this case is a consequence of flow through a system. There are two basic mechanisms of dispersion, and molecular diffusion is their basic modifier. Except in plug flow, i.e. simple translations, the first passage times across a system, between the entrance and exist planes, differ on different streamlines. In a random network (chaotic porous medium) streamtubes divide and rejoin repeatedly at the junctions of flow passages in the highly interconnected space. This causes tangling and divergence of streamlines which is accentuated by the variations in orientations of flow passages and coordinations of junctions. Thus wide variations result in both the length of streamlines and their downstream separation location. This is a kinematic mechanism of dispersion. The second mechanism is a dynamical one. The speed within flow passages depends on the conductance of the passage, its orientation, and the local pressure field. These wide variations in flow passage speeds alter the first passage time distribution of a colored tracer particle that is injected into the fluid flow. This first passage time distribution is a fundamental measure of dispersion in a porous medium (or equivalently a random network).

Traditional modeling of dispersion in a chaotic porous medium has been based on the macroscopic convective-diffusion equation

$$\frac{\partial C}{\partial t} + \bar{\underset{\sim}{V}} \cdot \nabla \underset{\sim}{C} = \nabla \cdot (\underset{\approx}{D} \cdot \nabla C) = D_L \frac{\partial^2 C}{\partial x^2} + D_T \nabla_2^2 C \, . \tag{9}$$

Here $\bar{\underset{\sim}{V}}$ is the macroscopic velocity: C is the macroscopic mean concentration: x is the mean flow direction; and ∇_2^2 is the Laplace operator in the transverse plane. If equation (9) holds, then dispersion is macroscopically diffusive. The idea is to account for dispersion with two effective diffusion or dispersion coefficients D_L and D_T, representing the longitudinal (mean flow) and transverse (perpendicular to mean flow) directions respectively. This is a purely phenomenological approach and does not provide any insight as to how D_L and D_T depend on topological and geometrical properties of a porous medium. It also presupposes that dispersion is a diffusive process, i.e. the concentration distributions caused by these mechanisms resemble a Gaussian one. One goal of the work reported here is to seek the conditions under which equation (9) is valid.

The two basic mechanisms of dispersion discussed above in no way depend on molecular diffusion which acts as a modifying factor by moving material from one streamline to another. But if molecular diffusion effects exist, they lie among "extreme" cases. One of these "extreme" cases is the Taylor-Aris mechanism as discussed above. In this case there is fast radial and axial molecular diffusion in a "pore throat," or a bond in network terminology, and no appreciable diffusion in "pore bodies" or sites or nodes of a network. In another "extreme" case there is no molecular diffusion in "pore throats," but overwhelming molecular diffusion, i.e. complete or at least partial mixing, in "pore bodies." As we shall discuss later, this picture is much closer to what takes place in a real world, as if the real world has converted to an extremist radical!

The case of complete mixing at nodes can be shown to be equivalent to a Markovian process. When we have only partial mixing at nodes, then the process is semi-Markovian. In either case if we consider continuous fluid as a continuum of very small fluid particles that retain their identities as they flow, just as do tracer particles added to the fluid, then the dispersion process can be treated by a continuous-time random walk (CTRW). This model was first developed by Montroll and Weiss[9] and was further extended by Scher and Lax[10] and Scher and Montroll[11] to investigate stochastic transport in disordered solids. (For a review of random walk theory see Barber and Ninham[12]; for further discussion of CTRW models see the review article by Weiss and Rubin[13].) The key probability density functions for a one dimensional CTRW are $P(\zeta - \zeta_{0,t})$, the probability density function for being at ζ at time t if one starts at $\zeta_{0,}Q(\zeta - \zeta_{0,t})$, the first passage time distribution for crossing the system at ζ for the first time starting at ζ_0, and $R(\zeta - \zeta_{0,}t$, which is that for a tracer particle arriving in the slice $(\zeta,\zeta + d\zeta)$ during the interval $(t, t + dt)$.

In general, in calculating a quantity such as $P(\zeta - \zeta_0,t)$ one has to consider all the possible paths from ζ_0 to ζ that the tracer can transverse in a time ξ with $\xi \leqslant t$ and a transition probability between nearest-neighbor sites is needed to describe transport along all the paths contributing to $P(\zeta - \zeta_0,t)$. The probability distribution for the length and direction of each displacement from a site to one of its allowed nearest-neighbor sites, and the waiting time for making the displacement are embodied in the probability density function $\Psi(\underset{\sim}{\zeta},t)$ generating individual steps of the walk; $\Psi(\underset{\sim}{\zeta},t)d\zeta dt$ is the probability of a displacement between $\underset{\sim}{\zeta}$ and $\underset{\sim}{\zeta} + d\underset{\sim}{\zeta}$ after the walker has waited some time between t and t + dt. In most of the previous works on CTRW's and their applications it has been assumed that the spatial and temporal distribution of each displacement can be independent of each other, i.e. $\Psi(\underset{\sim}{\zeta},t)$ can be written as

$$\Psi(\underset{\sim}{\zeta},t) = p(\underset{\sim}{\zeta})\rho(t) \ . \tag{10}$$

Shlesinger, Klafter and Wong[14] have noted that if the factorization (10) is impossible, a variety of novel cases are possible. For dispersion process studied here, we note that the time that a tracer particle needs to execute a displacement is exactly equivalent to the waiting time that a random walker needs in the usual CTRW models before performing an instantaneous jump. We also note that for dispersion processes we *cannot* decouple $\underset{\sim}{\zeta}$ and t, because the displacement time (or waiting time before making an instantaneous displacement) along a bond is related to its length, orientation, local flow field, etc. This distinguishes dispersion in an important way from the usual stochastic processes and the CTRW models. It also makes the development of a CTRW model of dispersion a more formidable task if one is to take into account the random topology of a random network. To derive an equation for the probability density function $\Psi(\underset{\sim}{\zeta},t$ one also needs to determine the flow field in the random network. The most attractive and also realistic way to do this is to use an effective-medium approximation. Percolation theory has been reviewed by several authors.[17,18] Most quantities of interest in percolation theory scaling laws near the percolation threshold

$$x \cong (p - p_c)^{\mu} \tag{11}$$

where p is the fraction of active bonds or sites, p_c the percolation threshold, and μ some universal constant which is believed to depend only on the dimensionality of the system. μ is called a critical exponent and its value can be either positive or negative depending on the quantity of interest x. We shall see how a scaling law such as (11) can shed light on a possible connection between dispersion processes in random percolating networks and percolation theory. Percolation thresholds of different lattices and the critical exponents are perhaps the most important quantities in percolation theory. Many analytical and numerical methods have been developed in the past few years for estimation of p_c's and μ's. These matters have been reviewed recently by Sahimi.[19] The relevance of percolation theory to flow in porous media was established by the papers of Larson, Scriven and Davis,[20] and that of Heiba, et. al.[21] However, percolation theory concepts and their relation to dispersion processes have not been previously used.

If there were no dispersion in steady, unidirectional flow in the x-direction in a macroscopically homogeneous random network, a tracer particle released at location (x_o,y_o,z_o) at time zero would be convected by the mean flow in time t to the location $x_o + \bar{v}t,y_o,z_o)$, where $\bar{v}$ is mean flow velocity. With dispersion, however, the tracer particle, which is convected through bonds and nodes of a random network, arrives at some location (x, y, z) after the same time. We define the squared-derivations of the actual particle position from the hypothetical one by the following equation

$$\sigma_x^2 = (x - x_o - \bar{v}t)^2,\ \sigma_y^2 = (y - y_o)^2,\ \sigma_z^2 = (z - z_o)^2 \ . \tag{12}$$

Then dispersion coefficients are defined as

$$D_\zeta = < \sigma_\zeta^2/2t > \ ; \ \zeta = x,y,z \tag{13}$$

Here the pointed brackets, $< \cdots >$, indicate an appropriate average or some population or concentration profile of injected tracer particles. Because x is the flow direction,D_x is also denoted by D_L, the longitudinal dispersion coefficient and D_y and D_z are denoted by D_T, the transverse dispersion coefficient. Since dispersion is flow driven, we expect a *priori* that $D_L \neq D_T$.

If the dispersion process is diffusive, then

$$P(\zeta - \zeta_0,t) = (4\pi D_\zeta t)^{-1/2} \exp[-(\zeta - \zeta_0 - \overline{V}_\zeta t)^2/4D_\zeta t] \ . \tag{14}$$

The two distributions $P(\zeta - \zeta_0,t)$ and $Q(\zeta - \zeta_0,t)$, the first passage time distribution, are related by [22]

$$P(\zeta - \zeta_0,t) = \int_0^t P(\zeta - \zeta_1,t - \xi)Q(\zeta_1 - \zeta_0,\xi)d\xi \ t \tag{15}$$

Thus for the case of *diffusive* dispersion

$$Q(\zeta - \zeta_0,t) = (\zeta - \zeta_o)\ (4\pi D_\zeta t^3)^{-1/2} \exp[-(\zeta - \zeta_0 - \overline{V}_\zeta t)^2/4D_\zeta t] \ . \tag{16}$$

Then it is easily shown that for such a process

$$D_\zeta = \int_{\infty}^{\infty} P(\zeta - \zeta_0,t)\ (\sigma_\zeta^2/2t)d\zeta = \int_0^\infty Q(\zeta - \zeta_o,t)\ (\sigma_\zeta^2/2t)dt \tag{17}$$

Equation (17) provides a very convenient way of calculating D_ζ in our Monte Carlo simulations. For example by fixing the location ζ one can measure the first passage time distribution from which D_ζ can be easily determined. One can also fix time and determine $P(\zeta - \zeta_o,t)$ and calculate D_ζ again. If the two D_ζ turn out to be equal then one may conclude that the dispersion process is diffusive.

To investigate dispersion, we inject a tracer particle into the network at a randomly chosen node in the plane x = o and monitor its motion through the network. We neglect the velocity distribution within each bond (the effect of this distribution will be discussed later). We assume that a particle arriving at a node leaves into one of the attached bonds with a conditional probability proportional to the fraction of the flow rate departing from the node through that bond. Thus, if q_i is volumetric flow rate in a bond i that transports fluid from the node, the conditional probability that the tracer will leave through bond i is $q_i / \sum_j q_j$, where $\sum_j q_j$ is the total volumetric flow rate leaving (and hence entering to) the node. Such an assumption is equivalent to complete mixing of converging streams at nodes. If there is only partial mixing at nodes, then the tracer has a memory. From the work of Montroll[23], one knows that so long as the tracers have a finite memory, the asymptotic regime in the partial mixing case will, qualitatively, be similar to that of a complete mixing case. Thus we do not consider the partial mixing case in this paper. To each bond of the network, we assign a random radius R, chosen from a Rayleigh distribution:

$$f(R) = 2Re^{-R^2} \tag{18}$$

This distribution mimics the pore size distribution of a real porous medium determined by several investigators, Thomas *el al.*[24] among others. We then assume that the bond conductance is given by

$$g = R^4 \tag{19}$$

as in Hagen-Poiseuille flow. The details of the Monte Carlo simulations procedure are given elsewhere.[25,26]

RESULTS FOR DISPERSION IN SINGLE-PHASE FLOW

We simulated dispersion in square and cubic networks. In Figure 1 we present the network size dependence of the dispersion coefficients. Based on these results and because of considerable simulation costs, we performed most of our calculations in 50 x 50 square networks and 10 x 10 x 10 cubic networks. For each study 15 different realizations of the networks were made. About 1500 tracers were injected into a square network before the averaged quantities in Equation (13) converged to some constant value. This number was about 800 tracers for the cubic network. For the case of dispersion in a percolating network near its percolation threshold (see below), the number of network realizations was considerably larger. For example, up to 50 different realizations of the square network were used. The number of injected tracers was also much larger. For example, we needed to inject about 3000 tracers into the square network before the quantities $\sigma_\xi^2/2t$ converged to a constant value. Then another averaging was made over the results of all realizations.

In Figures 2, 3, and 4 we present the longitudinal tracer particle distribution in the square network, the transverse tracer particle distribution in the cubic network, and the longitudinal first passage time distribution in the cubic network. Chi-squared test of goodness of fit showed that all these distributions fit their theoretical equations with more than 90% confidence. We thus conclude that the dispersion process simulated here is diffusive and therefore the dispersion coefficients can be identified as effective diffusion coefficients. The longitudinal dispersional coefficient was found to be about one order of magnitude larger than the transverse one. This lies comfortably in the range that this value has been observed experimentally, see for example Blackwell.[27]

In our Monte Carlo simulations if the overall pressure drop is changed, the pressure difference between the extremities of each bond of the network changes in proportion, and so also do the velocities and flow rates. Thus the first passage times of the tracer particles change also in proportion. Consequently the dispersion coefficients are proportional to the mean velocity in our simulations. Experimentally[27] it has been observed that

$$D_L \sim \overline{V}^{\beta_L} \tag{20}$$

$$D_T \sim \overline{V}^{\beta_T} \tag{21}$$

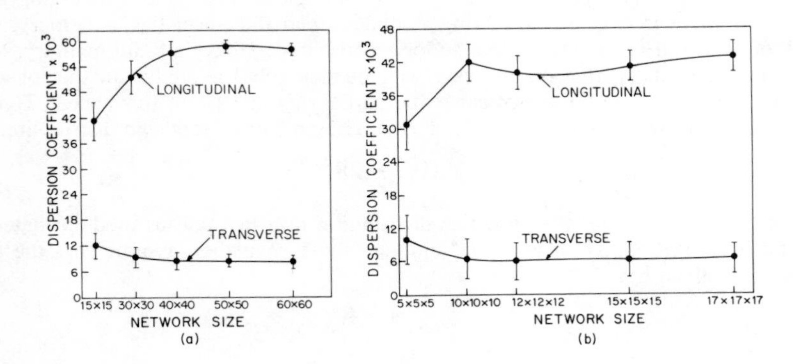

Fig. 1. Effect of network size on dispersion coefficients: (a) square network; (b) cubic network. Error bars denote one standard deviation.

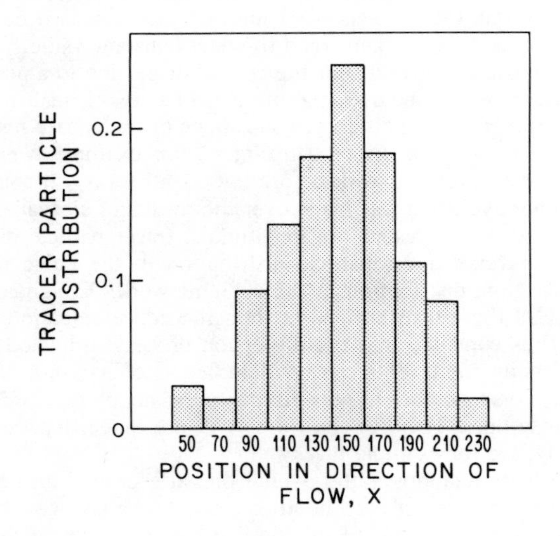

Fig. 2. Longitudinal tracer particle distribution at fixed time in the square network.

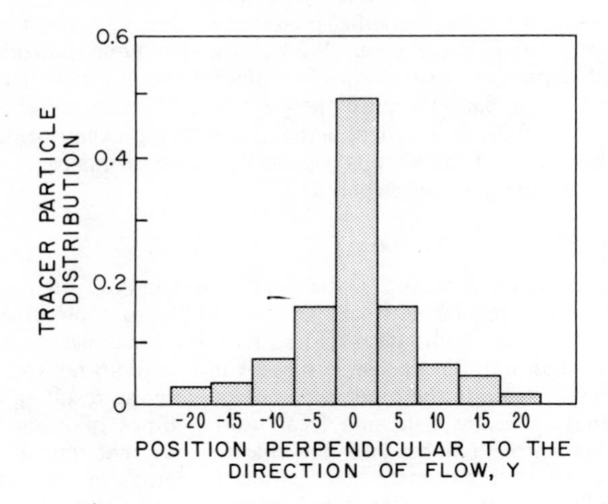

Fig. 3. Transverse tracer particle distribution in the cubic network.

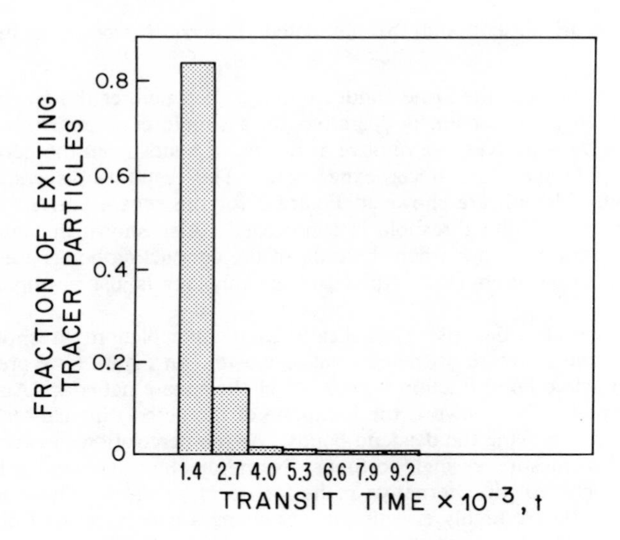

Fig. 4. First passage time distribution in the cubic network.

with $1.1 < \beta_L < 1.3$ and $\beta_T \cong 1$. Thus there is very good agreement between the predictions of our simulations of D_T and the experimental predictions. The agreement between our D_L and that of experimental data is not as good. We have modified our simulation procedure and have incorporated some more realistic mechanisms for the events within bonds and the redistribution of the tracer particles once they arrive at a node. We obtained $\beta_L \cong 1.27 \pm 0.1$ and $\beta_T \cong 0.95 \pm 0.07$, in excellent agreement with the experimental predictions. However, because this aspect of the work is beyond the scope of this paper, we will not discuss it any further. Details are given elsewhere[26].

DISPERSION IN A PERCOLATION NETWORK

The dispersion investigated above is caused by the random geometry of the pore space — the dynamical mechanism — in a regular network. One would like to know what effect, if any, a random topology has on the results discussed so far. For example, disperson is more complicated when more than one fluid phase is present in a random network. Each phase by its mere presence denies space to the other and consequently the resulting random topology alters the streamlines, pressure field and local waiting times of other phases. Dispersion in multiphase flow through a random network (porous medium) is a very important technological problem as it has many applications in petroleum and chemical industries. These phases are distributed in a complicated manner in the porous medium. A very important advancement is the percolation theory of two phase fluid distribution proposed by Heiba, et al.[21], which describes, for the first time, qualitatively as well as quantitatively the manner in which two immiscible fluids distribute themselves in a porous medium. We have investigated dispersion in multiphase flow in random networks results of which are presented elsewhere[28]. For our present purpose it suffices to investigate dispersion in a random walk whose bonds have a simple binary-percolation-like distribution

$$f(g) = (1 - p)\, \partial_t(g) + p\delta(g - g_o) \tag{22}$$

Thus geometrical variations are absent and the simulated dispersion process is purely kinematic.

If all bonds are conducting with the same conductivity g_o, then neither the kinematic nor the dynamical mechanism of dispersion is operative (in a simple cubic network in d dimensions) and thus $D_L = D_T = 0$. Next we remove a fraction of bonds from the network (assign $g\_ = 0$ to them) and repeat the tracer experiment. The resulting dispersivities $\alpha_\zeta = D_\zeta/\overline{V}$ which have unit of length are shown in Figure 5 for the square lattice. They increase dramatically as the percolation threshold is approached. Also shown in Figure 5 are the dispersivities of the same network when the radii of the conducting bonds are randomly distributed according to equation (18). Obviously the topology is just as important as the geometry.

The reason that the dispersivities rise dramatically as the percolation threshold is approached lies in the backbone structure of the percolating cluster. In Figure 6 we present the conducting bonds at an active bond fraction $x = 0.575$ of the square network. Among them many are deadend bonds. Also shown is the backbone of the percolating cluster at $x = 0.575$ which is obtained by removing the deadend bonds. As the percolation threshold is approached, the backbone becomes increasingly tortuous. Numerous blobs containing large number of loops appear which provide alternative paths for tracer particles. These blobs which have self-similar structure are highly effective in dispersing a concentration front of tracer particles. Moreover, the tortuosity of the backbone, which is defined as the ratio of the backbone fraction to the conductivity of the network, increases without bound as the percolation threshold is approached[29]. This increase in tortuosity conspires with the blobs and their associated loops to broaden the distribution of first passage times: the dispersivities grow larger as result. Most bonds in longitudinal direction are directed forward because

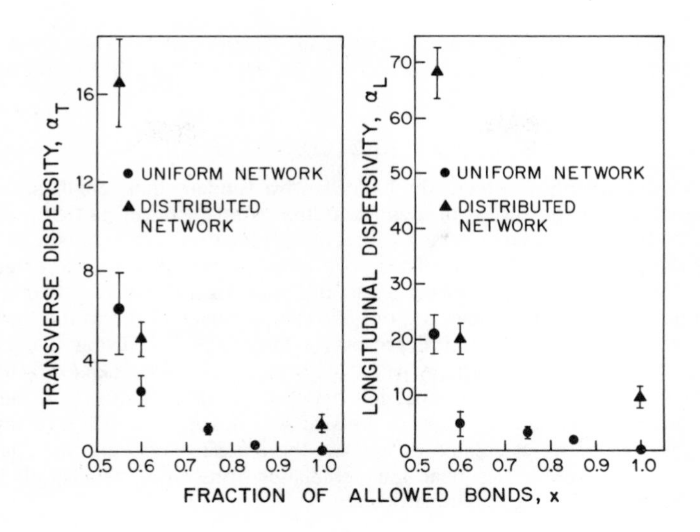

Fig. 5. Dispersion in a percolation square network. Error bars denote one standard revision.

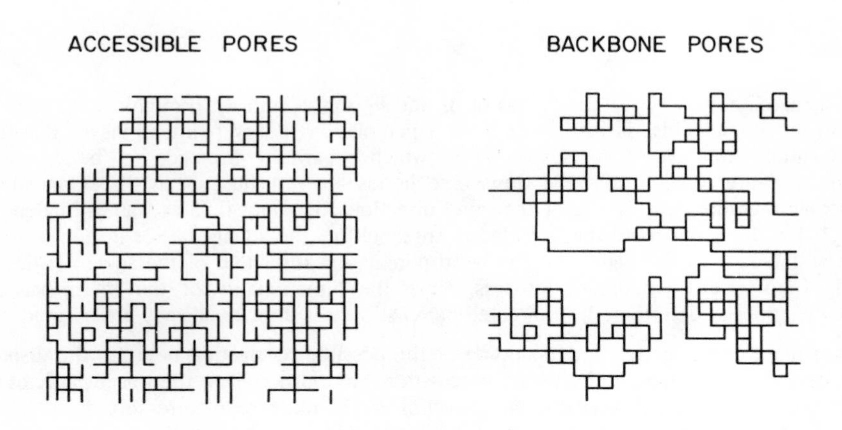

Fig. 6. A square network with a fraction 0.425 of bonds removed at random.

they are in the direction of macroscopic pressure drop. Thus a concentration front of tracer particles can spread more easily in this direction and therefore longitudinal dispersivity rises faster than the transverse one.

DISPERSION AND DIRECTED PERCOLATION

The rise of dispersivities, which are basically two fundamental length scales which measure the the extent of the spread of a concentration front of tracer particles, resembles the divergence of the correlation lengths in directed percolation (for a review of directed percolation see the paper by Redner[30]). These two lengths, too, measure the extent of the spread of a random walker who performs a random walk in a directed network. In our Monte Carlo simulations tracer particles are not allowed to move *relative* to the local flow in network bonds: the effect of the physical process of longitudinal diffusion is excluded by our simulation. Thus the tracer particle paths are *dynamically determined* directed paths. Thus one might hope that there is a connection between the dispersivities and the correlation lengths of directed percolation. Using finite-size scaling approach (for a review of this method see Nightingale[31]) we calculated d_L and d_T, the critical exponents that characterize the divergence of the dispersivities near the percolation threshold (details will be given elsewhere)[32]. The results for the square lattice are

$$d_L \cong 1.65 \tag{23}$$

$$d_T \cong 1.09 \tag{24}$$

These values are not incompatible with the values of ν_L and ν_T that characterize the divergence of longitudinal and transverse correlation lengths. Their values are given by Essam and De'Bell[33]

$$\nu_L \cong 1.733 \tag{25}$$

$$\nu_T \cong 1.098 \tag{26}$$

This close agreement between the values of d_L and ν_L and d_T and ν_T, though intuitively and physically understandable, is intriguing from a percolation theory point of view. Because the percolation threshold of the square lattice which is partially directed, i.e. bonds along only one axis are directed and the transverse bonds are not (just as in disperson when macroscopic pressure drop is one of the bond directions),is about 0.55 as was estimated by Redner.[34] On the other hand the percolation threshold of the square lattice in our simulations is still 1/2, the exact value for the bond percolation threshold of the square lattice is isotropic (undirected) percolation process, since the directionality of the tracer particle paths is dynamically determined; in directed percolation this directionality is preassigned.

In order to get a better understanding of the possible connection between the dispersion process simulated here and directed percolation, we did not allow the tracer particles to move in bonds that carry flow against the direction of the macroscopic pressure drop. Thus the square network became a partially directed network in the sense of directed percolation. The resulting dispersivities are presented in Figure 7. There are several differences between these results and those shown in Figure 5. The percolation threshold as shown in Figure 7 is closed to 0.55, the percolation threshold of the partially directed square lattice. This is because as the point $x = 0.55$ is approached an increasing fraction of bonds carry flow in the opposite direction of macroscopic pressure drop. Because the tracer particles are forbidden to move against the macroscopic pressure drop, they have to walk on a directed backbone of the infinite percolation cluster. The backbone of the infinite percolation cluster is the current-carrying part of it. If a tracer particle cannot find a directed path to move in the direction of macroscopic flow, then it becomes "trapped."

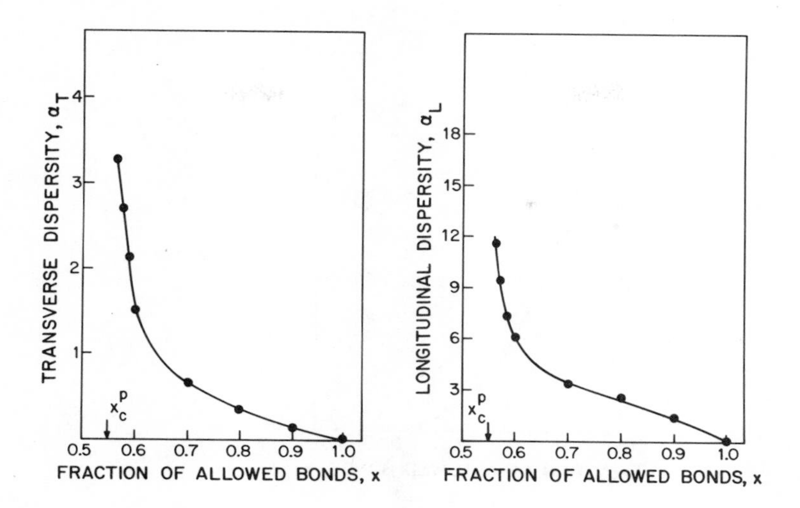

Fig. 7. Dispersion in partially directed square network. x_c^p denotes the percolation threshold of the network.

Figure 8 shows the fraction of tracer particles that get trapped as the percolation threshold is approached. At x = 0.55 all of the tracer particles are trapped and thus the process is stopped. The dispersivities that are shown in Figure 7 are also smaller than those of Figure 5. We attribute this to two reasons. The first one is the fact that a directed backbone is less tortuous than an isotropic (undirected) backbone. Consequently the first passage time distribution of the tracer particles is not as broad as in the isotropic case which results in smaller dispersivities. The second reason is that if the tracer particles are not allowed to move in bonds that carry flow against the direction of the macroscopic pressure drop, then they will not have access to many big blobs, the highly effective dispersers, that are connected to the rest of the backbone by bonds that carry flow in the opposite direction of the macroscopic pressure drop. Thus the first passage time distribution of the tracer particles is much narrower than in the isotropic case. This computer simulation of dispersion in a directed network manifests that an accurate structure of the backbone of the percolation infinite cluster is the one in which blobs with many complicated loops and with a self-similar structure are connected to each other by a few bonds which are called links (see Fig. 6). This structure of the backbone near the percolation threshold was first proposed by Stanley[35] and was further studied by Coniglio[37]. Our computer simulations of dispersion seems also to rule out other structures of the backbone. One of them was proposed by Skal and Shklovskii[37] and independently by de Gennes[38]. In their model the backbone is made of nodes which are connected by long chains of several bonds. Such a structure cannot give rise to the dramatic rise in dispersivities as the percolation threshold is approached. Another model was suggested by Gefen *et al.* [39] in which only blobs are present and the links ignored. This model which is the opposite extreme of the Skal-Shklovskii-de Gennes model cannot be accurate because the dispersivities in the directed case must be the same as those in the undirected case according to this model. The rise in dispersivities as the percolation threshold is approached has been observed experimentally.

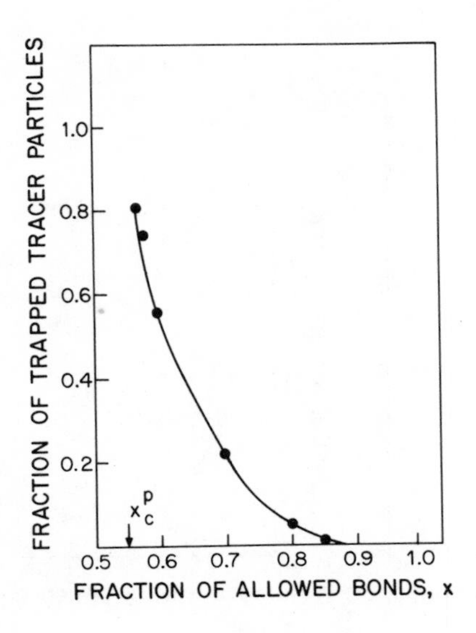

Fig. 8. Fraction of trapped tracer particles in dispersion in a partially directed square network. x_c^p denotes the percolation threshold of the network.

ACKNOWLEDGEMENTS

This work formed part of the author's thesis for the degree of Ph.D. at the University of Minnesota. The author would like to thank Professors H.T. Davis and L.E. Scriven for their constant encouragement and stimulating discussions. This work was supported in part by the U.S. Department of Energy and the University of Minnesota Computer Center.

REFERENCES

1. F.C. Fowler and G.G. Brown, Trans. AIChE, **39**. 491 (1943).
2. A. Griffiths, Proc. Phys. Soc. Lond., **23**. 190 (1911).
3. G.I. Taylor, Proc. Roy.Soc. Lond., **219A**, 186 (1953).
4. R. Aris, Proc. Roy. Soc. Lond., **235A**, 57 (1956).
5. J.J. Van Deemter, J.J. Broeder and H.A. Lauwerier, Appl. Sci. Res., **5A**, 374 (1956).
6. C. Van Den Broeck, Physica, **112A**, 343 (1982).
7. K.K. Mohanty, Ph.D. Thesis, The University of Minnesota, Minneapolis (1981).
8. M.H. Cohen and C. Lin, Lecture Notes in Physics, **154**, 74, (Springer-Verleg, 1982).
9. E.W. Montroll and G.H. Weiss, J. Math. Phys., **6**, 167 (1965).
10. H. Scher and M. Lax, Phys. Rev. **B7**, 4491; **B7**, 4502 (1973).
11. H. Scher and E.W. Montroll, Phys. Rev. **B12**, 2445 (1975).
12. M.N. Barber and B.W. Ninham, Random and Restricted Walks, Theory and Applications (Gordon and Breach, 1970).
13. G.H. Weiss and R.J. Rubin, Adv. Chem. Phys.,**52**, 363 (1982).
14. M.F. Shlesinger, J. Klafter and Y.M. Wong, J. Stat. Phys., **27**, 499 (1982).
15. S. Kirkpatrick, Rev. Mod. Phys., **45**, 574 (1973).
16. M. Sahimi, to be submitted to Journal of Fluid Mechanics.
17. D. Stauffer, Phys. Rep., **54**, 1 (1979).
18. J.W. Essam, Rep. Prog. Phys., **43**, 833 (1980).

19 M.Sahimi, in Proceedings of the Workshop on Mathematics and Physics of disordered Media, Lecture Notes in Mathematics, (Springer-Verlag, 1983).
20. R.G.Larson, L.E. Scriven and H.T. Davis, Chem. Engr. Sci., **36**, 57; **36**, 75 (1981).
21. A.A. Heiba, M. Sahimi, L.E. Scriven and H.T. Davis, Society of Petroleum Engineers preprint 11015 (1982).
22. E.W. Montroll and B.J. West, in Fluctuation Phenomena, edited by E.W. Montroll and J.L. Lebowitz, **61**, (North-Holland, 1979).
23. E.W. Montroll, J. Chem. Phys., **18**, 734 (1950).
24. G.H. Thomas, G.R. Countryman and I. Fatt, Soc. Pet. Eng. J., **3**, 189 (1963).
25. M. Sahimi, H.T. Davis and L.E. Scriven, Chem. Engr. Commu., in press (1983).
26. M. Sahimi, B.D.Hughes, A.A. Heiba, H.T. Davis and L.E. Scriven, submitted to Chem. Eng. Sci.
27. R.J. Blackwell, Soc. Pet. Eng. J., **2**. (1962).
28. M. Sahimi, A.A. Heiba, B.D. Hughes, L.E. Scriven and H.T. Davis, submitted to Chem. Eng. Sci.
29. R.G. Larson and H.T. Davis, J. Phys. C., **15**, 2327 (1982).
30. S. Redner, in Annals of Israel Physical Soc., A. Hilgen, Bristol, (1982).
31. P. Nightengale, J. Appl. Phys., **53**, 7927 (1982).
32. M. Sahimi, L.E. Scriven and H.T. Davis, to be submitted to J. Phys. A.
33. J.W. Essam and K. De'Bell, J. Phys. A., **181**, L459 (1981).
34. S. Redner, Phys. Rev., **B25**, 5656 (1982).
35. H.E. Stanley, J. Phys. A., **10**, L211 (1977).
36. A. Coniglio, J. Phys. A., **15**, 3829, (1982).
37. A.S. Skal and B.I. Shklovskii, Fiz. Tekh. Poluprov., **8**, 1586, (1974), (Sov. Phys.-Semicond., **8**, 1029 (1975)).
38. P.G. deGennes, J. Physique-Lett., **37**, L1 (1976).
39. Y. Gefen, A. Aharony, B.B. Mandlebrot, and S. Kirkpatrick, Phys. Rev. Lett., **47**, 1771 (1981).

A SELF-CONSISTENT MODE-COUPLING THEORY FOR THE ANDERSON LOCALIZATION

Shaul Mukamel
Department of Chemistry
University of Rochester
Rochester, New York 14627

ABSTRACT

Self consistent equations for the density response and the electrical conductivity of a fluid in a random potential are derived using projection operator techniques. The equations depend on the static density response function $\chi_\rho(q)$, the strength of the random potential U_k and the space dimensionality d, and are formally similar for classical and quantum fluids. The present formalism allows us to analyze directly the transport properties of the system without going into the fine details of its structure (i.e. the exact nature of the wavefunction in the quantum case).

I. INTRODUCTION

The motion of an electron gas in a random potential has drawn considerable theoretical attention in recent years since the pioneering works of Anderson and Mott.[1-5] Many theoretical efforts have been focussed on trying to understand the nature of the wavefunction (under what conditions is it localized or extended). The relevant experimental quantities are however the transport coefficients of the system such as the density response function or the electrical conductivity. One of the most intriguing and controversial aspects of the problem is the existence of a "mobility edge" as a function of the randomness and the spatial dimensionality, i.e. whether the DC conductivity undergoes a sharp phase transition.

In the present paper the problem will be attacked using a mode coupling approach, as is done in nonlinear hydrodynamics.[6-8] Our basic goal will be to derive a set of equations whose solution will directly lead to the quantities of interest, i.e. the electrical conductivity and the density response function, without having to go through a detailed analysis regarding the nature of the wavefunction. This paper is based on the mode-coupling formulation of Kawasaki[6] and Kadanoff and Swift[7] and we use the exact static density response function $\chi_\rho(p)$ as an input to the mode coupling calculations. This enables us to use various microscopic diagrammatic techniques to study the static properties of the system, or even treat them phenomenologically, and then the present equations may be used to calculate the dynamical response functions of interest. The present equations enable us to discuss in a convenient and unified way effects of randomness, energy, temperature and dimensionality. They also allow for a simple comparison of classical and quantum fluids.[9,10] In Section II we introduce the basic formal definitions to be used in this paper and in Section II we define the transport properties of interest and relate all of them to the kubo transformed density response function, $\hat{\chi}_{\rho\rho}(q,\omega)$. In Section IV we define the self consistent equations for the density and current response functions. Section V presents another set of equations derived using a different time-ordering scheme and finally, in Section VI we discuss our results and analyze their limiting behavior.

094-243X/84/1090205-20 $3.00

II. CORRELATION FUNCTIONS, RESPONSE FUNCTIONS AND KUBO TRANSFORMS

In this section we shall introduce the basic definitions and notation to be used later.[11] We consider a fluid consisting of N identical particles in a volume Ω. The coordinates and momenta of the j particle will be denoted r_j, p_j respectively and the set $\{r_j, p_j\}$ form a 6N dimensional phase space. Let A_j, B_j etc. be a set of single particle operators which depend on the j particle only. We then define in q space

$$A_q \equiv \frac{1}{2} \sum_j [\exp(iq \cdot r_j) A_j + A_j \exp(iq \cdot r_j)]. \tag{1}$$

Similarly we define B_q, C_q etc. We note that since A_j are Hermitian, then

$$A_q^\dagger = A_{-q} \tag{1a}$$

where $A_q^\dagger$ is the Hermitian conjugate of A_q.

We define the <u>correlation function</u> of A and B, $C_{AB}(q,\tau)$ as follows:

$$C_{AB}(q,\tau) \equiv \langle A_q(\tau) | B_q(0) \rangle \equiv \mathrm{Tr}(A_{-q}(\tau) B_q(0) \rho_{eq}), \tag{2}$$

where

$$A_q(\tau) \equiv \exp(iL\tau) A_q \equiv \exp(iH\tau) A_q \exp(-iH\tau), \tag{3}$$

and

$$\rho_{eq} \equiv \exp(-H/kT)/\mathrm{Tr} \exp(-H/kT). \tag{4}$$

Here H is the Hamiltonian of the system, L is the Liouville operator $L \equiv [H, \ ]$ and ρ_{eq} is the Canonical distribution function at temperature T. The <u>linear response function</u> (susceptibility) which measures the expectation value of A_q given an external field that couples B_q is[11]

$$\hbar \chi_{AB}(q,\tau) \equiv \langle [A_q(\tau), B_q] \rangle \equiv \langle A_q(\tau) | B_q(0) \rangle - \langle B_q(0) | A_q(\tau) \rangle. \tag{5}$$

The <u>Kubo transform</u> of A and B is defined as:

$$\begin{aligned} \hat{\chi}_{AB}(q,\tau) &\equiv \langle\langle A_q(\tau) | B_q \rangle\rangle \equiv \int_0^{1/kT} d\lambda \langle A_q(\tau - i\lambda) | B_q \rangle \\ &\equiv \int_0^{1/kT} d\lambda \ \mathrm{Tr}[\exp(\lambda H) A(\tau) \exp(-\lambda H) B \rho_{eq}]. \end{aligned} \tag{6}$$

In ω space we define the one sided (advanced) Fourier transform:

$$C_{AB}(q,\omega) = -i \int_0^\infty d\tau \exp(i\omega\tau) C_{AB}(q,\tau). \tag{7a}$$

and similarly for $\chi_{AB}(q,\omega)$ and $\hat{\chi}_{AB}(q,\omega)$. The inversion of (7a) is:

$$C_{AB}(q,\tau) = \frac{i}{2\pi} \int_{-\infty}^{\infty} d\omega \exp(-i\omega\tau) C_{AB}(q,\omega). \tag{7b}$$

Denoting the eigenstates of our Hamiltonian H (without the external field) by α and β with eigenvalues E_α and E_β and using Eqs. (2), (5), (6) and (7a) we have:

$$C_{AB}(q,\omega) = \sum_{\alpha,\beta} P(\alpha) \frac{(A_{-q})_{\alpha\beta}(B_q)_{\beta\alpha}}{\omega-\omega_{\beta\alpha}+i\varepsilon} \tag{8a}$$

$$\chi_{AB}(q,\omega) = \frac{1}{\hbar} \sum_{\alpha,\beta} [P(\alpha) - P(\beta)] \frac{(A_{-q})_{\alpha\beta}(B_q)_{\beta\alpha}}{\omega-\omega_{\beta\alpha}+i\varepsilon} , \tag{8b}$$

and

$$\hat{\chi}_{AB}(q,\omega) = \sum_{\alpha,\beta} \frac{P(\alpha)-P(\beta)}{\hbar\omega_{\beta\alpha}} \cdot \frac{(A_{-q})_{\alpha\beta}(B_q)_{\beta\alpha}}{\omega-\omega_{\beta\alpha}+i\varepsilon} , \tag{8c}$$

where $P(\alpha) \equiv \langle\alpha|\rho_{eq}|\alpha\rangle$, and

$$\omega_{\beta\alpha} = E_\beta - E_\alpha . \tag{8d}$$

Eqs. (8) are very useful for demonstrating the formal analytical properties of C, χ and $\hat{\chi}$, as well as for establishing various relations between them (see Eqs. (9)-(13)). However, the usage of Eqs. (8) for the actual evaluation of the correlation functions is quite limited since they contain the complicated many-body eigenstates α,β etc. of the total Hamiltonian, which are rarely known. The common methods for the evaluation of Eqs. (8) are diagrammatic techniques[12] or reduced equations of motion.[13-15] We further note that the three types of correlation functions defined here are not independent and any one of them determines the others,[11-16] however, for certain types of manipulations it is more convenient to adopt one of these definitions. As is clearly seen from Eqs. (8) we have:

$$\hbar\chi_{AB}(q,\omega) = C_{AB}(q,\omega) + C^*_{AB}(-q-\omega), \tag{9}$$

and

$$\hat{\chi}_{AB}(q,\omega) = \frac{1}{\omega} [\chi_{AB}(q,\omega) - \chi_{AB}(q,0)]. \tag{10}$$

The correlation functions may be decomposed into their real and imaginary parts as follows:

$$\Psi \equiv \Psi' - i\,\Psi'' \qquad \Psi = C, \chi, \hat{\chi} \text{ etc.} \tag{11}$$

Upon taking the immaginary part of Eq. (9) we get the fluctuation dissipation theorm:

$$\hbar\chi''_{AB}(q,\omega) = [1-\exp(-\hbar\omega/kT)]C''_{AB}(q,\omega). \tag{12}$$

Another useful relation which may be proved from Eqs. (8) is the Kramers-Kroning relation between the real and the imaginary parts of the correlation functions, i.e.

$$\Psi'(q,\omega) - \frac{1}{\pi} \mathrm{PP} \int d\omega' \frac{\Psi''(\omega')}{\omega'-\omega} \qquad \Psi = C, \chi, \hat{\chi}\,. \tag{13}$$

III. DENSITY AND CURRENT RESPONSE FUNCTIONS OF A FLUID

We consider a fluid of noninteracting particles which are moving in a random external potential. The Hamiltonian is assumed to have the form:

$$H = \sum_j [p_j^2/2m + U(r_j)] \tag{14}$$

Here U is a random potential defined by its correlation functions:

$$\overline{U(r)} = 0 \tag{14a}$$

$$\overline{U(r)U(r')} \equiv f(r-r') \tag{14b}$$

etc.

We shall now introduce the density and current operators ρ_q and J_q respectively defined as follows:

$$\rho_q = \sum_j \exp(iq\cdot r_j), \tag{15}$$

$$\dot{\rho}_q = \frac{i}{\hbar}[H,\rho_q] \equiv \frac{i}{\hbar} q\cdot J_q \quad . \tag{16a}$$

Using Eqs. (14) - (16) we have:

$$J_q = \frac{1}{2m}\sum_j [p_j \exp(iq\cdot r_j) + \exp(iq\cdot r_j)p_j], \tag{16b}$$

where m is the mass of our particles. We shall be interested in the electrical conductivity $\sigma(q,\omega)$ which is given by the Kubo transformed current response function, i.e.[11]

$$\sigma(q,\omega) = \frac{ie^2}{\hbar\Omega}\hat{\chi}_{JJ}(q,\omega) = \frac{ie^2}{\hbar\Omega\omega}[\chi_{JJ}(q,\omega) - \chi_{JJ}(q,0)] . \qquad (17)$$

Here e is the electrical charge of the particles and Ω is the volume of the system.

An important property of Kubo transforms $\hat{\chi}$ is that like the ordinary correlation functions C, they may be viewed as <u>scalar products</u> of two operators in Liouville space, i.e. the quantity $\langle\langle A_q|B_q\rangle\rangle$ defined in Eq. (6) (like $\langle A_q|B_q\rangle$ defined in Eq. (2)) satisfies all the requirements of a scalar product, namely

$$\text{(a)} \quad \langle\langle A_q|B_q\rangle\rangle^* = \langle\langle B_q|A_q\rangle\rangle, \qquad (18a)$$

$$\text{(b)} \quad \text{if } |B_q\rangle\rangle = c_1|B_q\rangle\rangle + c_2|\hat{B}_q\rangle\rangle \text{ then,}$$

$$\langle\langle A_q|B_q\rangle\rangle = c_1\langle\langle A_q|B_q\rangle\rangle + c_2\langle\langle A_q|\hat{B}_q\rangle\rangle \qquad (18b)$$

and

$$\text{(c)} \quad \langle\langle A_q|A_q\rangle\rangle \geqslant 0 \qquad (18c)$$

and the equality sign holds only when $|A_q\rangle\rangle \equiv 0$.

We note in passing that the randomness of the potential in our problem arises from other degrees of freedom and the trace operation in our scalar products involves therefore the averaging over the random potential as well.

Making use of Eqs. (18) we may define a Mori projection operator

$$\hat{P} \equiv |\rho_q\rangle\rangle\langle\langle\rho_q|\chi_\rho^{-1}(q) \quad , \qquad (19a)$$

and the complementary projection

$$\hat{Q} = 1 - \hat{P} . \qquad (19b)$$

Here:

$$\chi_\rho(q) \equiv \langle\langle\rho_q|\rho_q\rangle\rangle \equiv \sum_{\alpha,\beta} \frac{P(\alpha)-P(\beta)}{\hbar\omega_{\beta\alpha}} |(\rho_q)_{\alpha\beta}|^2 \qquad (20)$$

The significance of $\hat{P}$ is that when it operates on some arbitrary operator $|C\rangle\rangle$ is projects it onto $|\rho_q\rangle\rangle$, i.e.

$$\hat{P}|C\rangle\rangle \equiv |\rho_q\rangle\rangle\langle\langle\rho_q|C\rangle\rangle\cdot\chi_\rho^{-1}(q) \qquad (21)$$

$\chi_\rho(q)^{-1}$ is a normalization factor introduced to make sure that $\hat{P}^2=\hat{P}$.
Another quantity that we introduce is the tetradic T matrix :[17,18]

$$\Upsilon(\omega) \equiv L + L \frac{1}{\omega - L + i\varepsilon} L, \tag{22}$$

which may be alternatively written as

$$\Upsilon(\omega) = \omega^2\left[\frac{1}{\omega - L + i\varepsilon} - \frac{1}{\omega + i\varepsilon}\right] \quad . \tag{23}$$

The equivalence of Eqs. (22) and (23) may be easily shown as follows. (hereafter we shall omit the $+i\varepsilon$ factor and throughout this paper ω should be understood as $\omega+i\varepsilon$).

$$\begin{aligned}\Upsilon(\omega) &\equiv L+L \frac{1}{\omega-L} L = L+[(L-\omega)+\omega] \frac{1}{\omega-L} [(L-\omega)+\omega] = L-(L-\omega)+\omega^2 \frac{1}{\omega-L} - 2\omega \\ &= \omega^2\left(\frac{1}{\omega-L} - \frac{1}{\omega}\right)\end{aligned} \tag{24}$$

Upon multiplying Eq. (23) by a projection operator $\hat{P}$ from right and left we get:

$$\hat{P}\Upsilon\hat{P} = \omega^2\left[\hat{P}\frac{1}{\omega-L} \hat{P} - \frac{1}{\omega} \hat{P}\right] \tag{25}$$

when using the specific projection $\hat{P}$ (Eq. (19)) in Eq. (25) we get:

$$\hat{\chi}_{JJ} = \frac{\omega^2}{q^2} \left[\hat{\chi}_{\rho\rho} - \frac{1}{\omega} \chi_\rho(q)\right], \tag{26}$$

Eqs. (26) and (17) immediately result in:

$$\sigma(q,\omega) = \frac{ie^2\omega^2}{\hbar\Omega q^2} \left[\hat{\chi}_{\rho\rho}(q,\omega) - \frac{1}{\omega} \chi_\rho(q)\right]. \tag{27a}$$

We have now accomplished our goal to express $\sigma(q,\omega)$ in terms of $\hat{\chi}_{\rho\rho}(q,\omega)$. Other quantities of interest are the density response function $\chi_{\rho\rho}(q,\omega)$, the dielectric response function $\varepsilon(q,\omega)$ and the dynamic structure factor $S(q,\omega)$. They may all be expressed in terms of $\hat{\chi}_{\rho\rho}$,[19] i.e.

$$\chi_{\rho\rho}(q,\omega) = \omega \hat{\chi}_{\rho\rho}(q,\omega) + \chi_\rho(q) \tag{27b}$$

$$\frac{1}{\varepsilon(q,\omega)} = 1 + V_q\chi_{\rho\rho} = 1 + \frac{4\pi e^2}{q^2}[\omega\ \hat{\chi}_{\rho\rho}(q,\omega) + \chi_\rho(q)] \tag{27c}$$

$$S(q,\omega) \equiv C''_{\rho\rho}(q,\omega) = \hbar[1-\exp(\hbar\omega/kT)]^{-1}\chi''_{\rho\rho}\ . \tag{27d}$$

Here V_q is the q component of the interparticle interaction and is equal to $4\pi e^2/q^2$ for a coulomb force.[20] In the next section we shall develop a theory for $\hat{\chi}_{\rho\rho}$ which together with Eqs. (27) will yield $\sigma,\chi_{\rho\rho},\varepsilon$ and S. We should note that $\chi_\rho(q)\equiv\chi_{\rho\rho}(q,\omega=0)$ but is not necessarily equal to the low frequency limit of χ_{pp} (see Eqs. (83b) and (87b)).

IV. THE SELF CONSISTENT EQUATIONS FOR THE DENSITY AND CURRENT RESPONSE FUNCTIONS

In this section we shall derive the self consistent equations for the transport coefficients of our fluid. This will be done in two stages: we first consider the Kubo transformed density and current response functions $\chi_{\rho\rho}$ and χ_{JJ} and introduce reduced equations of motion (REM) which express them in terms of appropriate relaxation kernels R_ρ and R_J respectively (Eqs. (31) and (36)). We then derive a closed set of self consistent equations for these kernels in the long wavelength ($q\to 0$) limit. All the relevant transport properties ($\sigma,\chi_{\rho\rho},\varepsilon$ and S) may then be expressed in terms of R_ρ. Two comments of clarification should now be made:

1) The reason that we choose $\hat{\chi}_{\rho\rho}$ and $\hat{\chi}_{JJ}$ (and not $\chi_{\rho\rho}$ and χ_{JJ}) as our variables is that, as was shown in the previous section, the former (and not the latter) are scalar products in Liouville space and enable us to use standard projection operator techniques.

2) The usage of the kernels R_ρ and R_J instead of the response functions $\chi_{\rho\rho}$ and χ_{JJ} provides a partial ressumation of the appropriate series, i.e. a low order approximation to the kernels in some expansion parameter automatically yields an approximation for the response functions which contain certain types of terms to infinite order. The standard REM which $\hat{\chi}_{\rho\rho}$ satisfies is [13-15]

$$\frac{d\hat{\chi}_{\rho\rho}(q,t)}{dt} = -q^2 \int_0^t d\tau R_\rho(q,t-\tau)\hat{\chi}_{\rho\rho}(q,\tau) \tag{28}$$

where

$$R_\rho(q,\omega) = \langle\langle\rho_q|L\hat{Q}\frac{1}{\omega-LQ}L|\rho_q\rangle\rangle/[\chi_\rho(q)q^2] = \langle\langle J_q|\hat{Q}\frac{1}{\omega-L\hat{Q}}|J_q\rangle\rangle/\chi_\rho(q) \tag{29}$$

We shall denote the REM of the type (28), which involve a convolution in time as corresponding to the COP (chronological time ordering prescription). In Section V we shall consider another type of REM based on a different time ordering scheme.[16,21] Eq. (28) should be solved with the initial condition

$$\hat{\chi}_{\rho\rho}(q,t=0) \equiv \chi_\rho(q) \tag{30}$$

The solution (in ω space) is:

$$\hat{\chi}_{\rho\rho}(q,\omega) = \frac{1}{\omega - q^2 R_\rho(q,\omega)} \chi_\rho(q), \tag{31}$$

In a similar way we may define another projection operator onto the current operator $|J_q>>$ (Eq. (16)):

$$\tilde{P} \equiv |J_q>><<J_q|\chi_J^{-1}(q), \tag{32a}$$

and its complementary projection

$$\tilde{Q} \equiv 1 - \tilde{P} . \tag{32b}$$

where

$$\chi_J(q) = <<J_q|J_q>> = \sum_{\alpha,\beta} [P(\alpha)-P(\beta)] \frac{|(J_q)_{\alpha\beta}|^2}{\omega_{\alpha\beta}} . \tag{33}$$

Using the relation:

$$iq(J_q)_{\alpha\beta} = (\rho_q)_{\alpha\beta} = i\omega_{\alpha\beta}(\rho_q)_{\alpha\beta} , \tag{34}$$

we get

$$\chi_J \equiv \chi_J(q) = \frac{2}{q^2} \sum_{\alpha\beta} P(\alpha)|(\rho_q)_{\alpha\beta}|^2\omega_{\alpha\beta} = \frac{N}{m} . \tag{35}$$

This is the exact f sum rule.[19] Proceeding along the same lines of Eqs. (28) - (31) but this time with the new projection (32), we have

$$\hat{\chi}_{JJ}(q,\omega) = \frac{1}{\omega - R_J(q,\omega)} \cdot \chi_J , \tag{36}$$

where

$$R_J(q,\omega) = <<J_q|L\tilde{Q}\frac{1}{\omega - L\tilde{Q}} L|J_q>>/\chi_J . \tag{37}$$

Eqs. (31) and (36) relate the response functions $\hat{\chi}_{\rho\rho}$ and $\hat{\chi}_{JJ}$ to the corresponding kernels R_ρ and R_J respectively. Our goal now is to derive a set of self consistent equations for these two kernels. Once these are found we immediately have all our relevant quantities upon the substitution of Eq. (31) in Eqs. (27):

$$\sigma(q,\omega) = \frac{ie^2\omega}{\hbar\Omega} \frac{R_\rho(q,\omega)}{\omega - q^2 R_\rho(q,\omega)} \cdot \chi_\rho(q) \quad , \tag{38a}$$

$$\chi_{\rho\rho}(q,\omega) = \frac{R_\rho(q,\omega)q^2}{\omega - R_\rho(q,\omega)q^2} \chi_\rho(q) \quad , \tag{38b}$$

$$\frac{1}{\varepsilon(q,\omega)} = 1 + \frac{4\pi e^2}{q^2} \cdot \frac{R_\rho(q,\omega)q^2}{\omega - R_\rho(q,\omega)q^2} \chi_\rho(q) \quad , \tag{38c}$$

and

$$S(q,\omega) = \hbar[1 - \exp(-\hbar\omega/kT)]^{-1} \operatorname{Im} \frac{q^2 R_\rho(q,\omega)}{\omega - q^2 R_\rho(q,\omega)} \chi_\rho(q) \ . \tag{38d}$$

One relation between R_ρ and R_J is an exact relation which arises immediately by the substitution of Eqs. (31) and (36) in Eq. (26), i.e.

$$\frac{1}{\omega - R_J(q,\omega)} \cdot \chi_J = \frac{\omega}{q^2} \frac{R_\rho(q,\omega)q^2}{\omega - R_\rho(q,\omega)q^2} \chi_\rho(q) \ . \tag{39}$$

We shall now derive a second approximate relation between R_ρ and R_J which together with Eq. (39) will result in a closed set of nonlinear equations for these quantities. We start with Eq. (37) which may be written in the form:

$$R_J(q,\omega) \equiv \langle\langle \dot{J}_q | \tilde{Q} \frac{1}{\omega - L\tilde{Q}} | \dot{J}_q \rangle\rangle / \chi_J \ , \tag{40}$$

where

$$m\dot{J}_q = \frac{im}{\hbar} [H, J_q] = \sum_k U(q-k)(q-k)\rho_k + \frac{iq}{2} E_q + \frac{q^2}{2} J_q \ , \tag{41}$$

and where

$$E_q = \frac{1}{2m} \sum_j [\exp(iqr_j)p_j^2 + p_j^2 \exp(iqr_j)] \quad , \tag{42}$$

is the kinetic energy density. Hereafter we shall use units where m=1.

We shall now introduce three major approximations which allow us to express R_J (Eq. (40)) in terms of R_ρ.

(i) Long wavelength limit ($q \to 0$)

We are interested in the long wavelength limit of R_ρ and R_J. To that end we define

$$D(\omega) \equiv \lim_{q\to 0} R_\rho(q,\omega) \quad , \tag{43a}$$

and

$$M(\omega) \equiv \lim_{q\to 0} R_J(q,\omega) \quad . \tag{43b}$$

Using Eq. (41) we note that for $q\to 0$ we have:

$$\dot{J}_q \cong \sum_k U(k) k \rho_k \quad . \tag{44}$$

Substitution of Eq. (44) in Eq. (40) yields:

$$M(\omega) \cong \sum_{kk'} kk \langle\langle \rho_k U_k | \frac{1}{\omega - \hat{Q}L} \hat{Q} | U_{k'} \rho_{k'} \rangle\rangle / \chi_J \tag{45}$$

We recall that the scalar product $\langle\langle \cdots \rangle\rangle$ also includes averaging over the random potential.

(ii) Factorization

Since the critical behavior of the system is expected to come from collective long wavelength ($k\to 0$) modes, we shall split Eq. (45) into a high k part denoted $M_0(\omega)$ which may be evaluated using perturbation theory in U, and a long wavelength part. $M_0(\omega)$ does not undergo any drastic change near the Anderson transition. In the long wavelength part of the summation (45), we shall invoke a factorization approximation and factor out the $U_k U_k{}'$ terms in the averaging over the random potential. We thus get

$$M(\omega) = M_0(\omega) + \sum_k k^2 \overline{U_k^2} \langle\langle \rho_k | \frac{1}{\omega - \hat{Q}L} \hat{Q} | \rho_k \rangle\rangle \Big/ \chi_J \tag{46}$$

where the k summation now excludes the high k part included in $M_0(\omega)$. Eq. (46) is the most serious assumption made in the present deviation and it is a mean field type of approximation.

(iii) $\hat{Q} \cong 1$

Since the $\hat{P}$ space (Eq. (21)) is very small and the $\hat{Q}$ space is overwhelmingly bigger, it is reasonable to substitute $\hat{Q}=1 - \hat{P} \cong 1$ in Eq. (46), resulting in

$$\begin{aligned} M(\omega) &\cong M_0(\omega) + \sum_k k^2 \overline{U_k^2} \hat{\chi}_{\rho\rho}(k,\omega)/\chi_J \\ &\cong M_0(\omega) + \sum_k k^2 \overline{U_k^2} \frac{1}{\omega - k^2 R_\rho(k,\omega)} \chi_\rho(k)/\chi_J \end{aligned} \tag{47}$$

We have now accomplished our goal. Upon the substitution of Eqs. (43) in Eq. (47) and in Eq. (39) we get:

$$M(\omega) = M_0(\omega) + \sum_k k^2\overline{U_k^2}\,\frac{1}{\omega - k^2 D(\omega)}\,\chi_\rho(k)/\chi_J, \tag{48a}$$

and

$$D(\omega) = \frac{1}{\omega - M(\omega)} \cdot \frac{\chi_J}{\chi_\rho(q=0)} \tag{48b}$$

Eqs. (48a) and (48b) are our final self consistent equations for $D(\omega)$ and $M(\omega)$. The input information is U_k and $\chi_\rho(k)$. Once these are solved, we have, using Eqs. (38) and (43a), for the transport coefficients:

$$\sigma(\omega) \equiv \sigma(q\to 0,\omega) = \frac{ie^2}{\hbar\Omega} D(\omega)\cdot\chi_\rho(q=0), \tag{49a}$$

$$\chi_{\rho\rho}(q,\omega) = \frac{D(\omega)q^2}{\omega - D(\omega)q^2}\,\chi_\rho(q), \tag{49b}$$

$$\frac{1}{\varepsilon(q,\omega)} = 1 + 4\pi e^2\,\frac{D(\omega)}{\omega - D(\omega)q^2}\,\chi_\rho(q), \tag{49c}$$

and

$$S(q,\omega) = \hbar[1-\exp(-\hbar\omega/kT)]^{-1}\,\mathrm{Im}\,\frac{q^2 D(\omega)}{\omega - q^2 D(\omega)}\,\chi_\rho(q)\ . \tag{49d}$$

We should note at this point that for a free Fermi gas we have:

$$\frac{1}{\Omega}\chi_\rho(q=0) = N(E_F) \tag{50a}$$

whereas for a classical gas with n particles per unit volume

$$\frac{1}{\Omega}\chi_\rho(q=0) = \frac{n}{kT} \tag{50b}$$

Eqs. (48) together with (50a) were derived recently by Volldhardt and Wolfle[22, 23] for a Fermi gas at T=0.

V. THE SELF CONSISTENT EQUATIONS - REVISITED

In Section IV we have derived REM for R_ρ and R_J using the conventional COP equations (Eq. (28)). We shall now consider an alternative scheme based on different REM denoted POP (partial time-ordering prescription). A detailed study and comparison of the COP and POP equations was given elsewhere.[16,21] Suffice it to say here that both typ of REM are in principle exact, however once approximations are made they may yield very different results, especially in the non-Markovian limit (when the kernels are frequency dependent). The resulting self consisten equations for the density and current response kernels will be different in the POP scheme. The derivation of the new equations is parallel to that of the previous section, the only difference being the form of the REM. Therefore, we shall not go into the details and merely sketch the main steps of the derivation. The POP formulation is based on a convolutionless reduced equation of motion which is local in time.[21] Instead of Eq. (28), we write:

$$\frac{d\hat{\chi}_{\rho\rho}(q,t)}{dt} = -q^2 \int_0^t d\tau \tilde{R}_\rho(q,\tau)\cdot\hat{\chi}_{\rho\rho}(q,t), \tag{51}$$

where

$$\tilde{R}_\rho(q,\tau) = \frac{1}{q^2}\frac{d}{d\tau} \langle\langle\rho_q|L\exp(-iL\tau)|\rho_q\rangle\rangle\langle\langle\rho_q|\exp(-iL\tau)|\rho_q\rangle\rangle^{-1} \cdot \tag{52}$$

Using Eq. (16a) we may rearrange Eq. (52) in the form:

$$R(q,\tau) = \left\{\langle\langle J_q(\tau)|J_q\rangle\rangle\langle\langle\rho_q(\tau)|\rho_q\rangle\rangle - q^2\left[\int_0^\tau d\tau_1\langle\langle J_q(\tau_1)|J_q\rangle\rangle\right]^2\right\}\Big/ \\ \left[1+q^2\int^\tau d\tau_1(\tau-\tau_1)\langle\langle J_q(\tau_1)/J_q\rangle\rangle\right]^{-1}\chi_\rho^{-1}(q) \tag{53}$$

The solution of Eq. (51) with the initial condition Eq. (30) is:

$$\hat{\chi}_{\rho\rho}(q,t) = \exp\left[-q^2\int_0^t d\tau(t-\tau)\tilde{R}_\rho(q,\tau)\right]\cdot\chi_\rho(q) \tag{54}$$

Similarly using the POP scheme we have in analogy with (36)

$$\hat{\chi}_{JJ}(q,\tau) = \exp\left[-\int_0^\tau d\tau_1(\tau-\tau_1)\tilde{R}_J(q,\tau_1)\right]\cdot\chi_J \tag{55}$$

where

$$\tilde{R}_J(q,\tau) = \frac{d}{d\tau}\langle\langle J_q|L\exp(-iL\tau)|J_q\rangle\rangle\langle\langle J_q|\exp(-iL\tau)|J_q\rangle\rangle^{-1}. \tag{56}$$

We reiterate that Eqs. (54) and (55) are exact provided the kernels (52) and (56) are evaluated rigorously.

Upon substitution of Eq. (54) in Eqs. (27) we get:

$$\sigma(q,\omega) = \frac{ie^2}{\hbar\Omega}\frac{\omega^2}{q^2}\int_0^\infty d\tau\exp(i\omega\tau)\{\exp[-q^2K(q,\tau)]-1\}\chi_\rho(q), \tag{57a}$$

$$\chi_{\rho\rho}(q,\omega) = i\omega \int_0^\infty d\tau \exp(i\omega\tau)\{\exp[-q^2K(q,\tau)]-1\}\chi_\rho(q), \quad (57b)$$

$$\frac{1}{\varepsilon(q,\omega)} = 1 + \frac{i4\pi e^2\omega}{q^2} \int_0^\infty d\tau \exp(i\omega\tau)\{\exp[-q^2K(q,\tau)]-1\}\chi_\rho(q), \quad (57c)$$

$$S(q,\omega) = \frac{\hbar}{1-\exp(-\hbar\omega/kT} \mathrm{Im} \int_0^\infty d\tau \exp(i\omega\tau)\exp[-q^2K(q,\tau)]\chi_\rho(q), \quad (57d)$$

where

$$K(q,\tau) = \int_0^\tau dt_1(\tau-\tau_1)\tilde{R}_\rho(q,\tau_1). \quad (58)$$

The POP kernels $\tilde{R}_\rho$ and $\tilde{R}_J$ are related by the exact relation obtained by substituting Eqs. (54) and (55) in Eq. (26)

$$\exp[-\int_0^\tau d\tau_1(\tau-\tau_1)\tilde{R}_J(q,\tau_1)] =$$

$$\frac{\omega^2}{q^2} \int_0^\infty d\tau \exp(i\omega\tau)\{\exp[-q^2 \int_0^\tau d\tau_1(\tau-\tau_1)\tilde{R}_\rho(q,\tau)]-1\}\chi_\rho(q)/\chi_J \quad (59)$$

We now, as in Section IV, go to the long wave length limit and define

$$\tilde{M}(\tau) \equiv \lim_{q\to 0} R_J(q,\tau), \quad (60a)$$

$$\tilde{D}(\tau) = \lim_{q\to 0} R_\rho(q,\tau). \quad (60b)$$

Making assumptions (i) - (iii) of Section IV we finally get the self consistent POP equations:

$$\tilde{M}(\tau) = \tilde{M}_0(\tau) + \sum_k \overline{U_k^2}\, k^2 \exp[-k^2 \int_0^\tau d\tau_1(\tau-\tau_1)\tilde{D}(\tau_1)]\chi_\rho(q)/\chi_J, \quad (61a)$$

and

$$\tilde{D}(\tau) = \exp[-\int_0^\tau d\tau_1(\tau-\tau_1)\tilde{M}(\tau_1)]\cdot\chi_J/\chi_\rho(q=0). \quad (61b)$$

where $\tilde{M}_0(\tau)$ is a non-critical (high k) contribution to $M(\tau)$. Eqs. (61) are analogous to Eqs. (48) but in general they may yield different solutions since approximation (i) - (iii) when made within the POP scheme have a different significance. Once $M(\tau)$ and $D(\tau)$ are solved then the transport coefficients are obtained by the substitution of Eq. (54) in Eqs. (57):

$$\sigma(\omega) = \frac{ie^2}{\hbar\Omega}\,\omega^2 \int_0^\infty d\tau\, \tilde{K}(\tau)\exp(i\omega\tau)\cdot\chi_\rho(q=0), \tag{62a}$$

$$\chi_{\rho\rho}(q,\omega) = i\omega \int_0^\infty d\tau\, \exp(i\omega\tau)\{\exp[-q^2\tilde{K}(\tau)]-1\}\chi_\rho(q=0), \tag{62b}$$

$$\frac{1}{\varepsilon(q,\omega)} = 1 + i\,\frac{4\pi e^2\omega}{q^2} \int_0^\infty d\tau\, \exp(i\omega\tau)\{\exp[-q^2\tilde{K}(\tau)]-1\}\chi_\rho(q) \tag{62c}$$

$$S(q,\omega) = \frac{\hbar\omega}{[1-\exp(-\hbar\omega/kT)]}$$

$$\mathrm{Re}\int_0^\infty d\tau\, \exp(i\omega\tau)\{\exp[-q^2K(\tau)]-1\}\chi_\rho(q) \tag{62d}$$

where

$$\tilde{K}(\tau) = \int_0^\tau d\tau_1(\tau-\tau_1)\,\tilde{D}(\tau_1)\cdot \tag{62e}$$

VI. DISCUSSION

We have derived coupled self consistent equations which enable us to calculate the electrical conductivity, dielectric constant, the density response function and the dynamic structure factor of a fluid in a random potential. All these quantities are expressed in terms of the density response kernel $D(\omega)$ (Eqs. (49)) or $\tilde{D}(\tau)$ (Eqs. (62)). $D(\omega)$ or $\tilde{D}(\tau)$ may be obtained by solving the self consistent equations, Eqs. (48) and Eqs. (61), respectively. It should be noted that the density kernels $D(\omega)$ and $\tilde{D}(\tau)$ are simply related by a Fourier transformation, i.e.

$$D(\omega) = -i \int_0^\infty d\tau\, \exp(i\omega\tau)\tilde{D}(\tau)\cdot \tag{63}$$

This may easily be seen by noting that using Eq. (29) we have:

$$D(\omega) = \lim_{q\to 0} R_\rho\,(q,\omega) = \langle\langle J_0|\frac{1}{\omega-L}|J_0\rangle\rangle/\chi_J \tag{64a}$$

whereas when using Eq. (53) we get

$$\tilde{D}(\tau) = \lim_{q\to 0} R_\rho\,(q,\tau) = \langle\langle J_0|\exp(-iL\tau)|J_0\rangle\rangle/\chi_J\ . \tag{64b}$$

Eqs. (63) immediately follows from Eqs. (64). Despite the simple relation (63), the solutions of the self consistent Eqs. (48) and (61) do not necessarily satisfy Eq. (63). The reason is that the approximations made in the derivation of Eqs. (48) and (61) are different and we may get different results especially when $D(\omega)$ is strongly frequency dependent.[16,21] Only detailed study may reveal which form is to be preferred. It should be noted however that Eqs. (48) are easier to solve since they may be solved for each frequency independently whereas Eqs. (61) are solved for the entire function $D(\tau)$. At this stage we wish to clarify the connection between $D(\omega)$ and the ordinary diffusion coefficient $D_0(\omega)$. The latter is the kernel of the density correlation function, i.e.

$$C_{\rho\rho}(q,\omega) = \frac{1}{\omega - q^2 D_0(\omega)} S(q) \tag{65}$$

where $S(q)$ is the static structure factor

$$S(q) = \langle \rho_q | \rho_q \rangle = \sum_{\alpha,\beta} P(\alpha) |(\rho_q)_{\alpha\beta}|^2 \tag{66}$$

Using Eqs. (12), (49b) and (65) and taking the $q \to 0$ limit we have:

$$\frac{D''(\omega)}{D_0''(\omega)} = \frac{1-\exp(-\hbar\omega/kT)}{\hbar\omega} \cdot \frac{S(q=0)}{\chi_\rho(q=0)} = \frac{kT}{\hbar\omega}[1-\exp(-\hbar\omega/kT)] \tag{67}$$

In the limit $\omega \to 0$, $T \to 0$ but $h\omega \ll kT$ we thus have

$$\frac{D''(\omega)}{D_0''(\omega)} \xrightarrow{\omega \to 0} 1, \tag{68}$$

and utilizing the Kramers Kronig relation (13), we then have:

$$D(\omega) \cong D_0(\omega) \quad . \tag{69}$$

Eq. (69) holds for typical DC experiments in the Anderson localization problem. For T=1°K, e.g. it holds for $\omega < 10^9$ sec^{-1}.

We shall now focus on Eqs. (48) and simplify them further. Towards that end let us define the Laplace transform:

$$\hat{D}(s) = \int_0^\infty d\tau \, \tilde{D}(\tau) \exp(-s\tau) \tag{70}$$

It is clear from Eq. (63) that

$$D(\omega) = -i\hat{D}(-i\omega) \tag{71a}$$

Similarly we define $\hat{M}(s)$ such that

$$M(\omega) = -i\hat{M}(-i\omega) \tag{71b}$$

As is clearly seen from eqs. (64) and (71) D and M are real and this simplifies our analysis.

Upon multiplying Eq. (48a) by D(s) and substituting Eq. (48b), we get:

$$\hat{D}(s)M(s) = \frac{\chi_J}{\chi_\rho(q=0)} - \hat{D}(s)\cdot s \ ,$$

$$= \hat{M}_0(s)\hat{D}(s) + \sum_k \frac{\overline{U_k^2} k^2 \hat{D}(s)}{s+k^2\hat{D}(s)} \frac{\chi_\rho(k)}{\chi_J} \ , \tag{72}$$

i.e.

$$\frac{\hat{D}(s)}{D_0(s)} = 1 - \frac{\chi_\rho(q=0)}{\chi_J} \sum_k \frac{\overline{U_k^2} k^2 \hat{D}(s)}{s+k^2\hat{D}(s)} \frac{\chi_\rho(k)}{\chi_J} \ , \tag{73}$$

where

$$\hat{D}_0(s) = \frac{1}{s+M_0(s)} \frac{\chi_J}{\chi_\rho(q=0)} \tag{74}$$

is a microscopic non-critical diffusion kernel which does not vary significantly in the vicinity of the transition. Assuming that U_k and $\chi_\rho(k)$ are weakly dependent on k for small values of k, we then get:

$$\frac{\hat{D}(s)}{D_0(s)} = 1 - \lambda \left(\frac{s}{\hat{D}(s) k_0^2} \right)^{\frac{d}{2}} \int_0^{\sqrt{Dk_0^2/s}} dy \frac{y^{d+1}}{1+y^2} \tag{75}$$

Here we have changed the k summation to an integration

$$\sum_k \rightarrow \frac{\Omega}{(2\pi)^3} \int_0^{k_0} k^{d-1} dk \tag{76}$$

where k_0 is an upper cutoff (of the order of the lattice spacing), and

$$\lambda = \frac{\Omega}{(2\pi)^3} \overline{U_0^2} \left(\frac{\chi_\rho(q=0)}{\chi_J} \right)^2 k_0^d \tag{77}$$

is a dimensionless coupling strength.

Eq. (75) is our final equation for the diffusion kernel. The input information that enters this equation is the dimensionless coupling strength λ, the microscopic diffusion coefficient $D_0(s)$ and the microscopic cutoff wavevector k_0.

Let us see what do we expect from our kernel D(s) for the Anderson localization. In the metallic side of the transition there is a finite DC conductivity whereas in the insulating side it vanishes like ω^β. Assuming that the polarizability $\alpha(\omega)=\sigma(\omega)/\omega$ remains finite in the insulating phase[22,23] we have $\beta=1$. We thus get:

$$\sigma(\omega) \underset{\omega\to 0}{\to} \begin{cases} \text{const.} & \text{conducting phase} \\ \omega & \text{insulating phase} \end{cases}, \tag{78}$$

This implies using Eq. (49a):

$$\hat{D}(s) \underset{s\to 0}{\to} \begin{cases} \text{const.} & \text{conducting phase} \\ \xi^2 s & \text{insulating phase} \end{cases} \tag{79}$$

Making use of a Tauberian theorm,[24] which relates the small frequency behavior of D(s) to the long time behavior of its Fourier transform $D(\tau)$, and of Eq. (62e), we get also:

$$\tilde{K}(t) \underset{t\to\infty}{\to} \begin{cases} t & \text{conducting phase} \\ \xi^2 & \text{insulating phase} \end{cases} \tag{80}$$

The existence of the Anderson transition may thus be investigated using Eq. (75) by looking for a value of $\lambda=\lambda_c$ such that for $\lambda<\lambda_c$ D(s) ~ const. and for $\lambda>\lambda_c$ $D(s) = \xi^2 s$ at low frequencies. By virtue of Eqs. (63e), (63) and (69) it is clear that K(t) is simply the second moment of a density fluctuation in the medium created initially at r=0. Eq. (80) shows that for $\lambda<\lambda_c$ the second moment diverges linearly at long times, whereas for $\lambda>\lambda_c$ the second moment assumes asymptotically a finite value, namely the correlation length ξ. Let us investigate the existence of λ_c in 3d. To that end we rewrite Eq. (75) for d=3 as:

$$\frac{\hat{D}(s)}{D_0(s)} = 1 - \frac{1}{3}\lambda + \lambda\frac{s}{\hat{D}k_0^2} - \lambda\left(\frac{s}{\hat{D}k_0^2}\right)^{3/2} \operatorname{arctg}\sqrt{\frac{\hat{D}k_0^2}{s}} \tag{81}$$

A simple inspection of Eq. (80) shows that $\lambda_c=3$ and we have:

$$\hat{D}(s) = \hat{D}_0(s)\,[1 - \frac{1}{3}\lambda\,] \qquad s\to 0 \qquad \lambda < 3 \tag{82a}$$

$$\hat{D}(s) = 0 \qquad s\to 0 \qquad \lambda > 3 \tag{82b}$$

Vollhardt and Wolfle[23] have recently derived similar equations. However, their derivation is made specifically for a Fermi gas at T=0. The preser derivation is less restrictive and we show that the general structure of the equations is insensitive to the details of the model (classical or quantum fluid, temperature, space dimensionality, etc.) This enables us connect the Anderson problem with the classical phenomena of long time tails in nonlinear hydrodynamics.[6,9,10] After a lengthy diagramatic calculation Vollhardt and Wolfle got an equation similar to Eq. (75) but with d replaced by d-2. They have shown that their equation predicts a phase transition for d>2 as it should, whereas the present equation predicts such a phase transition for all d>0. This arises due to our me field approximation (Eq. (46)) which certainly fails for low dimensionalities. It is interesting to note however that the structure the equation is the same for a classical and a quantum fluid. We also se that for d>3 there is no difference between our result and that of Vollhardt and Wolfle.

Using the present derivation we may conclude that some of the featu of the Anderson localization in 3d are not at all quantum mechanical and maybe the detailed study of the nature of the wave functions, whether th are localized or extended is superfluous in the sense that the transport coefficients are not so sensitive to these details, and the transition exists also in a classical fluid.[10]

Finally let us see what we do predict for the density response function (Eq. (27b)) near the Anderson transition. Upon substituting (7 in (49b) we get:

$$\chi_{\rho\rho}(q,\omega) = \begin{cases} \chi_\rho\ (q) & \text{conducting phase} \quad (83a \\ \dfrac{\xi^2 q^2}{1-\xi^2 q^2} \cdot \chi_\rho\ (q) & \text{insulating phase} \quad (83b \end{cases}$$

since $\chi_\rho(q) \equiv \chi_{\rho\rho}(q,\omega=0)$ this means that for the insulating phase

$$\lim_{\omega\to 0} \chi_{\rho\rho}(q,\omega) \neq \chi_{\rho\rho}(q,\omega=0) \quad (84$$

In fact, in r space we have for the insulating phase

$$\chi_{\rho\rho}(r,\omega=0) = \int \chi_\rho(r-r')F(r') \quad (85$$

where

$$F(r') = \delta(r') + \exp(-r'/\xi)/(r')^{d/2} \quad (86$$

is a dynamic contribution to $\chi_{\rho\rho}$. Eq. (86) was derived recently by Vollhardt and Wolfle.[23] Within the POP formalism, substitution of Eq. (in (62b) yields:

$$\lim_{\omega\to 0} \chi_{\rho\rho}(q,\omega) = \begin{cases} \chi_\rho(q) & \text{conducting phase} \quad (87a \\ [\exp(-\xi^2 q^2)-1]\chi_\rho(q) & \text{insulating phase} \quad (87b \end{cases}$$

So that for the insulating phase we have again Eq. (85) where

$$F(r') = \exp(-r'^2/\xi^2) + \delta(r') \tag{88}$$

We note that in the metallic phase where $D(\omega)$=const. and $\tilde{K}(t)\sim t$ both formulations(83a) or (87a) yield the same result. This is a general feature of the POP and COP schemes that in the Markovian limit they are identical.[16,21] However the difference between Eqs. (86) and (88) is a manifestation of the difference in predictions of both formulations in the non-Markovian limit, where $D(\omega)$ is frequency dependent.

REFERENCES

1. a) P. W. Anderson, Phys. Rev. 109, 1492 (1958).
   b) E. Abrahams, P. W. Anderson, D. C. Licciardello and T. V. Ramakrishnan, Phys. REv. Lett. 42, 673 (1979).
2. a) N. F. Mott, Metal Insulator Transition, Taylor and Francis, New York (1974).
   b) N. F. Mott and G. A. Davis, Electronic Processes in Non-Crystalline Materials, Clarendon (London) (1979).
   c) N. F. Mott, Phil. Mag. B44, 265 (1981).
   d) N. F. Mott and M. Kaveh, J. Phys. C14, L659 (1981).
3. D. J. Thouless, Phys. Rep. 13C, 93 (1974).
4. J. M. Ziman, Models of Disorder, Cambridge, London (1979).
5. F. J. Wegner, Zeit. Phys. B25, (1976); ibit 36, 209 (1980).
6. a) K. Kawasaki, Ann. Phys. (N.Y.) 61, 1 (1970).
   b) K. Kawasaki in "Phase Transitions and Critical Phenomena" V. 5A, C. Domb and M. S. Green, eds. Academic N.Y. (1971), p. 166.
7. L. P. Kadanoff and J. Swift, Phys. Rev. 166, 89 (1968).
8. a) T. Keyes and I. Oppenheim, Phys. Rev. A7, 1384 (1973); A8, 937 (1973).
   b) T. Keyes in "Statistical Mechanics", V.6B, B. J. Berne, Ed., Plenum, N.Y. (1977), p. 259.
9. B. J. Alder and T. E. Wainwright, Phys. Rev. Lett. 18, 988 (1967).
10. H. H. H. Yuan and I. Oppenheim, Physica 90A, 1 (1978) ibid 90A, 21 (1978).
11. a) L. P. Kadanoff and P. C. Martin, Ann. of Phys. 24, 419 (1963).
    b) P. C. Martin, Measurements and Correlation Functions, Gordon and Breach, New York (1968).
12. A. A. Abrikosov, L. P. Gorkov and D. I. Dzyaloshinski, Methods of Quantum Field Theory in Statistical Physics, Dover, N.Y. (1963).
13. H. Mori, Prog. Theor. Phys. 33, 423 (1965).
14. R. Zwanzig, Physica 30, 1109 (1964).
15. I. Oppenheim in "Topics in Statistical Mechanics and Biophysics", R. A. Piccirelli, Ed. (A.I.P. New York) 1976, p. 111.
16. S. Mukamel, J. Stat. Phys. 27, 317 (1982).
17. A. Ben-Reuven and S. Mukamel, J. Phys. A8, 1313 (1975).
18. S. Mukamel, Phys. Reports 93, 1 (1982).
19. D. Pines and P. Nozieres, The Theory of Quantum Liquids, V. I. Benjamin, N.Y. (1966).
20. Eq. (27c) is to lowest order V_q since $\chi_{\rho\rho}$ is evaluated for a noninteracting fluid described by the Hamiltonian Eq. (14).

21. S. Mukamel, Adv. Chem. Phys. 47, 509 (1981); Phys. Rev. A 26, 617 (1982); Phys. Rev. B B25, 830 (1982); J. Stat. Phys. 30, 179 (1983).
22. a) W. Gotze and P. Wolfle, Phys. Rev. B6, 1226 (1972).
b) W. Gotze, Sol. St. Comm. 27, 1393 (1978); J. Phys. C12, 1279 (1979); Phil. Mag. B43, 219 (1981).
23. D. Vollhardt and P. Wolfle - Phys. Rev. B, 22, 4666 (1980).
24. W. Feller, An Introduction to Probability Theory and its Applications, Wiley, N.Y. (1968), Vol. II. The Tauberian theorm states that when $D(s) \sim s^{\beta}$ then $\hat{D}(\tau) \sim \tau^{-1-\beta}$ and $K(\tau)$ (eq. (62e)) behaves therefore as $\tau^{1-\beta}$.

KINETIC ASPECTS OF ION TRANSPORT THROUGH NEURAL MEMBRANE

John R. Clay
Section of Neural Membranes, Laboratory of Biophysics, IRP, NINCDS,
National Institutes of Health at the Marine Biological Laboratory,
Woods Hole, MA 02543

Michael F. Shlesinger
La Jolla Institute, P.O. Box 1434, La Jolla, CA 92038
and
Institute for Physical Science and Technology, University of Maryland,
College Park, MD 20742

ABSTRACT

The Hodgkin-Huxley model for the nerve membrane action potential is reviewed. The model can be described in terms of channels and gates. That is, ions permeate the nerve membrane through narrow ion specific pores, or channels, which are modulated by a voltage dependent gating process. The Hodgkin-Huxley model provides a detailed kinetic scheme for channel gating, but not for channel permeation. Radioactive tracer flux experiments suggest that permeation occurs via single file motion of ions through a channel. We have modeled this process as a random walk with internal states. The theory leads to expressions for one way fluxes which can be compared with experimental tracer flux data. Recent experiments on channel gating have indicated that the Hodgkin-Huxley model of the gating process requires certain modifications. We present a class of modifications involving temporal memory of gating and interactions between gating particles of any single channel. These models can also be described in terms of random walks.

I. THE HODGKIN-HUXLEY EXPERIMENTS AND MODEL: ACTION POTENTIALS, CHANNELS AND GATES

1. Membrane Excitability

We will be concerned with the properties of excitable membranes, primarily the membranes of nerve axons. We will emphasize results from the giant axon of the common squid. This preparation has been the source of much of the detailed information physiologists have acquired about the process of excitability during the past few decades. The axon is the primary branching process, in general, of most nerve cells. Its physiological role is to transmit electrical signals, called action potentials, from the nerve cell body to other nerve, or muscle cells. An example of a recording of an action potential from a squid axon in response to current pulse stimulation is shown in Figure 1. The action potential has a characteristic triangular shape consisting of a rapid depolarization phase from rest followed by a somewhat slower return, or repolarization phase, which usually undershoots the resting potential. The membrane then gradually returns to

0094-243X/84/1090225-19 $3.00

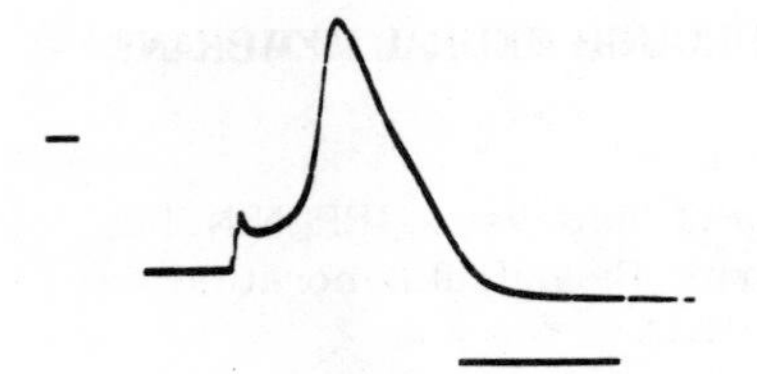

Figure 1. Action potential recoding from squid giant axon. Axon initially at rest (-50 mV). Brief current pulse stimulus (not shown) applied to depolarize axon membrane to theshold. The horizontal line below the action potential represents 5 ms. The line to the left indicates 0 mV. T=7° C.

rest (The term potential or voltage used throughout this article refers to the potential difference between the inside and the outside of the membrane.) The mechanisms underlying this event are differential gradients between the cellular interior and the external environment for sodium and potassium ions, and differential, voltage-dependent permeabilities of the membrane to these ions. The gradients are maintained by a metabolically active pumping process which will not be discussed in this article. The voltage dependent permeabilities can be understood in terms of the Hodgkin-Huxley (hereafter termed HH) equations, which relate macroscopic ion current flow across the membrane to the membrane potential.
We stress the fact that the microscopic processes of how specific ions cross the membrane are still not completely understood. In addition, recent experiments call for modification of the HH equations. However, the basic ideas of their model are still a useful framework for any discussion of membrane excitability.

In the remainder of this section we will present experimental results similar to those on which the HH model was originally based. We will then give a brief description of the model itself. In Section II a specific theory for the ion transport process will be presented in detail and applied to calculate one-way tracer fluxes. These results are compared with recent experiments. A short discussion of the voltage dependent effects of other ions, such as cesium, rubidium, and sodium, on potassium ion current then ensues. In Section III particular aspects of voltage dependent permeability, or channel gating, are examined in more detail. Modifications of the HH model of channel gating are proposed.

2. Voltage Clamp Experiments and the Hodgkin-Huxley Model

It is necessary to introduce several concepts prior to a discussion of the HH equations. In a macroscopic picture the steady current, I, flowing across a membrane due to a concentration gradient of, for example, K ions and an electrostatic potential, φ, is given by

$$I = -u(x)\,\rho(x)\left[F^2\,\frac{d\varphi(x)}{dx} + \frac{FRT}{\rho(x)}\,\frac{d\rho(x)}{dx}\right], \tag{1.1}$$

where x is the distance measured from the outer surface normally into the membrane, u is the K ion mobility, ρ is the K^+ concentration, F is the Faraday (the charge on one mole of K^+), R is the universal gas constant, and T is the absolute temperature. Dividing Eq. (1.1) by $u\rho F^2$ and integrating from the inner to the outer surface of the membrane, we have

$$I \int_{in}^{out} \frac{dx}{u(x)\rho(x)F^2} = - \int_{in}^{out} \nabla\varphi(x)dx - \frac{RT}{F}\log\left[\frac{\rho_{out}}{\rho_{in}}\right]. \tag{1.2}$$

The first term on the RHS of Eq. (1.2) is the membrane potential V, the second term is the Nernst potential, and the integral on the LHS is the membrane resistance (or reciprocal of the conductivity g). We rewrite Eq. (1.2) as

$$I_K = g_K (V - V_K), \tag{1.3}$$

where the subscript K reminds us that so far we have considered only the K^+ current. Na^+ is the other major contributor to nerve membrane ionic current. I_{Na} is given by an equation similar to Eq. (1.3) with K replaced by Na. In equilibrium

$$I_K + I_{Na} = 0, \tag{1.4}$$

which allows us to solve for the equilibrium (resting) membrane voltage V_{REST}. The result of this calculation is

$$V_{REST} = \left(\frac{g_K^{REST}}{g_K^{REST} + g_{Na}^{REST}} \right) V_K + \left(\frac{g_{Na}^{REST}}{g_K^{REST} + g_{Na}^{REST}} \right) V_{Na} . \tag{1.5}$$

In nerve membrane the Nernst potentials V_K and V_{Na} are similar in magnitude but different in sign. (There is a higher concentration of K^+ inside than outside, and vice versa for Na^+). However, g_K is about one hundred times greater than g_{Na} at rest, so

$$V_{REST} \simeq V_K. \tag{1.6}$$

A typical value of V_{REST} is -60 mV.

The above analysis is modified when the voltage changes with time, because the conductances are functions of potential, i.e., $g_K = g_K(V)$ and $g_{Na} = g_{Na}(V)$, and the membrane has a capacitative property described by a DC capacitance C. Consequently, Eq. (1.4) is replaced by

$$C\, dV/dt + g_K(V)(V - V_K) + g_{Na}(V)(V - V_{Na}) = 0 . \tag{1.7}$$

The sole independent variable in Eq. (1.7) is V. Consequently, the best way to experimentally determine $g_K(V)$ and $g_{Na}(V)$ is to examine the relaxation of membrane current following a step change in potential. This procedure also effectively eliminates the capacitative current, except for a transient at the beginning of the step. Experiments of this type (now classic) were carried out by HH using the "voltage clamp" technique, invented by Cole[2], which controls V by a negative feedback circuit while monitoring the net ionic current across the membrane. HH deduced from their experimental results that K^+ and Na^+ were the major contributors to the net ionic current and that these respective components were independent of each other. The above analysis implies this result, *a priori*. The primary evidence for the independence of Na^+ and K^+ currents was that the time constant of the change in g_{Na} following a depolarizing step was approximately an order of magnitude faster than that of g_K. Moreover, when a step change in V was made to V_{Na}, i.e. I_{Na} = 0, the remaining current was unchanged when Na was replaced by impermanent ions. Examples of voltage clamp records from a squid axon in normal ionic conditions are shown in Figure 2, upper left panel. These results, which are similar to the ones originally obtained by HH, are superimposed, digital recordings, of membrane current following a step change in V from a holding potential of -80 mV to -40, -20, 0, ... +120 mV, respectively. The step to -40 mV produces a relatively slow increase of inward (negative) current which reaches a peak value several ms following the voltage step and then relaxes back to baseline. The steps to -20 and 0 mV show a similar result, but with more rapid kinetics. The steps to more positive potentials also show an early inward current followed by a clear increase of outward current activated with a slower time course than activation of g_{Na}. The Na^+ current at the two most positive potentials is outward, which is consistent with the fact that these potentials (+100 and +120 mV) are more positive than V_{Na}. Similar results enabled HH to deduce the mechanism underlying the action potential. Depolarization of the membrane leads to an increase in g_{Na}, which causes further depolarization and, hence, a further increase in g_{Na}. That is, the initial depolarizing phase of the action potential is due to a positive feedback mechanism inherent in I_{Na}. During the latter phase of depolarization, the effect of I_{Na} is less pronounced, because the driving force, $(V-V_{Na})$ is less. Moreover, I_K begins to be activated, which brings the membrane back to rest with a somewhat slower time course than that of the initial rising phase, since the I_K kinetics are slower than the I_{Na} kinetics. The membrane undershoots the resting level, since V_K is less than V_{REST}. The membrane subsequently returns to rest, as I_K deactivates. Experiments more recent than HH provided further support for this mechanism.

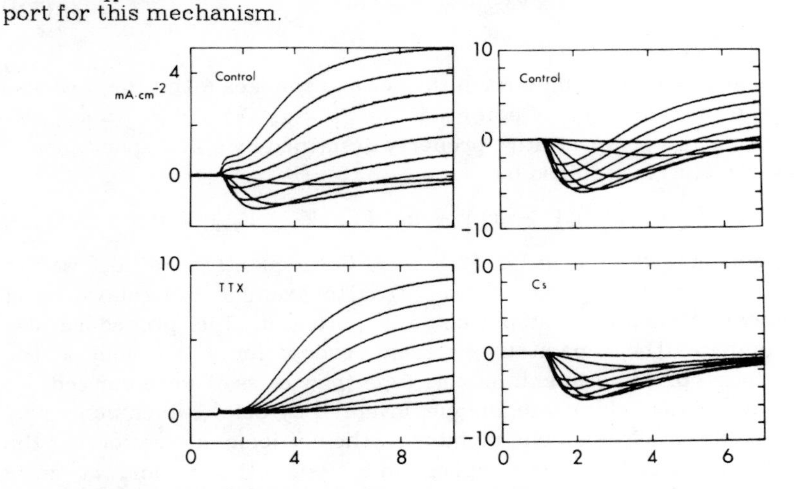

Figure 2. Voltage clamp records of squid axon membrane currents as described in the text. T = 7° C. The abscissa is in units of ms. Further experimental details provided elsewhere.[3,24]

Specific chemical agents and toxins were found to selectively block either the I_K or the I_{Na} component. For example, tetrodotoxin (TTX), a naturally occurring substance found in the ovaries of female Japanese puffer fish selectively blocks I_{Na} when placed outside of an axon,[4] as illustrated by the lower left panel of Figure 2. This experiment describes superimposed records from an axon in TTX (different from the axon in the upper left panel of Figure 2) following voltage steps from holding potential to -20, 0, +20, ... +100 mV. The outward I_K component is present; I_{Na} is not. Other agents, such as tetraethylamonium and cesium ions, are able to block I_K selectively, when placed inside an axon,[5,6] as the records in the upper and lower right panels of Figure 2 illustrate. Both sets of records are from the same axon. The control results for steps to -40, -30, ... +40 mV are qualitatively similar to the results in the upper left of Figure 2. The lower right panel shows the effect of internal cesium ions. Outward current is eliminated; the inward I_{Na} component remains. (The slight reduction of I_{Na} in the cesium results was due to a change in the internal Na^+ concentration during this particular experiment.) Moreover, an essential difference between I_{Na} and I_K kinetics is revealed: I_{Na} inactivates with time following a voltage step; I_K does not, at least not during the course of 10 ms. The I_K component does partially inactivate, but the time constant of this process is several *seconds*, rather than several milliseconds[7]. Recent experiments have shown that I_{Na} inactivation can be separated kinetically from I_{Na} activation, if appropriate chemical agents are used.[8]

The results in Figure 1 suggested a model to HH consisting essentially of first order differential equations with voltage dependent parameters. The model is described by

$$g_K(V) = \bar{g}_K n^4 \tag{1.8}$$

$$g_{Na}(V) = \bar{g}_{Na} m^3 h \tag{1.9}$$

where $\bar{g}_K$ and $\bar{g}_{Na}$ are constants, and h, m, and n all satisfy linear differential equations of the form

$$dy/dt = (y_\infty - y)/\tau_y \tag{1.10}$$

where y is either h, m or n and $y_\infty = y(t=\infty)$. HH found it necessary to raise m and n to a power in order to describe the sigmoidal activation of both I_{Na} and I_K (Figure 2). They determined τ_y and y_∞ for each process, experimentally, by fitting the model to membrane current records. They represented the voltage dependence of these parameters by empirical mathematial expressions inspired by chemical rate theory. They then used the model to calculate the shape of the action potential following current pulse stimulation. For this work HH received the Nobel Prize in Physiology in 1963.

One interpretation of h, m, and n is that they represent charged structures, whose orientation, or configuration, depends on V. There would be four such particles which act as gates for a channel through which K^+ flows. All four gates must be open for a K ion to be allowed into the channel. Therefore, the probability that a channel is open is given by n^4. The more positive V is, the closer n approaches unity. For the Na^+ channel the three m gates are open with a higher probability when V is made more positive, but the single h, or inactivation gate closes for the same conditions. The probability that any single Na channel is open is m^3h. As we indicate below, the h, m, and n scheme requires revision. However, the basic ideas of channels and gates, which follow almost directly from the original HH model, have withstood all experimental tests. Indeed, they have been confirmed rather dramatically in recent years by the invention of the patch voltage clamp technique. This method has enabled investigators for the first time to directly observe currents from single channels in excitable membranes.[9,10]

II. A RANDOM WALK MODEL OF ONE-WAY TRACER FLUXES THROUGH POTASSIUM CHANNELS.

1. The Hodgkin-Keynes experiment: Single File Motion.

As pointed out in Section I, the HH model does not elucidate the microscopic mechanisms of ion transport across the nerve membrane. One of the first experiments which provided information on this process was carried out by Hodgkin and Keynes,[11] hereafter referred to as HK. They measured the exchange of radioactive ions across the membrane at various different potentials. They found that ion influx and efflux were both considerably less than that predicted by independent movement of K ions. The ratio of one-way fluxes should be

$$\frac{EFFLUX}{INFLUX} = \frac{[K^+_{in}]}{[K^+_{out}]}\exp\left(eV/k_BT\right) = \exp\left[e\left(V-V_K\right)/k_BT\right] \qquad (2.1)$$

for ion independence, where $[K^+_{in}]$ and $[K^+_{out}]$ refer to the internal and external K^+ concentrations, respectively, e is the unit electrical charge, and k_B is the Boltzmann constant. Eq. (2.1) is called the Ussing flux ratio.[12] In contrast, HK found that

$$\frac{EFFLUX}{INFLUX} = \exp\left[se\left(V-V_K\right)/k_BT\right] \qquad (2.2)$$

with $s \simeq 2.5$. The interpretation which HK offered for their result was that K ions interact with each other within the membrane. Specifically, HK assumed that the membrane contained channels with s sites (potential wells) in a *single file* through which ions must pass. It is easy to see that single file motion would decrease the amount of ion exchange, because an ion which entered a channel would not necessarily cross from one side to the other. An ion may enter a channel, progress part way to the other side, and then reverse direction and exit the channel on the same side from which it entered. The longer the channel, the more difficult would be the exchange of ions across it.

The HK experiment has inspired a number of theoretical formulations of single file transport.[13-16]. The model we present below is a formulation in terms of random walks.

2. A Random Walk Model

A. Description of the Model

We present a specific model of a channel with s sites, where s = 1, 2, ..., n with n finite, in which each site is always occupied by an ion. Collisions of K ions from either the internal or external bathing fluid with the channel cause movement of the row of s ions. We assume that all ions in a channel move simultaneously in the same direction when a collision occurs having sufficient energy to overcome the static forces which tend to keep the ions in place in the channel. Such a collision is termed successful. The rate of successful collisions from either side of the membrane depends upon the membrane potential and the concentration of permeant ions in both the internal and external solutions. A successful collision will knock one K ion out of the membrane. We assume that another K ion immediately enters the channel to fill the vacant site. We let p_+ and p_- be the probabilities that the channel ions move toward either the external or the internal solution, respectively, following a successful collision. A reasonable assumption to make regarding p_+ and p_- is that their ratio is equal to the Ussing ratio

$$p_+/p_- = \exp\left[e(V-V_K)/k_B T\right]. \tag{2.3}$$

Moreover, $p_+ + p_- = 1$, since the row must move in one direction or the other following a successful collision. Consequently, we find that

$$p_+ = \left[1+\exp(-e(V-V_K)/k_B T)\right]^{-1}. \tag{2.4}$$

We let $\bar{t}$ be the mean time between successful collisions. Consequently, the net flux through a single channel is given by

$$J_1 = (p_+ - p_-)/\bar{t} = \bar{t}^{-1}\tanh\left[e(V-V_K)/2k_B T\right]. \tag{2.5}$$

The mean collision time $\bar{t}$ is undoubtedly a function of $|(V-V_K)|$, since the probability that any collision is successful will increase as V moves away from V_K in either the hyperpolarizing or the depolarizing direction. Since the total number of collisions per unit time, either successful or unsuccessful, is finite, $\bar{t}$ must have a minimum value and J_1 must ultimately saturate as $|V-V_K| \to \infty$. The exact dependence of $\bar{t}$ on membrane potential is unknown. We have simply assumed that $\bar{t}$ increases with $|V-V_K|$ so that J_1 is a linear function of $(V-V_K)$, which is consistent with the HH equations.

The total K ion current of the above model is given by

$$I_K = (eNn^4\bar{t}^{-1})\tanh\left[e(V-V_K)/2\,k_B T\right], \tag{2.6}$$

where N is the density of K channels and n^4 is the HH gating variable. Eq. (2.6) is to be contrasted with the HH model for K channel current, which is given by

$$I_K = \bar{g}_K n^4 (V-V_K). \tag{2.7}$$

When $e|V-V_K| \simeq 2k_BT$, a comparison of Eq. (2.6) with Eq. (2.7) yields $\bar{g}_K = eN/(2k_B T\bar{t})$. Using $\bar{g}_K = 36$ mS.cm^{-2} from HH, N = 60 channels μm^{-2} from noise measurements[17], and $k_BT/e = 25$ mV, we find that $\bar{t} \simeq 0.5$ μs, which is consistent with the diffusion time for an ion to traverse a distance equal to the width of a membrane in an aqueous solution.

The formulation of ion flux in Eq. (2.6) can be readily adapted to the analysis of tracer flux measurements. The primary modification concerns the fact that isotopically labeled K^+ are distinguishable from unlabeled K^+. Hence, the physical state of the channel will depend upon the configuration of labeled (hot) and unlabeled (cold) ions at the s ion sites, with the number of different states given by 2^s. Almost all aspects of the model can be demonstrated by a two site channel which we will consider in detail before presenting our results for an arbitrary number of sites. We let $P_{cc}, P_{ch}, P_{hc}, P_{hh}$ be the probabilities of the channel being in one of its four states: c represents a cold ion, h represents a hot ion, and the left hand subscript represents the channel ion site adjacent to the axon interior. Throughout the derivation of our theoretical unidirectional flux expressions, we assume that the channel gates are fully open. That is, $n^4 = 1$.

We consider first the loss of label from an axon into a non-radioactive solution. We let f be the fraction of label inside the axon and assume that ions which leave a channel are immediately removed from the vicinity of the nerve fiber. That is, there is no backflux of label into the axon. The differential equations describing the four probability functions may be derived by considering all possible interstate transitions. For example, if the channel is in state hc at time t, it can make a transition to state hh in time (t, t+ dt) if a labeled ion strikes the channel from the axon interior and the row moves to the right. The probability of this event is $p_+ f\, dt/\bar{t}$. The complete set of equations in matrix form is given by $\bar{t}\, d\vec{P}/dt = \vec{A}\vec{P}$ in which

$$\vec{A} = \begin{pmatrix} -p_+f & p_- & p_+(1-f) & 0 \\ p_+f & -1 & p_-+p_+f & p_- \\ 0 & p_+(1-f) & -1 & p_+(1-f) \\ 0 & p_+f & 0 & -(1-p_+f) \end{pmatrix} \tag{2.8}$$

and the vector $\vec{P} = (P_{cc}, P_{hc}, P_{ch}, P_{hh})$. The approach of $\vec{P}$ to its steady-state value occurs in a time approximately equal to $\bar{t}$. Because the time constant of the K^+ channel gates in nerve is a few ms and $\bar{t} \simeq 1$ μs, we can ignore $d\vec{P}/dt$ and use the steady state value of $\vec{P}$ when computing K fluxes during either an action potential or voltage clamp steps. Consequently, the rate at which labeled ions are lost from an axon is given by

$$dk_h/dt = -\pi d N\, p_+\left(P_{ch}^{ss} + P_{hh}^{ss}\right)/\bar{t}\ , \tag{2.9}$$

in which k_h is the number of hot ions per cm of axon, d is the axon diameter, and N is the number of K^+ channels per unit area. We note that πdN is the total number of channels per unit length of axon. The solution of $\vec{A}\vec{P}^{ss} = 0$ with the condition that $P_{cc}+P_{ch}+P_{hc}+P_{hh} = 1$ yields

$$P^{ss}_{ch} + P^{ss}_{hh} = p_+^2 f / \left[1 - p_+ + p_+^2 \right] \tag{2.10}$$

By expanding the denominator, we show later that Eq. (2.10) represents the probability that a labeled ion enters the channel from the internal fluid and then goes back and forth between the two sites any number of times before stopping at the site adjacent to the external fluid. Since $k_h = fk_i$, where k_i is the total number of ions in the axon interior per unit length, both hot and cold, we obtain from Eqs. (2.9) and (2.10) the result

$$\dot{f}/f = -4N p_+^3 / \left[\left(d\, N_A [K_i] \bar{t} \right) (1 - p_+ + p_+^2) \right] = -\lambda \tag{2.11}$$

in which N_A is Avogadro's number and $[K_i]$ is the internal K concentration in moles.

B. Recursive Enumeration of the Random Walk Paths

The rate constant λ may be solved for any number of sites s in a similar way, although the amount of algebraic manipulation is forbidding for $s \geq 3$. We now present a novel recursion method for determining λ for arbitrary s that circumvents this computational problem.

In general, the rate of loss of label is given by

$$\dot{f} = -4Np_+ P_s^h / (dN_A [K_i] \bar{t}) \tag{2.12}$$

in which P_s^h is the probability that a labeled ion is in the outermost ion site regardless of the state of the rest of the channel. The simplest way to calculate P_s^h to find all the ways that an ion can reach the outermost site and weight each way by its probability of occurrence. Consider the three site case. A labeled ion enters the channel with probability $p_+ f$. It must move to site 2, which occurs with probability p_+, or else it gets knocked back into the internal solution. The labeled ion can go back and forth between sites 1 and 2 n times, for any n, with probability $(p_- p_+)^n$. At this stage of the random walk we have included all transitions back to site 1. The next transition is to site 3 which occurs with probability p_+. At this point the ion can leave the channel, or it can go back to site 2, and we simply repeat the analysis appropriate for that site. All of the properly weighed paths the ion may take to reach site 3 are given by

$$P_3^h = (p_+ f)(p_+) \left[\sum_{n=0}^{\infty} (p_- p_+)^n \right] (p_+) [1 + p_- \left[\sum_{n=0}^{\infty} (p_- p_+)^n p_+ [1 + \ldots \right. \tag{2.13}$$

let

$$\gamma_1 = \sum_{n=0}^{\infty} (p_- p_+)^n = (1 - p_+ + p_+^2)^{-1} \tag{2.14}$$

so that

$$P_3^h \;=\; p_+^3 f\,\gamma_1[1 + p_-\gamma_1 p_+\;[1 + p_-\gamma_1\, p_+[1 + \;\cdots \tag{2.15}$$

or

$$P_3^h \;=\; p_+^3 f\,\gamma_1 \sum_{n=0}^{\infty}(p_-\gamma_1\, p_+)^n \;=\; p_+^3 f\,\gamma_1\gamma_2 \;, \tag{2.16}$$

in which

$$\gamma_2 \;=\; \sum_{n=0}^{\infty}\,(p_-\gamma_1\, p_+)^n \tag{2.17}$$

Eq. (2.15) is in the form of a recursion relationship, because

$$P_3^h \;=\; p_3^+ f\,\gamma_1\gamma_2 \;=\; P_2^h\, p_+\,\gamma_2 \;=\; p_+^3 f\,/\,(1-2p_+ + 2p_+^2)\,. \tag{2.18}$$

In general,

$$P_s^h \;=\; P_{s-1}^h\; p_+\gamma_{s-1} \;=\; P_1^h\; p_+^{s-1}\prod_{k=1}^{s-1}\;\gamma_k;\;\; s>1 \tag{2.19}$$

in which

$$\gamma_k \;=\; \sum_{n=0}^{\infty}\;(p_-\gamma_{k-1}\, p_+)^n \tag{2.20}$$

and

$$P_1^h \;=\; p_+ f \;\;. \tag{2.21}$$

Combining Eqs. (2.12) and (2.19) we obtain

$$\dot{f}\,/\,f \;\;=\;\; -\,4N\;p_+^{s+1}\;\prod_{k=1}^{s-1}\;\gamma_k/\left[dN_A[K_i]\bar{t}\,\right]\;;\;\; s>1 \tag{2.22}$$

or

$$\dot{f}\,/\,f \;\;=\;\; -4\chi_s N/\left[dN_A[K_i]\bar{t}\,\right];\;\; s>1\;\;. \tag{2.23}$$

The problem of influx of tracer ions into an axon is a simple extension of the preceding analysis. The quantity obtained from influx experiments is the initial influx. That is, initially there are no labeled ions inside and only labeled ions outside, which is the reverse of the initial condition for the efflux experiment. Consequently, the initial rate of entry of labeled ions into the axon may be obtained from Eq. (2.23) with $f=1$ and with the role of right and left hand sides of the channel reversed in the quantity χ_s. That is, p_+ and p_- are exchanged with one another. The quantity γ_k is unaffected by this procedure since it contains products of p_+ and p_-, whereas p_+^{s+1} now becomes p_-^{s+1}. Consequently, the initial influx Φ_i is

$$\Phi_i \;\;=\;\; 4\,(p_-/\,p_+)^{s+1}\chi_s N/\,(dN_A[K_i]\bar{t}\;)\;\;, \tag{2.24}$$

and

$$\Phi_i \;/\; \Phi_e \;=\; (p_-/p_+)^{s+1} \;=\; \exp\left[(s+1)(V-V_K)\;/\;k_B T)\right] \;, \qquad (2.25)$$

which is similar to the flux ratio relationship in Eq. (2.2).

C. Application to Tracer Flux Data

(i) Flux During an Action Potential

The one-way fluxes at any time during an action potential are given by

$$\Phi_e \;=\; n^4(t)\lambda'\chi_s, \qquad (2.26)$$

and

$$\Phi_i \;=\; (p_-/p_+)^{s+1}\, n^4(t)\lambda'\chi_s, \qquad (2.27)$$

in which n(t) is the HH gating parameter and

$$\lambda' \;=\; 4N/\left[dN_A[K_i]\bar{t}\right]. \qquad (2.28)$$

The flux to be expected at rest can also be calculated and subtracted from Φ_e andΦ_i to give the fluxes (extra fluxes) associated purely with firing of the axon. We calculate $N/\bar{t}$ so that the current voltage relation is linear, i.e.,

$$N/\bar{t} \;=\; g_k(V-V_K)/\left[e\tanh[e(V-V_K)/2k_B T]\right] \qquad (2.29)$$

in which g_K = 36 $mS.cm^{-2}$ and V_K = -12 mV with respect to the rest potential. A membrane action potential was generated from the HH model using a standard Runga-Kutta iteration technique. The extra fluxes were calculated from temperatures between 5.5 deg and 27 deg C. with a Q_{10} of 3 for h, m, and n.[15] Q_{10} is a temperature coefficient which multiplies the rates in Eq. (3.1). The absolute extra efflux is plotted against temperature in Figure 3 for $0 \leq s \leq 4$. The efflux data are clearly inconsistent with independent ion movement, i.e. s=0. Unfortunately, the scatter in the data makes it impossible to ascertain the appropriate value of s from these experiments. The situation is more promising for the influx data as shown in Fig. 4. Either s=2 or s=3 appears to provide a fairly good representation of these results, which HK also inferred from their earlier flux ratio measurements.

(ii) Flux During Voltage Clamp Steps.

A more direct way to measure one-way fluxes is to use the voltage clamp technique, which effectively was the procedure used by HK. Begenisich and DeWeer[21] repeated the original HK experiment a few years ago using improved techniques. They measured the one way flux of perfused axons with K_{in} = 400 mV and various different levels of external K (K_{out}), and with the voltage V clamped at various different potentials. Their results are shown in Figure 5. For V = -50 mV and -40 mV. These results are plotted on a relative scale. The appropriate theoretical result for comparison with these data is

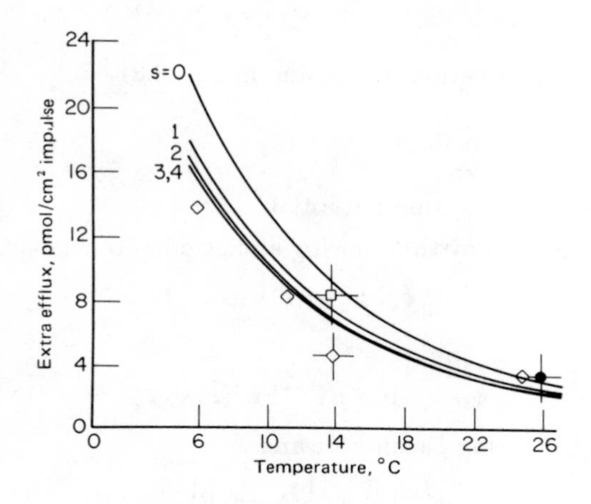

Figure 3. Extra efflux of K^+ from experimental axons during an action potential.[18-20] The solid lines were calculated as described in the text. Reproduced from Reference 15 with permission.

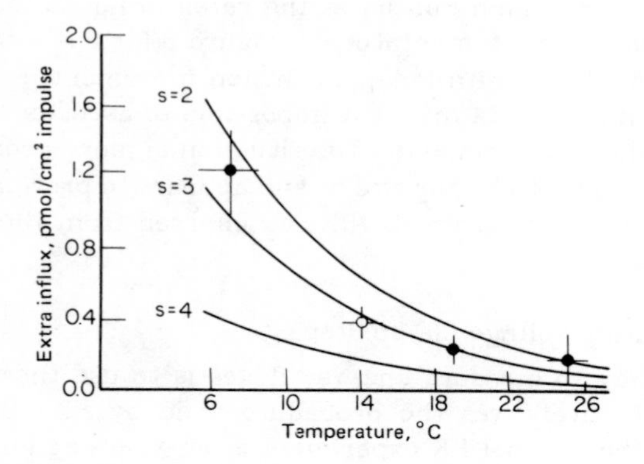

Figure 4. Comparison of theory with extra influx experimental results.[18-20] Reproduced from Ref. 15 with permission.

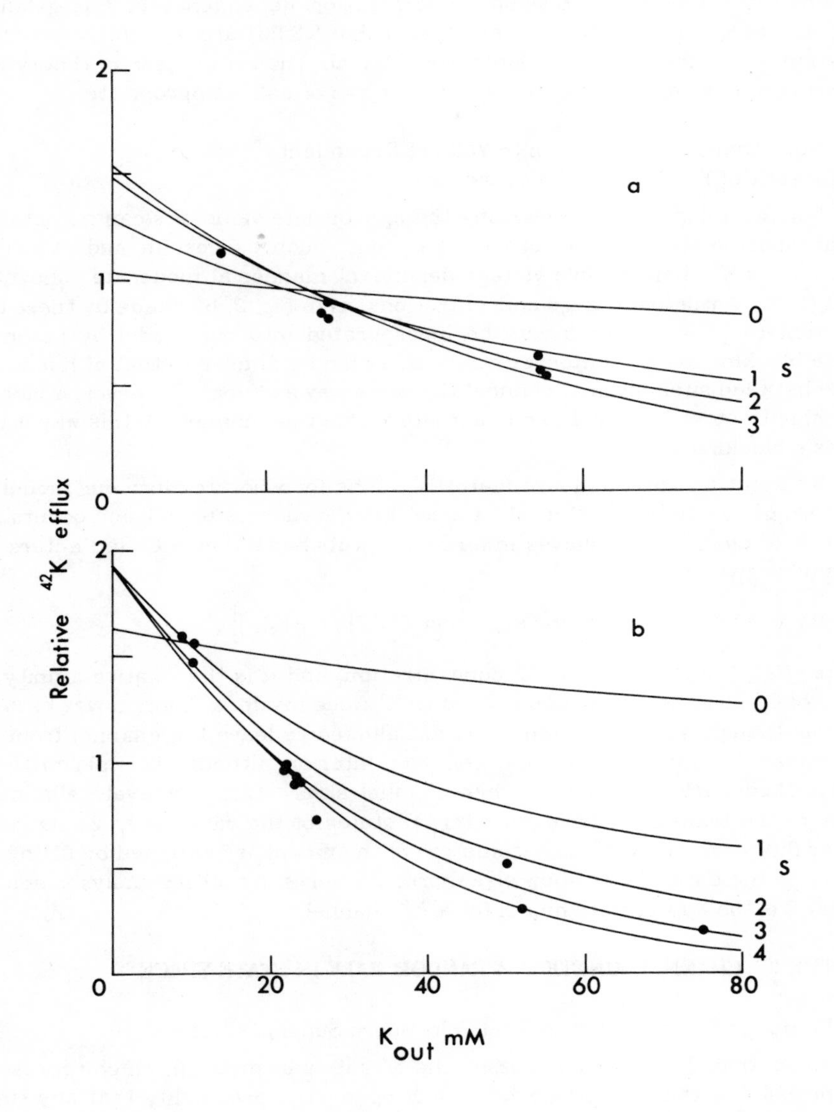

Figure 5. Potassium efflux from squid axon vs. K_{out} with K_{in} = 400 mm. (a.) V clamped at -50 mV (b.) V clamped at -40 mV.

$$\Phi_e \approx p_+^{s+1} / \bar{t} \prod_{k=1}^{s-1} \gamma_k \tag{2.30}$$

with γ_k given by Eq. (2.20). A straightforward calculation with our model[22] demonstrates that the potassium concentration dependence of $\bar{t}$ is given by $\bar{t} \approx ([K_{in}] + [K_{out}])^{-1}$ The free parameters in Eq. (2.30) are an arbitrary scaling constant and s, the number of sites per channel. The comparison of theory with experiment once again indicates that either s=2 or s=3 is appropriate.

3. A Note About Generalization to Voltage Dependent Blockade of K^+ Channels by Other Ions

Further information concerning the appropriate value of s can be obtained from experiments involving other alkali ions, such as cesium and rubidium, which block K^+ channels in a voltage dependent manner at moderate concentrations.[23,24] At relatively large concentrations, as in Fig. 2, blockade by these ions is complete. These effects can be incorporated into our model by assuming that a blocking ion can enter a channel in a manner similar to that of K ions and move between sites within a channel the same way as K ions. However, a blocker is unable to move beyond a particular site within the channel. In this way it produces a blocking effect.

We describe how the above features can be incorporated into our model for the case of blockade by external Cs ions. When a successful collision occurs and the row of channel ions moves inward, the probability that a Cs ion enters the channel is given by

$$r = \alpha[Cs_{out}^+] / \left(\alpha[Cs_{out}^+] + [K_{out}^+]\right), \tag{2.31}$$

Where $[Cs_{out}^+]$ is the external C_s^+ concentration, and α is the relative affinity for entry of C_s^+ into the channel compared to K^+ Once inside, a C_s ion moves between sites as though it were a K ion. It is not allowed to leave the channel from the innermost site and thereby enter the axon interior, although it is permitted to be knocked back toward the other channel sites. This effectively eliminates many of the transitions between internal states of the random walk. As in the tracer flux work, the various parameters of the model are obtained by fitting the theory to the data. As we show elsewhere,[24] the results of this analysis indicate that s=2 is the appropriate model for a K^+ channel.

III. THE K^+ CHANNEL KINETICS: A RANDOM WALK IN STATE SPACE

1. The Hodgkin-Huxley Picture and Cole-Moore Superposition

In Sections I and II we discussed the K^+ gate parameter n, which represents the probability that any single K^+ gate is open. The probability that any single K^+ channel is open is n^4. Consider for now a single n gate which flips back and forth between its open and closed configurations with rates α (V) (closed to open) and β (V) (open to closed). The probability n(V,t) satisfies the following master (kinetic) equation.

$$dn(V,t)/dt = -\beta(V)n(V,t) + \alpha(V)\left[1-n(V,t)\right], \quad (3.1)$$

which has the solution

$$n(V,t) = \frac{\alpha(V)}{\alpha(V)+\beta(V)} + \left[n_0 - \frac{\alpha(V)}{\alpha(V)+\beta(V)}\right] e^{-(\alpha(V)+\beta(V))t} \quad (3.2)$$

where n_0 is the initial condition of the gate. When Eq. (3.2) is compared, with Eq. (1.10) we find that

$$n_\infty = \alpha(V)/\left[\alpha(V)+\beta(V)\right]$$

and

$$\tau_n = 1/(\alpha(V)+\beta(V)). \quad (3.3)$$

Eq. (3.2) can be written as

$$n(t) = n_\infty\left[1 - \frac{n_\infty - n_0}{n_\infty}\exp\left(-t/\tau_n\right)\right] \quad (3.4)$$

$$= n_\infty\left[1 - \exp\left(-(t+t_0)/\tau_n\right)\right]$$

where $t_0 = \tau_n \ln[n_\infty/(n_\infty - n_0)]$, and $n_0 = n(t=0)$ where $0 \le n_0 < n_\infty$ is satisfied. The kinetics of all four gates can be further understood by considering the following diagram

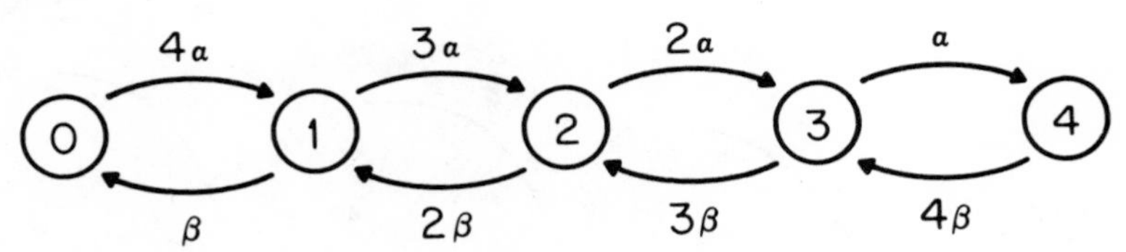

in which the number in each circle indicates the number of individual gates which are open. The channel itself is open only when it is in the right hand most state of the diagram. The rate constants in the above picture are obtained by considering the number of open and closed gates in each state. For example, no particles are open in the left hand state. Each gate has the rate $\alpha(V)$ for transition to its open condition. Therefore, the total rate constant for the channel to move to a state in which one gate is open is $4\alpha(V)$.

One of the earliest tests of the n^4 scheme was carried out by Cole and Moore,[25] who hyperpolarized the membrane to various levels prior to depolarization. They found that the resulting delay, or sigmoidicity, of the membrane response was greater than that predicted by HH. For all hyperpolarizing initial conditions where n_0=D the HH equations predict the same time dependence upon depolarizing to the same n_∞, regardless of the initial condition. Experimentally, a delay in activation of potassium current is observed under these conditions. Moreover, the currents essentially superimpose when the control current response is translated along the time axis to correspond to the delay in the response which was preceded by hyperpolarization. Exact superposition imposes very tight constraints, on channel gating models, as Hill and Chen[26] have shown. However, recent experiments indicate that superposition is not exact for all conditions.[3,27] The records in Figure 6 (top panel) illustrate this point for squid axons. These results correspond to potassium current kinetics for depolarizations to 0, +50, and +100 mV (left to right, respectively) from a starting potential of either -80 mV or -240 mV. The record with the starting potential of -80 mV has been translated in time until maximum superposition is achieved with the -240 mV record. The difference in starting time of the capacitative transient gives the amount of time translation. The records for 0 mV essentially superimpose. The other two sets of records do not. They show a slight lack of complete superposition.

2. Modified Kinetics for K^+ Channels with a Delay

The results in the lower panel of Figure 6 represented by (•) are the same as in the top panel without the time shift. The solid lines correspond to an alternative model of channel gating in which a delay, or a memory effect, is incorporated into α (V). That is, α is a function of both time and voltage, represented by α (V,t), where

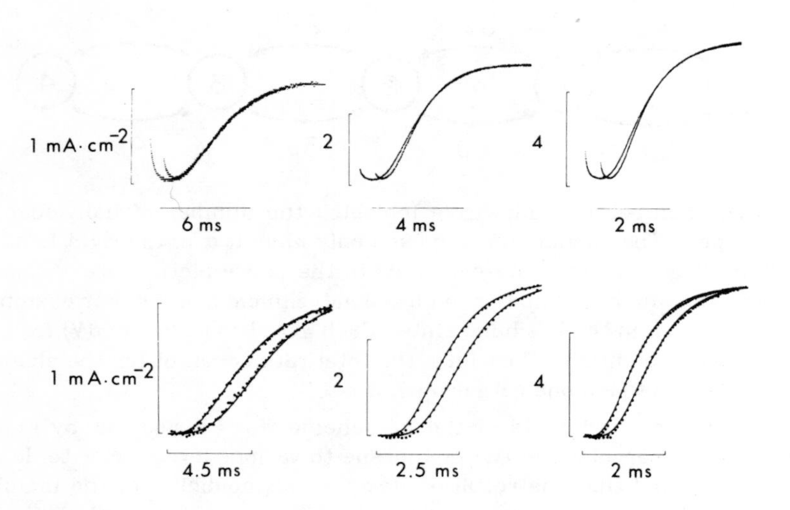

Figure 6. Potassium kinetics from squid axon, as described in the text. Reprinted with permission from the *Biophysical Journal* 1982, **37**, 677-680 by copyright permission of the Biophysical Society.

$$\alpha(V,t) = \alpha(V_2) + \left[\alpha(V_1) - \alpha(V_2)\right]e^{-t/\tau_D} \qquad (3.5)$$

where $\alpha(V_1)$ and $\alpha(V_2)$ are the HH values of α corresponding to the starting and the final potentials, V_1 and V_2, respectively, and τ_D is the time constant for the change of α from $\alpha(V_1)$ to $\alpha(V_2)$ following a voltage step. Eq. (3.2) is now modified to

$$dn(V,t)/dt = -\left[\alpha(V,t) + \beta(V)\right]n(V,t) + \alpha(V,t) \qquad (3.6)$$

with α (V,t) given by Eq. (3.5). Eq. (3.6) can be solved exactly to give a solution for n(V,t). The model now has the additional parameter τ_D, which was obtained by fitting $n^4(t)$ with the modified form of n(t) in Eq. (3.6) to the data of Figure 6. The results of this procedure are represented by the solid lines in Figure 6 (bottom panel).

3. A Word About Modifying the Hodgkin-Huxley Na^+ Gate Kinetics

The Na^+ gate kinetics within the HH model are described by independent m and h parameters with I_{Na} proportional to m^3h. The m and h parameters are chosen to satisfy the same kinetic equation as n(t) but with the rates α and β suitably modified through the voltage dependencies of m_∞, h_∞, τ_m and τ_h. The m^3h kinetics can be represented by the following diagram. The numbers in the circles represent the number of open m gates. The upper tier of states corresponds to a condition in which the h gate is closed. The lower tier of states corresponds to an open h gate. Only state 3 in the lower tier is conducting.

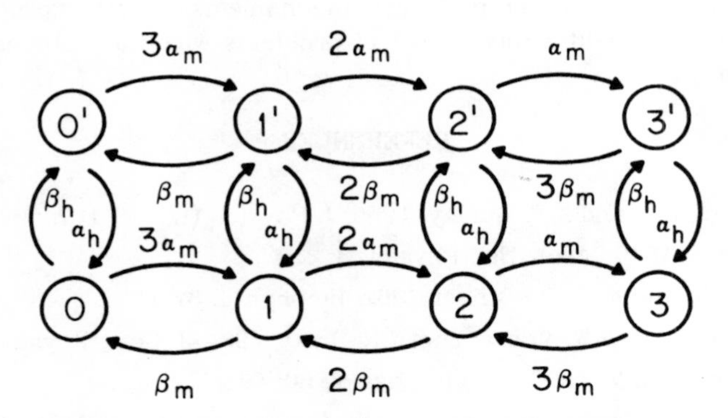

Recent experiments cast doubt on the assumption that the m and h parameters can be viewed as non-interacting.[28,29] For example, there is a delay in closing the h gate following a depolarization as if the inactivated state can be reached during a depolarization only by the system first passing through the conducting state. One possible kinetic diagram with this feature is exhibited below. The bottom tier of states corresponds to the HH m^3 kinetics. The upper state labeled I represents an inactivation state, which is reached in a counterclockwise manner during a depolarization by the system first passing through the conducting state 3. In the HH model diagram, shown above for m^3h, the system can always switch between the upper and lower tiers without having to pass through state 3. Other than this point, we have chosen the rates in the modified model to be in accordance with HH as much as possible. For example, the steady state probability in the above diagram of being in the conducting state is m^3h, the steady state probability of being in the inactivated state I is (1-h), and the action potential (with standard n^4 kinetics) is similar to that of the HH model.[30] The differences in the gates kinetics will show up dramatically in power spectrum measurements of current fluctuations.[31,32]

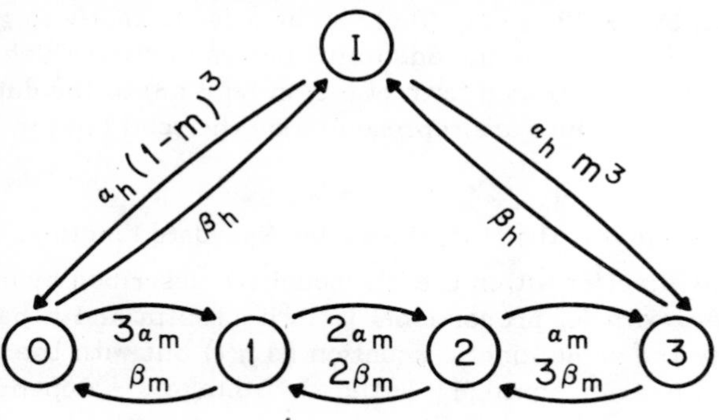

IV. SUMMARY

We have shown that the problem of membrane excitability has a number of issues which can be addressed from the viewpoint of random walks. The primary challenge which remains in this field is to understand the concepts of channels and gates in terms of detailed molecular mechanisms.. Further progress in this direction may well yield a further set of problems which are amenable to random walk analysis.

REFERENCES

1. Hodgkin, A.L. and A.F. Huxley. 1955. J. Physiol. (Lond.) **117:** 500.
2. Cole, K.S. 1949. Arch. Sci. Physiol. **3:** 253.
3. Clay, J.R. and M.F. Shlesinger 1982. Biophys. J. **37**: 677.
4. Narahashi, T., J.W. Moore, and W.R. Scott, 1964. J. Gen. Physiol. **47:** 965.
5. Armstrong, C.M. 1966. J. Gen. Physiol. **50:** 491.
6. Adelman,W., Jr. In: *Biophysicis* and *Physiology of Excitable Membranes.* Ch. 15 (Van Nostrand Reinhold, W.J. Adelman, Jr. Ed.)
7. Ehrenstein, G. and D.L. Gilbert, 1966. Biophys. J. **6:** 553.
8. Oxford, G.S. 1981. J. Gen. Physiol. **77:** 1.
9. Conti, F. and E. Neher. 1980. Nature. **285:** 140.
10. Sigworth, F.J. and E. Neher. 1980. Nature. **287** 447.

11 Hodgkin, A.L. and R.D. Keynes. 1955. J. Physiol. (Lond.) **128:** 61.

12. Ussing, H.H. 1949. Physiol. Rev. **29:** 127.

13. In: *Funktionelle und Morphologische Organization per Zelle.* (Springer-Verlag, P. Krlson, Ed.) p. 241.

14. Hladky, S.D. and J.D. Harris, 1967. Biophys. J. **7:** 535.

15. Clay, J.R. and M.F. Shlesinger. 1977. P.N.A.S. (U.S.A.) **74:** 409.

16. Hille, B. and W. Schwarz. 1978. J. Gen. Physiol. **72:** 409.

17. Conti, F., L.J. DeFelice, and E. Wanke, 1975. J. Physiol. (Lond.) **248:** 45.

18. Landowne, D. and V. Scruggs. 1976. J. Physiol. (Lond.) **259:** 145.

19. Brinley, F.J. and L.J. Muillins. 1965. J. Neurophysiol. **28:** 526.

20. Keynes, R.D. 1951. J. Physiol. (Lond.) **114:** 119.

21. Begenesich, T. and P. DeWeer 1977. Nature. **269:** 710.

22. Clay, J.R. and M.F. Shlesinger, 1976. Biophys. J. **16:** 121.

23. Adelman, W.J., Jr. and R.J. French. 1978. J. Physiol. (Lond.) **276:** 13.

24. Clay, J.R. and M.F. Shlesinger. 1983. Biophys. J. In press.

25. Cole, K.S. and J.W. Moore, 1960. Biophys. J. **1** 1.

26. Hill, T.L. and Y.D. Chen. 1972. Biophys. J. **12:** 960.

27. Begenesich, T. 1979. Biophys. J. **27:** 257.

28. Bezenilla, F. and C.M. Armstrong. 1977. J. Gen. Physiol. **70:** 549.

29. Gilly, W.J. and C.M. Armstrong. 1979. J. Gen. Physiol. **74:** 691.

30. E. Jakobsson. Personal Communication

31. Clay, J.R. 1977. J. Membrane Biol. **42:** 215.

32. Conti, F., B. Neumcke, W. Nonner, and R. Stamphi. 1980. J. Physiol. **308:** 217.

AIP Conference Proceedings

		L.C. Number	ISBN
No.1	Feedback and Dynamic Control of Plasmas	70-141596	0-88318-100-2
No.2	Particles and Fields - 1971 (Rochester)	71-184662	0-88318-101-0
No.3	Thermal Expansion - 1971 (Corning)	72-76970	0-88318-102-9
No.4	Superconductivity in d-and f-Band Metals (Rochester, 1971)	74-18879	0-88318-103-7
No.5	Magnetism and Magnetic Materials - 1971 (2 parts) (Chicago)	59-2468	0-88318-104-5
No.6	Particle Physics (Irvine, 1971)	72-81239	0-88318-105-3
No.7	Exploring the History of Nuclear Physics	72-81883	0-88318-106-1
No.8	Experimental Meson Spectroscopy - 1972	72-88226	0-88318-107-X
No.9	Cyclotrons - 1972 (Vancouver)	72-92798	0-88318-108-8
No.10	Magnetism and Magnetic Materials - 1972	72-623469	0-88318-109-6
No.11	Transport Phenomena - 1973 (Brown University Conference)	73-80682	0-88318-110-X
No.12	Experiments on High Energy Particle Collisions - 1973 (Vanderbilt Conference)	73-81705	0-88318-111-8
No.13	π-π Scattering - 1973 (Tallahassee Conference)	73-81704	0-88318-112-6
No.14	Particles and Fields - 1973 (APS/DPF Berkeley)	73-91923	0-88318-113-4
No.15	High Energy Collisions - 1973 (Stony Brook)	73-92324	0-88318-114-2
No.16	Causality and Physical Theories (Wayne State University, 1973)	73-93420	0-88318-115-0
No.17	Thermal Expansion - 1973 (lake of the Ozarks)	73-94415	0-88318-116-9
No.18	Magnetism and Magnetic Materials - 1973 (2 parts) (Boston)	59-2468	0-88318-117-7
No.19	Physics and the Energy Problem - 1974 (APS Chicago)	73-94416	0-88318-118-5
No.20	Tetrahedrally Bonded Amorphous Semiconductors (Yorktown Heights, 1974)	74-80145	0-88318-119-3
No.21	Experimental Meson Spectroscopy - 1974 (Boston)	74-82628	0-88318-120-7
No.22	Neutrinos - 1974 (Philadelphia)	74-82413	0-88318-121-5
No.23	Particles and Fields - 1974 (APS/DPF Williamsburg)	74-27575	0-88318-122-3
No.24	Magnetism and Magnetic Materials - 1974 (20th Annual Conference, San Francisco)	75-2647	0-88318-123-1
No.25	Efficient Use of Energy (The APS Studies on the Technical Aspects of the More Efficient Use of Energy)	75-18227	0-88318-124-X

No.	Title	LC No.	ISBN
No.26	High-Energy Physics and Nuclear Structure - 1975 (Santa Fe and Los Alamos)	75-26411	0-88318-125-8
No.27	Topics in Statistical Mechanics and Biophysics: A Memorial to Julius L. Jackson (Wayne State University, 1975)	75-36309	0-88318-126-6
No.28	Physics and Our World: A Symposium in Honor of Victor F. Weisskopf (M.I.T., 1974)	76-7207	0-88318-127-4
No.29	Magnetism and Magnetic Materials - 1975 (21st Annual Conference, Philadelphia)	76-10931	0-88318-128-2
No.30	Particle Searches and Discoveries - 1976 (Vanderbilt Conference)	76-19949	0-88318-129-0
No.31	Structure and Excitations of Amorphous Solids (Williamsburg, VA., 1976)	76-22279	0-88318-130-4
No.32	Materials Technology - 1976 (APS New York Meeting)	76-27967	0-88318-131-2
No.33	Meson-Nuclear Physics - 1976 (Carnegie-Mellon Conference)	76-26811	0-88318-132-0
No.34	Magnetism and Magnetic Materials - 1976 (Joint MMM-Intermag Conference, Pittsburgh)	76-47106	0-88318-133-9
No.35	High Energy Physics with Polarized Beams and Targets (Argonne, 1976)	76-50181	0-88318-134-7
No.36	Momentum Wave Functions - 1976 (Indiana University)	77-82145	0-88318-135-5
No.37	Weak Interaction Physics - 1977 (Indiana University)	77-83344	0-88318-136-3
No.38	Workshop on New Directions in Mossbauer Spectroscopy (Argonne, 1977)	77-90635	0-88318-137-1
No.39	Physics Careers, Employment and Education (Penn State, 1977)	77-94053	0-88318-138-X
No.40	Electrical Transport and Optical Properties of Inhomogeneous Media (Ohio State University, 1977)	78-54319	0-88318-139-8
No.41	Nucleon-Nucleon Interactions - 1977 (Vancouver)	78-54249	0-88318-140-1
No.42	Higher Energy Polarized Proton Beams (Ann Arbor, 1977)	78-55682	0-88318-141-X
No.43	Particles and Fields - 1977 (APS/DPF, Argonne)	78-55683	0-88318-142-8
No.44	Future Trends in Superconductive Electronics (Charlottesville, 1978)	77-9240	0-88318-143-6
No.45	New Results in High Energy Physics - 1978 (Vanderbilt Conference)	78-67196	0-88318-144-4
No.46	Topics in Nonlinear Dynamics (La Jolla Institute)	78-057870	0-88318-145-2
No.47	Clustering Aspects of Nuclear Structure and Nuclear Reactions (Winnepeg, 1978)	78-64942	0-88318-146-0
No.48	Current Trends in the Theory of Fields (Tallahassee, 1978)	78-72948	0-88318-147-9
No.49	Cosmic Rays and Particle Physics - 1978 (Bartol Conference)	79-50489	0-88318-148-7

No.	Title	LC Number	ISBN
.50	Laser-Solid Interactions and Laser Processing - 1978 (Boston)	79-51564	0-88318-149-5
.51	High Energy Physics with Polarized Beams and Polarized Targets (Argonne, 1978)	79-64565	0-88318-150-9
.52	Long-Distance Neutrino Detection - 1978 (C.L. Cowan Memorial Symposium)	79-52078	0-88318-151-7
.53	Modulated Structures - 1979 (Kailua Kona, Hawaii)	79-53846	0-88318-152-5
.54	Meson-Nuclear Physics - 1979 (Houston)	79-53978	0-88318-153-3
.55	Quantum Chromodynamics (La Jolla, 1978)	79-54969	0-88318-154-1
.56	Particle Acceleration Mechanisms in Astrophysics (La Jolla, 1979)	79-55844	0-88318-155-X
. 57	Nonlinear Dynamics and the Beam-Beam Interaction (Brookhaven, 1979)	79-57341	0-88318-156-8
. 58	Inhomogeneous Superconductors - 1979 (Berkeley Springs, W.V.)	79-57620	0-88318-157-6
. 59	Particles and Fields - 1979 (APS/DPF Montreal)	80-66631	0-88318-158-4
. 60	History of the ZGS (Argonne, 1979)	80-67694	0-88318-159-2
. 61	Aspects of the Kinetics and Dynamics of Surface Reactions (La Jolla Institute, 1979)	80-68004	0-88318-160-6
. 62	High Energy e^+e^- Interactions (Vanderbilt , 1980)	80-53377	0-88318-161-4
. 63	Supernovae Spectra (La Jolla, 1980)	80-70019	0-88318-162-2
. 64	Laboratory EXAFS Facilities - 1980 (Univ. of Washington)	80-70579	0-88318-163-0
. 65	Optics in Four Dimensions - 1980 (ICO, Ensenada)	80-70771	0-88318-164-9
. 66	Physics in the Automotive Industry - 1980 (APS/AAPT Topical Conference)	80-70987	0-88318-165-7
. 67	Experimental Meson Spectroscopy - 1980 (Sixth International Conference , Brookhaven)	80-71123	0-88318-166-5
68	High Energy Physics - 1980 (XX International Conference, Madison)	81-65032	0-88318-167-3
69	Polarization Phenomena in Nuclear Physics - 1980 (Fifth International Symposium, Santa Fe)	81-65107	0-88318-168-1
70	Chemistry and Physics of Coal Utilization - 1980 (APS, Morgantown)	81-65106	0-88318-169-X
71	Group Theory and its Applications in Physics - 1980 (Latin American School of Physics, Mexico City)	81-66132	0-88318-170-3
72	Weak Interactions as a Probe of Unification (Virginia Polytechnic Institute - 1980)	81-67184	0-88318-171-1
73	Tetrahedrally Bonded Amorphous Semiconductors (Carefree, Arizona, 1981)	81-67419	0-88318-172-X
4	Perturbative Quantum Chromodynamics (Tallahassee, 1981)	81-70372	0-88318-173-8

No. 75	Low Energy X-ray Diagnostics-1981 (Monterey)	81-69841	0-88318-174
No. 76	Nonlinear Properties of Internal Waves (La Jolla Institute, 1981)	81-71062	0-88318-175
No. 77	Gamma Ray Transients and Related Astrophysical Phenomena (La Jolla Institute, 1981)	81-71543	0-88318-176
No. 78	Shock Waves in Condensed Matter - 1981 (Menlo Park)	82-70014	0-88318-177
No. 79	Pion Production and Absorption in Nuclei - 1981 (Indiana University Cyclotron Facility)	82-70678	0-88318-178
No. 80	Polarized Proton Ion Sources (Ann Arbor, 1981)	82-71025	0-88318-179
No. 81	Particles and Fields - 1981: Testing the Standard Model (APS/DPF, Santa Cruz)	82-71156	0-88318-180
No. 82	Interpretation of Climate and Photochemical Models, Ozone and Temperature Measurements (La Jolla Institute, 1981)	82-071345	0-88318-181
No. 83	The Galactic Center (Cal. Inst. of Tech., 1982)	82-071635	0-88318-182-
No. 84	Physics in the Steel Industry (APS.AISI, Lehigh University, 1981)	82-072033	0-88318-183-
No. 85	Proton-Antiproton Collider Physics - 1981 (Madison, Wisconsin)	82-072141	0-88318-184-
No. 86	Momentum Wave Functions - 1982 (Adelaide, Australia)	82-072375	0-88318-185-
No. 87	Physics of High Energy Particle Accelerators (Fermilab Summer School, 1981)	82-072421	0-88318-186-
No. 88	Mathematical Methods in Hydrodynamics and Integrability in Dynamical Systems (La Jolla Institute, 1981)	82-072462	0-88318-187-
No. 89	Neutron Scattering - 1981 (Argonne National Laboratory)	82-073094	0-88318-188
No. 90	Laser Techniques for Extreme Ultraviolt Spectroscopy (Boulder, 1982)	82-073205	0-88318-189-
No. 91	Laser Acceleration of Particles (Los Alamos, 1982)	82-073361	0-88318-190-
No. 92	The State of Particle Accelerators and High Energy Physics(Fermilab, 1981)	82-073861	0-88318-191-
No. 93	Novel Results in Particle Physics (Vanderbilt, 1982)	82-73954	0-88318-192-
No. 94	X-Ray and Atomic Inner-Shell Physics-1982 (International Conference, U. of Oregon)	82-74075	0-88318-193-
No. 95	High Energy Spin Physics - 1982 (Brookhaven National Laboratory)	83-70154	0-88318-194-
No. 96	Science Underground (Los Alamos, 1982)	83-70377	0-88318-195-

97	The Interaction Between Medium Energy Nucleons in Nuclei-1982 (Indiana University)	83-70649	0-88318-196-7
98	Particles and Fields - 1982 (APS/DPF University of Maryland)	83-70807	0-88318-197-5
99	Neutrino Mass and Gauge Structure of Weak Interactions (Telemark, 1982)	83-71072	0-88318-198-3
100	Excimer Lasers - 1983 (OSA, Lake Tahoe, Nevada)	83-71437	0-88318-199-1
101	Positron-Electron Pairs in Astrophysics (Goddard Space Flight Center, 1983)	83-71926	0-88318-200-9
102	Intense Medium Energy Sources of Strangeness (UC-Santa Cruz, 1983)	83-72261	0-88318-201-7
103	Quantum Fluids and Solids - 1983 (Sanibel Island, Florida)	83-72440	0-88318-202-5
104	Physics,Technology and the Nuclear Arms Race (APS Baltimore-1983)	83-72533	0-88318-203-3
105	Physics of High Energy Particle Accelerators (SLAC Summer School, 1982)	83-72986	0-88318-304-8
106	Predictability of Fluid Motions (La Jolla Institute, 1983)	83-73641	0-88318-305-6
107	Physics and Chemistry of Porous Media (Schlumberger-Doll Research, 1983)	83-73640	0-88318-306-4
108	The Time Projection Chamber (TRIUMF, Vancouver, 1983)	83-83445	0-88318-307-2
109	Random Walks and Their Applications in the Physical and Biological Sciences (NBS/La Jolla Institute, 1982)	84-70208	0-88318-308-0
110	Hadron Substructure in Nuclear Physics (Indiana University, 1983)	84-70165	0-88318-309-9